T0269613

CAMBRIDGE LIBRARY COLLECTION

Books of enduring scholarly value

Zoology

Until the nineteenth century, the investigation of natural phenomena, plants and animals was considered either the preserve of elite scholars or a pastime for the leisured upper classes. As increasing academic rigour and systematisation was brought to the study of 'natural history', its subdisciplines were adopted into university curricula, and learned societies (such as the London Zoological Society, founded in 1826) were established to support research in these areas. These developments are reflected in the books reissued in this series, which describe the anatomy and characteristics of animals ranging from invertebrates to polar bears, fish to birds, in habitats from Arctic North America to the tropical forests of Malaysia. By the middle of the nineteenth century, this work and developments in research on fossils had resulted in the formulation of the theory of evolution.

Evenings at the Microscope

English zoologist Philip Henry Gosse (1810–88) spent several years studying the biodiversity of habitats in North America and the Caribbean. His *Naturalist's Sojourn in Jamaica* (1851) is reissued in this series. When he settled on the Devonshire coast, the area proved equally rich for research. In this 1859 publication, the deeply religious Gosse considers the 'Divine mechanics' of animal body parts and microorganisms seen through the lens of a microscope. He leads the reader through a selection of specimens ranging from a hog's bristle to the shoe-like protist Paramecium. Gosse's writing style, enlivened with anecdotes and literary references, earned him considerable appreciation among Victorian audiences. His entertaining text is complemented by more than 100 illustrations which showcase his draughtsmanship. While the work shares its year of publication with Darwin's groundbreaking *Origin of Species*, Gosse's religious views firmly shaped his interpretation of the specimens on show.

Cambridge University Press has long been a pioneer in the reissuing of out-of-print titles from its own backlist, producing digital reprints of books that are still sought after by scholars and students but could not be reprinted economically using traditional technology. The Cambridge Library Collection extends this activity to a wider range of books which are still of importance to researchers and professionals, either for the source material they contain, or as landmarks in the history of their academic discipline.

Drawing from the world-renowned collections in the Cambridge University Library and other partner libraries, and guided by the advice of experts in each subject area, Cambridge University Press is using state-of-the-art scanning machines in its own Printing House to capture the content of each book selected for inclusion. The files are processed to give a consistently clear, crisp image, and the books finished to the high quality standard for which the Press is recognised around the world. The latest print-on-demand technology ensures that the books will remain available indefinitely, and that orders for single or multiple copies can quickly be supplied.

The Cambridge Library Collection brings back to life books of enduring scholarly value (including out-of-copyright works originally issued by other publishers) across a wide range of disciplines in the humanities and social sciences and in science and technology.

Evenings at the Microscope

Or, Researches among the Minuter Organs and Forms of Animal Life

Philip Henry Gosse

CAMBRIDGE
UNIVERSITY PRESS

University Printing House, Cambridge, CB2 8BS, United Kingdom

Cambridge University Press is part of the University of Cambridge.
It furthers the University's mission by disseminating knowledge in the pursuit of education, learning and research at the highest international levels of excellence.

www.cambridge.org
Information on this title: www.cambridge.org/9781108081269

This edition first published 1859
This digitally printed version 2015

ISBN 978-1-108-08126-9 Paperback

EVENINGS AT THE MICROSCOPE;

OR,

RESEARCHES

AMONG THE MINUTER ORGANS AND FORMS OF

ANIMAL LIFE.

BY

PHILIP HENRY GOSSE, F.R.S.

PUBLISHED UNDER THE DIRECTION OF
THE COMMITTEE OF GENERAL LITERATURE AND EDUCATION,
APPOINTED BY THE SOCIETY FOR PROMOTING
CHRISTIAN KNOWLEDGE.

LONDON:
SOCIETY FOR PROMOTING CHRISTIAN KNOWLEDGE;
SOLD AT THE DEPOSITORIES:
GREAT QUEEN STREET, LINCOLN'S INN FIELDS,
4, ROYAL EXCHANGE; 16, HANOVER STREET, HANOVER SQUARE;
AND BY THE BOOKSELLERS.

PREFACE.

To open the path to the myriad wonders of creation, which, altogether unseen by the unassisted eye, are made cognisable to sight by the aid of the Microscope, is the aim and scope of this volume. Great and gorgeous as is the display of Divine power and wisdom in the things that are seen of all, it may safely be affirmed that a far more extensive prospect of these glories lay unheeded and unknown till the optician's art revealed it. Like the work of some mighty genie of Oriental fable, the brazen tube is the key that unlocks a world of wonder and beauty before invisible, which one who has once gazed upon it can never forget, and never cease to admire.

This volume contains but a gleaning: the author has swept rapidly across the vast field of marvels, snatching up a gem here and there, and culling one and another of the brilliant blossoms of this flowery region, to weave a specimen chaplet, a sample coronal, which may tell of the good things behind. Yet the selection has been so made as to leave untouched no considerable area of the great field of Zoology which is under the control of the Microscope; so that the student who shall have verified for himself the observations here detailed, will be no longer a tyro in microscopic science, and will be well prepared to extend his independent researches, without any other limit than that which the finite, though vast, sphere of study itself presents to him.

The staple of the work now offered to the public consists of original observation. The author is far from thinking lightly of the labours of others in this ample field; but, still, it is true that, respecting very many of the subjects that came under his notice, he found, in endeavouring to reproduce and verify published statements, so much perplexity and difficulty, that he was thrown back upon himself and nature, compelled to observe *de novo*, and to set down simply what he himself could see. The ever accumulating stock of observed and recorded facts is the common property of science; and the author has not scrupled to reproduce, to amplify, or to abridge his own observations which have already appeared in his published works and scientific memoirs, as freely as he would have cited those of any other observer, in which he had confidence, and which were germane to his purpose. Yet in almost all cases the observations so used have been subjected to renewed scrutiny, and have been verified afresh, or corrected where found defective.

In order to relieve as much as possible the dryness of technical description, a colloquial and familiar style has been given to the work; which has been thrown into the form of a series of imaginary *conversaziones*, or microscopical *soirées*, in which the author is supposed to act as the provider of scientific entertainment and instruction to a circle of friends. It is proper to add, however, that the precision essential to science has never been consciously sacrificed. A master may be easy and familiar without being loose or vague.

A considerable amount of information will be found incidentally scattèred throughout the work, on microscopic manipulation,—the selecting, securing, and preparing objects for examination;—an important matter, and one which presents a good deal of practical difficulty to the beginner. Not a little help will be afforded to him, also, on the power to observe and to discriminate what he has under his eye. In almost every instance, the objects selected for illustration are common things, such as any one placed in tolerably favourable circumstances, with access to sea-shore and country-side, may reasonably expect to meet with in a twelvemonth's round of research.

The pictorial illustrations are almost co-extensive with the descriptions; they are one hundred and thirteen in number; all, with the exception of eighteen, productions of the author's own pencil, the great majority having been drawn on the wood direct from the Microscope, at the same time as the respective descriptions were written. He ventures to hope that they will be found accurate delineations of the objects represented.*

TORQUAY, *February*, 1859.

* The subjects on pp. 51, 58, 118, 120 (the lower figures), and 184, have been copied, under the courteous permission of the publisher, from Dr. Carpenter's valuable work, "The Microscope, and its Revelations." (Churchill, London.)

LIST OF ILLUSTRATIONS.

EVENINGS

AT THE MICROSCOPE.

CHAPTER I.

HAIRS, FEATHERS, AND SCALES.

NOT many years ago an eminent microscopist received a communication inquiring whether, if a minute portion of dried skin were submitted to him, he could determine it to be *human* skin or not. He replied, that he thought he could. Accordingly a very minute fragment was forwarded to him, somewhat resembling what might be torn from the surface of an old trunk, with all the hair rubbed off.

The professor brought his microscope to bear upon it, and presently found some fine hairs scattered over the surface; after carefully examining which, he pronounced with confidence that they were *human* hairs, and such as grew on the naked parts of the body; and still further, that the person who had owned them was of a fair complexion.

This was a very interesting decision, because the fragment of skin was taken from the door of an old church in Yorkshire;* in the vicinity of which a tradition is preserved, that about a thousand years ago a Danish robber had violated this church, and having been taken, was condemned to be flayed, and his skin nailed to the church-door, as a terror to evil-doers. The action of the weather and other causes had long ago removed all traces of the stretched and dried skin, except that from under the edges of the broad-headed nails, with which the door was studded, fragments still peeped out. It was one of these atoms, obtained by drawing one of the old nails, that was subjected to microscopical scrutiny; and it was interesting to find that the wonder-showing tube could confirm the tradition with the utmost certainty; not only in the general fact, that it was really the skin of man, but in the special one of the race to which that man belonged, viz. one with fair complexion and light hair, such as the Danes are well known to possess.

It is evident from this anecdote that the human hair presents characters so indelible that centuries of exposure have not availed to obliterate them, and which readily distinguish it from the hair of any other creature. Let us then begin our evening's entertainment by an examination of a human hair, and a comparison of it with that which belongs to various animals.

Here, then, is a hair from my own head. I cut off

* I am writing from memory, having no means of referring to the original record, which will be found in the first (or second) volume of the "Transactions of the Microscopical Society" of London. The general facts, however, may be depended on.

about half an inch of its length, and, laying it between two plates of glass, put it upon the stage of the microscope. I now apply a power of 600 diameters; that is, the apparent increase of size is the same as if six hundred of these hairs were placed side by side. Now, with this eye-piece micrometer, we will first of all measure its diameter.

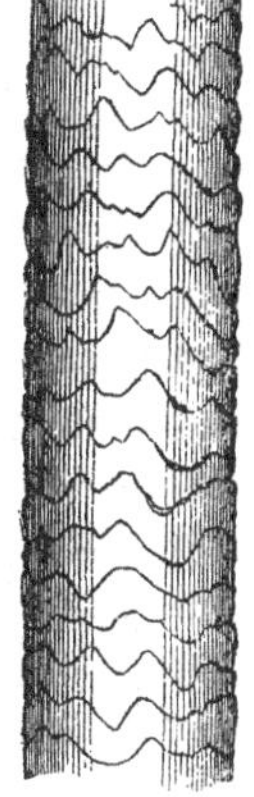
HUMAN HAIR.

You see, crossing the bright circular field of view, a semi-pellucid cylindrical object; that is the hair. You see also a number of fine lines drawn parallel to each other, exactly like those on an ivory rule or scale, with every fifth line longer than the rest, and every tenth longer still. This is the micrometer, or scale by which we measure objects; and the difference in the length of the lines, you will readily guess, is merely a device to facilitate the counting of them. By moving the stage up or down, or to either side, we easily get the hair to be exactly in the centre of the field; and now, by adjusting the eye-piece, we make the scale to lie directly across the hair, at right angles with its length. Thus we see that its diameter covers just thirty of the fine lines; and as, with this magnifying power, each line represents 1-10,000th of an inch, the hair is 30-10,000ths, $= \frac{1}{333}$rd of an inch, in diameter.

In all branches of natural history, but perhaps pre-eminently in microscopic natural history,—owing to its greater liability to error from illusory appearances,—we gain much information on any given structure by comparing it with parallel or analogous structures in other forms. Thus we shall find that our understanding of the structure of this hair will be much increased when we have seen, under the same magnifying power, specimens of the hair of other animals. In order, however, to explain it, I must anticipate those observations.

What we see, then, is a perfectly translucent cylinder, having a light brown tinge, and marked with a great number of delicate lines, having a general transverse direction, but very irregularly sinuous in their individual courses. These lines we perceive to be on the surface; because, if we slowly turn the adjustment-screw, the lines grow dim on the central part of the cylinder, while those parts that lie near the edges (speaking according to the optical appearance) come into distinctness. Presently the edges of the cylinder become sharply defined, and are seen to be cut into exceedingly shallow saw-like teeth, about as far apart as the lines; these, however, are so slight that they can be seen only by very delicate adjustment. We go on turning the screw, and presently another series of transverse lines, having the same characters as the former, but differing from them individually, come into view, at the sides first, and presently in the middle, and then, as we still turn, become dim, and the whole is confused. In fact, our eye has travelled, in this process, from the nearer surface of the hair, right through its transparent substance, to the farther surface; and we have seen that it is surrounded by these sinuous lines, which the edges—or those portions of the hair which would be the edges, if it were split through the middle (for, optically, this is the same thing)—show to be successive coats of the surface, suddenly terminated. If we suppose a cylinder to be formed of very thin paper, rolled up, and then, with a turning-lathe, this cylinder to be tapered into a very lengthened cone, the whole would be surrounded by lines marking the cut-through edges of the successive layers of paper; and, owing to the thickness of the paper not being mathematically equal in every part, these edges would be sinuous; exactly as we see in these lines upon the hair. The effect and the cause are the same in the two cases.

A hair is closely analogous to the stem of a plant; inasmuch as it grows from a root, by continual additions of cells to the lower parts, which, as they lengthen, push forward the ever-lengthening tip. Indeed, in some of the hairs which we shall presently look at, there is the most curious resemblance to the stem of a palm, with the projections produced by the successive growth and sloughing of leaf-bases around the central cylinder. Internally, too, the resemblance is remarkable; for, if we split a human hair, and especially if we macerate it in weak muriatic acid, we shall find it composed of (1) a thin but dense kind of bark, forming the successive overlapping scales just described; (2) a fibrous substance, extending from the bulb to the point of the hair. By soaking the hair in hot sulphuric acid, this fibrous substance resolves itself into an immense number of very long cells, pointed at each end, and squeezed by mutual pressure into various angular forms. "A human hair, of one-tenth of a line in thickness,* has about 250 fibrils in its mere diameter, and about 50,000 in its entire calibre: so that these ultimate fibrils are finer than those of almost any other known tissue, from the great elongation and narrowing of their constituent cells as they are drawn out into the shaft of the hair during growth; and hence the expanded bulb of the hair, where the cells are yet spherical and soft."† (3) Running through the very centre of the fibrous portion may be sometimes discerned a dark slender line, which is a sort of pith (*medulla*), composed of minute roundish cells, filled with air, and arranged in two or three rows.

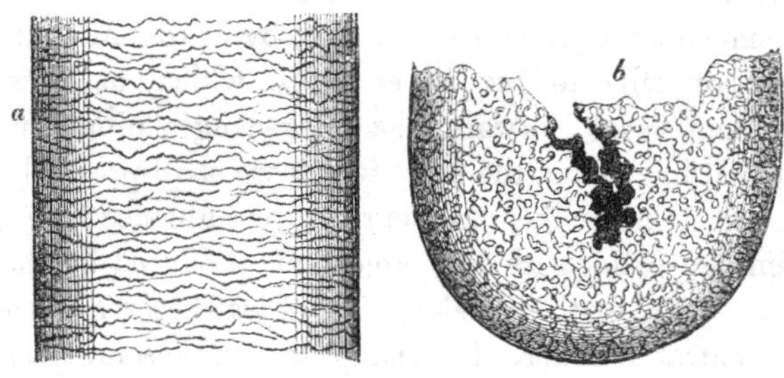

HOG'S BRISTLE.

The bristles of the Hog bear much resemblance to the human hair. On this slide is one, which you perceive is just thrice as thick as the hair that we have been examining, or $\frac{1}{100}$th of an inch in diameter. The sinuous lines across the surface are proportionally

* This is nearly thrice as great as the diameter I had given above, which was the result of several careful admeasurements of different hairs, taken from childhood and adult age.

† Grant, Outl. Comp. Anat. 647.

far finer and closer together, and no saw-teeth are visible at the edge, the most delicate adjustment showing only a minute undulation in the outline; that is to say, the overlapping scales are far thinner, and therefore their terminations are nearer together, in the hair of the Swine than in that of Man. I will now show you a transverse section of a similar bristle, which I will obtain thus: I take this old brush, and with a razor cut off one of the bundles of bristles, close to the wood; then I take off as thin a shaving as I can cut, wood, bristles, and all: I repeat the same operation two or three times. Now, picking out the shavings of wood, I take up a few of the dust-like atoms with the point of my penknife, and scatter them on this plate (or slide) of glass, and these I cover with another plate of thin glass; for this dust is composed of thin transverse slices of the bristles, and as I scatter them, some will fall upon their cut ends, so that we shall look through them endwise.

Here is one, very suitable for examination,—since it is not a whole section, the razor having passed somewhat obliquely across it, coming out beyond the middle, where it thins away to an edge. The outline is not circular, but elliptical; that is, the hair is not round, but flattened. There is no separable *cortex*, or bark, and the whole substance appears made up of excessively fine fibres, of which we see the ends cut across. A rough dark line occupies the middle of the slice, in the plane of the greater diameter; but at the edge of the slice we are able to see that this is not a solid core, as has been sometimes supposed, but a cavity passing up through the hair. It is surrounded by a layer of medullary cells, which appear black, because they are filled with air.

The finer hairs of the Horse and the Ass, such as those selected from the cheeks, have the sinuous edges of the plates about as close as in human hair. But they are distinguished at once by the conspicuousness of the medullary portion, which is thick, and quite opaque, and is broken up (especially towards each extremity of the hair) into separate longitudinal irregular masses.

The fine wool of the Sheep is clothed with imbrications, proportionally much fewer than those of human hair, while the diameter is also much less. Thus these examples, selected from fine flannel and from coarse worsted, vary in diameter from $\frac{1}{2000}$th to $\frac{1}{700}$th of an inch; and there are, upon an average, about two imbrications in a space equal to the diameter. No colour is perceptible in these specimens; they are as transparent and colourless as glass. The imbricated plates project here considerably more than in either of the examples we before examined; the "teeth," however, form an obtuse angle.

FIBRE OF SHEEP'S WOOL.

We shall presently see the importance of this imbricate structure; but we will first look at a few more examples, in which we shall find it still more strongly developed, in conjunction with some other peculiarities. All the hairs that we have looked at are what I have called fibrous in their interior texture, but those of many animals are more distinctly cellular.

Thus, in these specimens, plucked from the fur of the Cat that lies coiled up on the hearthrug, we see, first, that the imbrications are short, being about equal to the diameter in length, but are very strongly marked; though, like those of the Sheep's wool, obtuse. Hence, the contour is extremely like that of the stipe of an old rough palm-tree. There is a distinct bark (*cortex*), which is thick, and marked with longitudinal lines, which add to the resemblance just alluded to. The interior is clear, marked off at pretty regular intervals by the broad flattened medullary cells, in single series, each cell occupying, for the most part, the whole breadth of the interior. These cells are transparent and apparently empty; but their walls appear opaque and almost black,—an optical illusion, dependent on the absorption of the light by their surfaces at certain angles with the eye of the beholder. The fibrous portion is here almost displaced by the great development of the medullary cells.

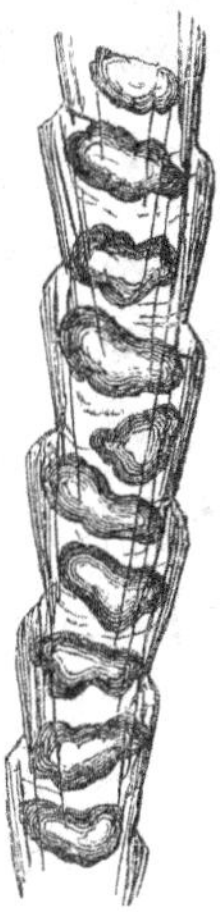

HAIR OF CAT.

In the larger hairs of the Mole, which we will now look at, the bark is very thin; and though the surface is marked with sinuous lines, these do not project into teeth. The pith here again forms the greater portion of the hair, the cells of which it is composed being placed in single series, which, for the most part, extend all across the body of the hair, though they are somewhat irregular both in size and shape. They are rather flattened, and appear perfectly black (that is, opaque) by transmitted light, their surfaces absorbing all the rays of light. The small hairs of the same animal, however, are very different in form: they are flattened, so as to appear twice as broad in one aspect as in another at right angles to it; and, what is curious, the scales of the bark project into strongly-marked imbrications on one side, and are scarcely perceptible on the other. Here, as in the larger hairs, there is a single row of oval transverse cells, perfectly opaque.

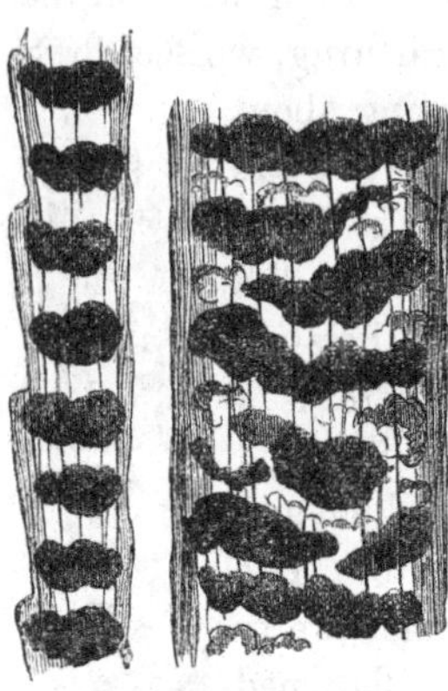

HAIRS OF MOLE.

The hair of many of the smaller Mammalia shows considerable diversity of form, according to the part which we select for observation. Thus, if we take a long hair out of this Sable tippet, and examine it near the base, we see that it is very slender, transparent, and colourless, covered with strongly-marked imbrications, which are not obtuse teeth, but long, pointed, overlapping scales, about ten of which complete a whorl. The fibrous portion is moderately thick; inclosing a wide pith of roundish cells, set in two rows, that allow the rays of light to be transmitted through their central parts.

As we trace the hair upwards, by moving the stage of the microscope, by and by it swells and rapidly increases in thickness; the imbrications are scarcely perceptible; while the pith-cells have greatly augmented in number and in breadth. These are arranged in confused, close-set, transverse rows, and are nearly opaque.

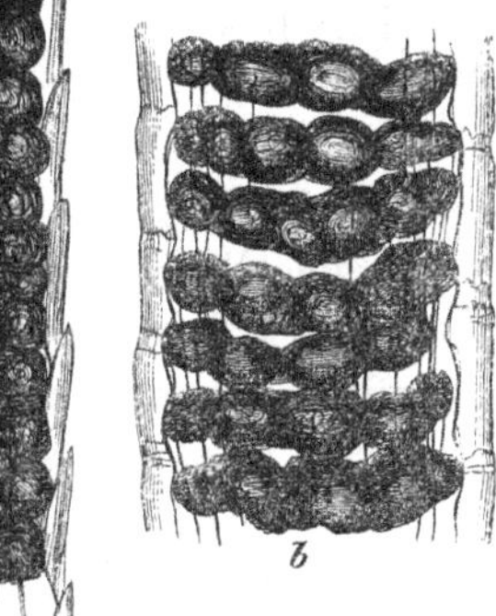

HAIR OF SABLE.

Still tracing up the same hair, as we approach the tip, the bark and fibrous part become very thin; the cells are fewer and fewer till they cease altogether, and a long slender point, of a clear yellow tinge, without cells, presents transverse wavy lines of imbrication scarcely projecting.

The hair of the common Mouse is a pretty and interesting object. In the larger specimens the fibrous portion is reduced almost to nothing. The imbrications project very little, but careful observation reveals slanting lines proceeding from the "teeth;" which show that the whole surface is clothed with large pointed scales, which are excessively thin, and lie close. The pith consists of large flattened cells, arranged thus; one row passes up through the centre, and other similar

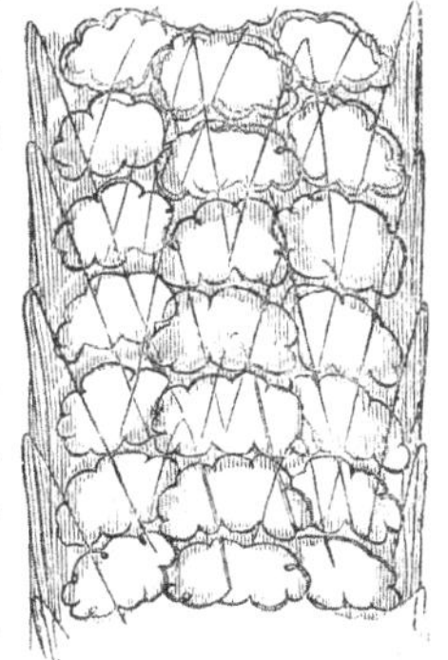
HAIR OF MOUSE.

ones are set in a circle around it, so that a longitudinal section would show three parallel rows. These cells are translucent, and some of them are either wholly or partially lined with a clear yellow pigment.

The smaller hairs from the same little animal are scarcely distinguishable from those of the Cat, already described, except that the imbrications are proportionally larger. In all the extremity is drawn out to a lengthened fine point, and is occupied with clear yellow cells, except the very tip, which is colourless, and imbricated with sinuous whorls, each consisting of a single scale.

But it is in the Bats that the imbricated character attains its greatest development. On this slide is a number of hairs from the fur of one of our English Bats, in which it is far more conspicuous than in any example we have yet seen. In the middle portion of each hair the scales lie close, embracing their successors to the very edges, or nearly; but the lower part, which is more slender, resembles a multitude of trumpet-shaped flowers formed into a chain, each being inserted into the throat of another. The lip of the "flower" is generally oblique, and here and there we can perceive that each is formed of two half-encircling scales; for one scale occasionally springs from the level of its fellow, so as to make the imbrication alternate.

TIP OF SMALL HAIR OF MOUSE.

HAIR OF BAT.

Even this, however, is far excelled by a species of Bat from India, of whose hair I have now specimens on the stage. The trumpet-like cups are here very thin and transparent, but very expansive; the diameter of the lip being, in some parts of the hair, fully thrice as great as that of the stem itself. The margin of each cup appears to be undivided, but very irregularly notched and cut. In the middle portion of the hair, the cups are far more crowded than in the basal part, more brush-like, and less elegant; and this structure is continued to the very extremity, which is not drawn out to so attenuated a point as the hair of the Mouse, though it is of a needle-like sharpness. The trumpet-shaped scales are, it seems, liable to be removed by accident; for in these dozen hairs there are several, in which we see one or more cups rubbed off, and in one the stem is destitute of them for a considerable space. The stem so denuded closely resembles the basal part of a Mouse's hair in its normal condition.

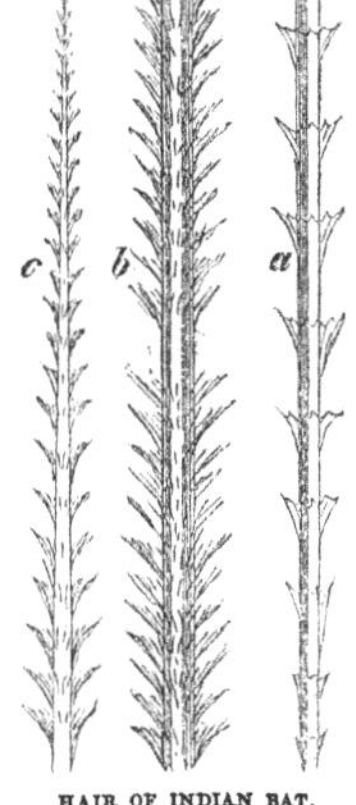

HAIR OF INDIAN BAT.

This character of being clothed with overlapping scales, each growing out of its predecessor, is common, then, to the hairs of the Mammalia, though it exists in different degrees of development. It may be readily detected by the unaided sense, even when the eye, though assisted by the microscope, fails to recognise it. Almost every schoolboy is familiar with the mode by which the tip of any hair may be distinguished from its base; and even of the least fragment, the terminal end from the basal end. The initiated lad assembles a few younger ones, and says, "Now you may make a mark with ink on one end of a white horse-hair, and I'll tell you, by feeling it, which end you have marked." He does, infallibly. He rubs it to and fro between his thumb and finger, and the hair regularly travels through in the direction of its base: one or two rubs of course determine this, and the verdict is given oracularly. Now you see the cause of this property lies in the imbricate structure; the scales may be excessively thin and close, but still they project sufficiently in any specimen to present a barrier to motion in the terminal direction when pressed between two surfaces, such as the fingers, while they very readily move in the opposite.

But more than the success of a schoolboy's magic depends on the imbricate surface of hairs. England's time-honoured manufacture, that which affords the highest seat in her most august assembly, depends on it. The hat on your head, the coat on your back, the flannel waistcoat that shields your chest, the double hose that comfort your ankles, the carpet under your feet, and hundreds of other necessaries of life, are what they are, because mammalian hairs are covered with sheathing scales.

It is owing to this structure that those hairs which possess it in an appreciable degree, are endowed with the property of *felting;* that is, of being, especially under the combined action of heat, moisture, motion, and pressure, so interlaced and entangled as to become inseparable, and of gradually forming a dense and cloth-like texture. The "body," or substance of the best sort of men's hats, is made of lamb's wool and rabbit's fur, not interwoven, but simply beaten, pressed, and worked together, between damp cloths. The same property enables woven woollen tissues to become close and thick: every one knows that worsted stockings shrink in their dimensions, but become much thicker and firmer after they have been worn and washed a little; and the "stout broad-cloth," which has been the characteristic covering of Englishmen for ages, would be but a poor open flimsy texture, but for the intimate union of the felted wool-fibres, which accrues from the various processes to which the fabric has been subjected.

In a commercial view, the excellence of wool is tested by the closeness of its imbrications. When first the wool-fibre was submitted to microscopical examination, the experiment was made on a specimen of Merino; it presented 2,400 serratures in an inch. Then a fibre of Saxon wool, finer than the former, and known to possess a superior felting power, was tried: there were 2,720 serratures in an inch. Next a specimen of South-Down wool, acknowledged to be inferior to either of the former, was examimed, and gave 2,080 serratures. Finally, the Leicester wool, whose felting property is feebler still, yielded only 1850 serratures per inch. And this connexion of good felting quality with the number and sharpness of the sheathing scales, is found to be invariable.

The hairs of many Insects are curious and interesting. Here you may see the head of the hive-bee, which is moderately clothed with hair; each hair is slender and pointed, and is beset with a multitude of subordinate short hairs, which project from the main stem, and stand out at an angle: these are set on in a spiral order. Here, again, is one of the hinder legs of the same bee: the yellow hair, which you can see with the naked eye, consists of strong, horny, curved spines, each of which is scored obliquely, like a butcher's steel. These legs are used, as you are well aware, to brush off the pollen from the anthers of flowers, wherewith the substance called bee-bread, the food of the grubs, is made; and in this specimen, you may see hundreds of the beautiful oval pollen-grains entangled among these formidable-looking spines.

These rusty hairs are from a large caterpillar (that of the Oak Egger Moth, I believe); they appear, when highly magnified, like stout horny rods drawn out to an acute point, and sending forth alternate short pointed spines, which scarcely project from the line of the axis.

TIP OF HAIR OF DERMESTES.

But there is scarcely any hair more curious than that of a troublesome grub in museums and cabinets, the larva of *Dermestes lardarius*, which lives upon furskins, and any dried animal substances. It has a cylindrical shaft, which is covered with whorls of large close-set spines, four or five in each whorl, closely succeeding each other; the upper part of the shaft is surrounded by a whorl of larger and more knotted spines, and the extremity is furnished with six or seven large filaments, which appear to have a knob-like hinge in the middle, by which they are bent up on themselves.

The feathers of Birds are essentially hairs. That

shrivelled membrane which we pull out of the interior of a quill when we make a pen, is the medullary portion, dried. There is a beautiful contrivance in the barbs of most feathers, which I will illustrate by this feather from the body-plumage of the domestic fowl. Every one must have observed the regular arrangement of the vane of a feather, and the exquisite manner in which the beards of which it is composed are connected together. This is specially observable in the wing-feathers,—a goose-quill, for example; where the vane, though very light and thin, forms an exceedingly firm resisting medium, the individual beards maintaining their union with great tenacity, and resuming it immediately, when they have been violently separated.

BARB OF CLOTHING-FEATHER OF FOWL.

Now this property is of high importance in the economy of the bird. It is essential that with great lightness and buoyancy—for the bird is a *flying* creature — there be power to strike the air with a broad resisting surface. The wide vanes of the quill-feathers afford these two requisites, strength and lightness; the latter depending on the material employed, which is very cellular, and the former on the mode in which the individual barbs, set edgewise to the direction of the stroke, take a firm hold on each other.

Now, in the body-feather which is under the microscope, we see that the central stem carries on each side a row of barbs, which interlock with each other. The magnifying power shows us that these barbs are not simple filaments, but are themselves doubly bearded in the same fashion; and further, that these barbules of the second series are furnished with a third series. It is in this third series of filaments that the tenacity in question resides. If we isolate one of the primary beards, by stripping away a few on each side of it, and again put it on the stage, we see that the secondary barbules of one side are armed differently from those of the other side. Those of the lower side carry short and simple barbulets, whereas those of the side which looks towards the point of the feather bear much longer ones; and, moreover, many of them are abruptly hooked backwards. Now, whenever the primary beards are brought into contact, some of these hooks catch on the barbule next above, and, slipping into the angles formed by the barbulets, hold there, and thus the two contiguous beards are firmly locked together.

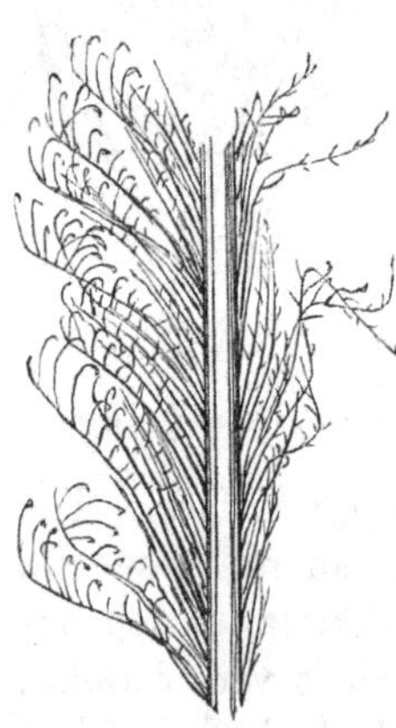
BARB FROM GOOSE QUILL.

In the beard of the goose-quill, the structure is essentially the same, but the barbulets are far more numerous and more closely set; they are also proportionally much larger,—both those which are hooked and those which are simple. Indeed, the latter manifest a tendency to the hooked form, and by all these peculiarities the interlocking power is augmented. It is interesting to observe the great dilatation of the beard in a direction towards the inferior surface of the feather—towards the stroke, as I just now observed. This is to increase the resisting power, as a thin board set edgewise will bear a great weight without bending or breaking, provided it can be kept from yielding laterally. The barbules are arranged only on the very edge—the upper edge—of the beard.

We will now examine some specimens of scales of Fishes, all of which are very interesting and beautiful objects under low powers of the microscope; though higher powers are requisite to resolve their structure. We will use both.

The scales of almost all the Fishes with which we are familiar, fall under two kinds, which have been named *ctenoid* (or comb-like), and *cycloid* (or roundish). The Perch affords us good examples of the former kind. On this slide are three scales from the body of this fish: the one on the left side is taken from the back (fig. *a*); the middle one from the lateral line (*b*); and the one on the right from the belly (*c*). In order to understand these objects, we must remember that the scales of fishes are horny or bony plates, developed in the substance of the proper skin, with a layer of which they are always covered. In most cases (as, for example, the Perch), the hinder end of each scale projects, carrying with it the thin layer of skin with which it is invested; and thus the scales overlay one another, like the tiles of a house, or like the feathers of a bird, and that for a like purpose. For as the rain, falling on the house-top, has a tendency to flow downwards, from gravitation; and as the slope of the roof is in that direction, the current passing over each tile is deposited from its bottom-edge on the middle of the next one, whence it still flows down to the free edge of this one, and so in succession. So the motion of the bird through the air, and of the fish through the water, produces the very same effect as if these fluids were in motion and the animals were still; and therefore the bodies of the latter are, as it were, *tiled* with feathers or scales, the free edges of which, looking in the opposite direction to the coming of the current (that is, the same direction as its flow), deposit the successive particles of the moving fluid in the midst of the successive feathers or scales. Thus two results ensue, both essential to the comfort of the animal: first, the air or water does not run upward between the feathers or scales to the skin; and secondly, the surface presents no impediment to free motion. This latter advantage will be appreci-

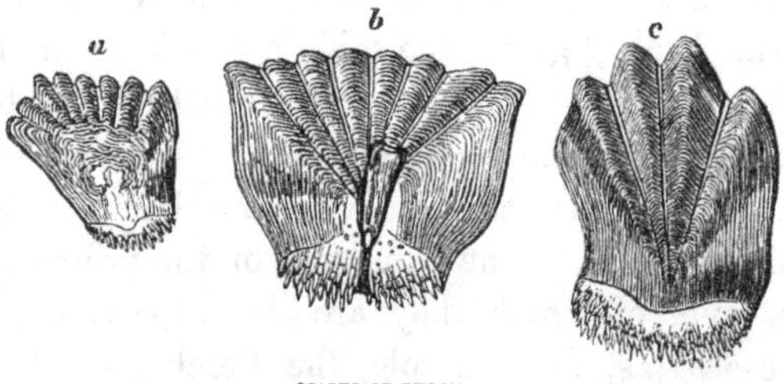

SCALES OF PERCH.

ated, if you take hold of a dead bird by the legs, and push it rapidly through the air tail-foremost: the feathers will instantly rise and ruffle up, presenting a powerful resistance to movement in that direction.

These scales of the Perch have their hinder, or free edge, set with fine crystalline points, arranged in successive rows, and overlapping. Their front side is cut with a scolloped pattern, the extremities of undulations of the surface that radiate from a common point behind the centre. These undulations are separated by narrow furrows, across which, contrary to the ordinary rule, the close-set concentric lines that follow the sinuosities of the outline are not visible. Under the microscope they look as if they had been split in these radiating lines, after the whole number of layers had been completed, and the fissures had then been filled with new transparent substance.

The middle scale is, as I have said, from the lateral line. Along each side, in most fishes, may be observed a line, known as the lateral line, formed by scales of peculiar form. They are commonly more bony than the other scales, and are pierced by a tubular orifice for the escape (as is generally supposed, though this has been denied) of a mucous secretion, which is poured out from glands beneath, and thus flows over the body for the double purpose of protecting the skin from the macerating influence of the surrounding water, and of diminishing friction in swimming.

Let us now look at some scales of the cycloid kind. The great majority of our fishes are clothed with such as are of this description. This dead Gold-fish shall give us examples. The three scales in the upper row are from the lateral line, the left-hand one (*a*) taken just behind the head, the second (*b*) near the middle of the body, and the right-hand one (*c*) near the tail. Of the lower row, the first (*d*) is from the back, the second (*e*) from the middle of the belly, and the last (*f*) from the throat. Thus we see there is considerable variety in form presented by the scales even of the same individual fish. They all, however, differ

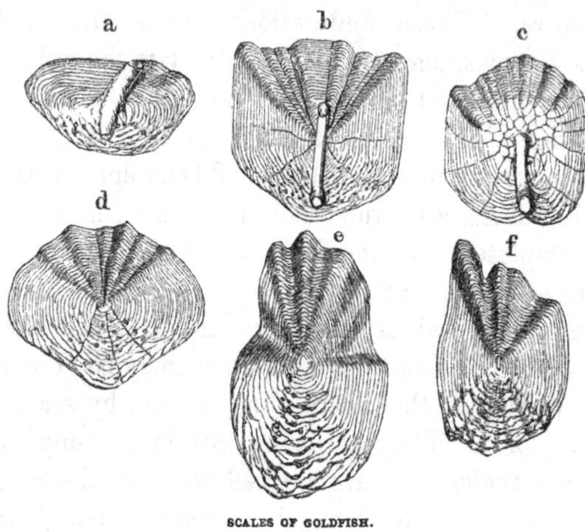

SCALES OF GOLDFISH.

from those of the Perch, in this respect;—that their free overlapping edges are entire, or destitute of the crystalline points which we saw in the former examples; while they agree in having the front edges, by which they are during life imbedded in the skin, cut into waves or sinuosities. The lower part, as we now look at them, is the free portion of each, which alone is visible in the living fish, the other parts being concealed by the three neighbouring scales that overlap it,—above, in front, and below.

In those from the lateral line, the tube already referred to is seen to pervade each, running through it longitudinally, so that it opens posteriorly on the outer surface, and anteriorly on the inner or under surface of the scale. In the scales near the front of the line, just behind the head, the tube is large and prominent (*a*), while in the scales at the opposite extremity it becomes slender; diminishing, in the very last scale, viz. at the commencement of the tail-fin, to a mere groove.

The whole surface of each scale, when viewed under a lens of low power, is seen to be covered with concentric lines, following the irregular sinuosities of the outline. These lines are the edges of the successive layers of which the scale is believed to be composed, each layer being added in the process of growth to the under surface, and each being a little larger every way than its predecessor; thus the scale is a very depressed cone, of which the centre is the apex. There is a marked difference (indicated in the figures) between that part of the surface which is exposed, and that which is covered by the other scales; the concentric marks in the former are much coarser and less regular, often being interrupted, and seeming to run into each other, and frequently swelling into oval scars. This may, perhaps, be owing to the surface having been partially worn down by rubbing against the gravel of the bottom, or against other objects in the water. Besides the concentric lines, there are seen on many of the scales, especially those of the lateral line, radiating lines varying in number from one to twenty, or more, diverging from the centre towards the circumference, and frequently connected by cross lines forming a sort of net-work around the centre (see *c*). Under the microscope, these lines appear to be elevated ridges, dividing the concentric lines; but of their use I am ignorant.

What I have just stated is the ordinary explanation of these fine concentric lines; but a careful examination of the structure with much higher powers than we have been using, induces me to doubt its correctness. Reverting to the scales of the Perch, let us notice the clear diverging bands, which look as if the whole scale had been split in several places, and the openings thus made filled with uniform clear substance. The same structure is seen in many other scales, as in this *cycloid* one from the Flounder, which, being coarsely lined, shows the structure well; or in these from the Green Wrasse. I will now apply to one of these a power of 600 diameters, concentrating the light thrown through the scale from the mirror by the achromatic condenser, and examine the scale anew. You now see two distinct layers; the upper one which bears the concentric lines, and a lower clear one which not only fills the radiating bands, but underlies the whole of the lined parts. The concentric lines of the upper layer do not now appear to be edges of successive plates, but irregular

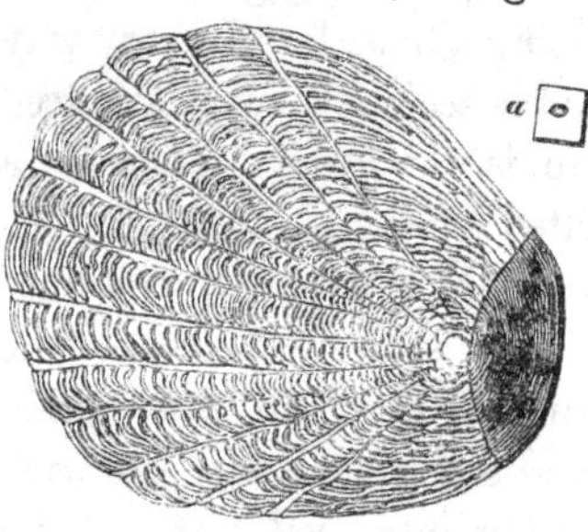

SCALE OF FLOUNDER.
a. Natural size.

canals running through the solid substance. This, however, is illusory: for, by delicate focussing, we perceive that each portion marked by these lines is really in a different plane from the others, that the highest is at the centre of radiation of the scale, and that each is successively lower till we reach the margin. But now, if with very sharp scissors we cut one of these scales longitudinally through the centre, and examine the cut edge, we find that each of these lines forms a distinct ridge. On the other hand, the under layer of clear substance is quite smooth, and always a little exceeds the margin of the concentrically lined portion. The clear substance that fills the radiating slits agrees both in texture and level with this lower layer, and is manifestly continuous with it.

Hence, I think that, in these slit scales, the upper layer is formed, as commonly believed, by successive deposits from beneath; but that, after a few have been deposited, they begin to slit, probably by contraction in becoming solid; that the lower layer is formed after each upper one is hardened, exceeding its length by a little, and filling up the slit; that this lower layer becomes the upper layer of the next course, slitting, and turning up its terminal edge as it hardens; that then the lower layer is deposited on this, filling up the slit as before; and that this process goes on as long as the fish lives.

It is curious that, in the scales of the Pike, the portions thus separated by slitting, instead of expanding and leaving spaces to be filled up, actually close over each other, the divided parts overlapping considerably, as you may see in these specimens,

The left hand scale (*a*) is from the back; the central one (*b*), which has only a deep narrow incision instead

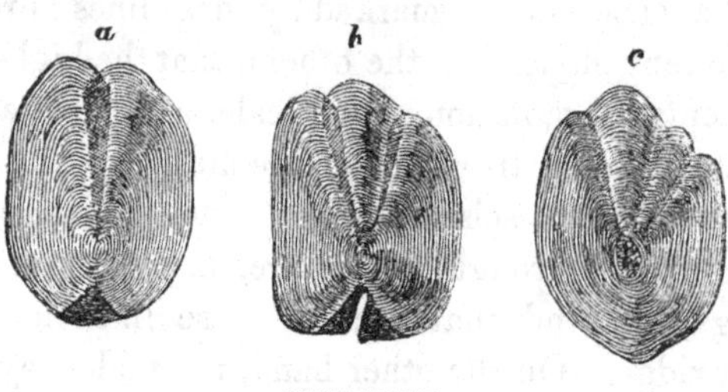

SCALES OF PIKE

of a tube, is from the lateral line; and the third (*c*) is from the belly of the fish.

Let us return now to the scales of our Gold-fish, and examine a highly interesting structure connected with them. The brilliant golden or silvery reflection that constitutes the beauty of these lovely fishes, depends not on the scales themselves, but on a soft layer of pigment spread over their inner surface, and seen through their translucent substance. On carefully detaching a scale, we see on the under side, opposite to that portion only which was exposed (all the concealed parts being colourless), a layer of soft gleaming substance, easily separable, either silvery or golden, according to the hue of the fish. If now we remove a small portion of this substance with a fine needle, and spread it on a plate of thin glass, we shall find, by the aid of the microscope, that it consists of two distinct substances; the one giving the colour, the other the metallic lustre. With a power of 300 diameters, the former is seen to be a layer of loose membranous cells of an orange colour, in what are properly called the Gold-fishes, and whitish or pellucid in the Silver-fishes. If we now add a minute drop of water to the mass, and gently agitate it with the point of a needle, and again submit it to the microscope, we shall have a beautiful and interesting spectacle. The water around the mass is seen to be full of an infinite number of flat spicula or crystals, varying much in size, but of very constant form, a flat oblong prism with angular ends (as represented in the accompanying engraving). By transmitted light they are so transparent and filmy as to be only just discernible; but by reflected light, and especially under the sun's rays, they flash like plates of polished steel. But what appears most singular, is that each spiculum is perpetually vibrating and quivering with a motion apparently quite spontaneous, but probably to be referred to slight vibrations of the water in which they float; and each independently of the rest, so as to convey the impression to the observer that each is animated with life, though the scale be taken from a fish some days dead. Owing to this irregular motion, and consequent change of position, each spiculum, as it assumes or leaves the reflecting angle, is momentarily brightening or waning, flashing out or retiring into darkness, producing a magic effect on the admiring observer. To this property, I suppose, is to be attributed the beautiful pearly play of light that marks these lovely fishes, as distinguished from the light reflected by an uniformly polished surface.

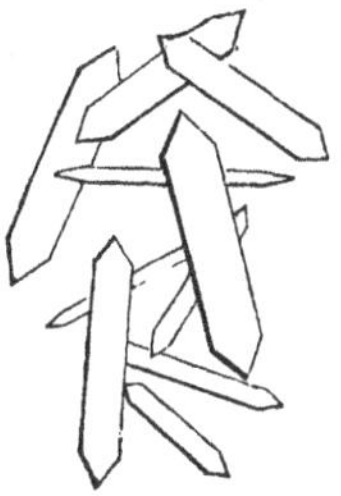

SPICULA OF GOLDFISH'S SCALE.

I have found the pearly pigment of the scales to be provided with similar spicula in fishes widely differing in size, structure, and habits; as the Gudgeon and Minnow, the Pike and the Marine Bream. The spicula of these fishes agree in general form with those of the Gold-fish; and also in size, with the exception of trifling variations in the comparative length and breadth. The colouring matter is lodged in lengthened cylindrical cells, arranged side by side, and running across the scale; that is, in a direction at right angles to the lateral line.

CHAPTER II.

BLOOD.

THE microscope is daily becoming a more and more important aid to legal investigation. An illustration of this occurred not long ago, in which a murder was brought home to the criminal by means of this instrument. Much circumstantial evidence had been adduced against him, among which was the fact, that a knife in his possession was smeared with blood, which had dried both on the blade and on the handle. The prisoner strove to turn aside the force of this circumstance by asserting that he had cut some raw beef with the knife, and had omitted to wipe it.

The knife was submitted to an eminent professor of microscopy, who immediately discovered the following facts:—1. The stain was certainly blood. 2. It was not the blood of a piece of dead flesh, but that of a living body; for it had coagulated where it was found. 3. It was not the blood of an ox, sheep, or hog. 4. It was human blood. Besides these facts, however, other important ones were revealed by the same mode of investigation. 5. Among the blood were found some vegetable fibres. 6. These were proved to be *cotton* fibres,—agreeing with those of the murdered man's shirt and neck-kerchief. 7. There were present also numerous tesselated epithelial cells. In order to understand the meaning and the bearing of this last fact, I must explain that the whole of the internal surface of the body is lined with a delicate membrane (a continuation of the external skin), which discharges mucus, and is hence termed mucous membrane. Now this is composed of loose cells, which very easily separate, called epithelial cells; they are in fact constantly in process of being detached (in which state they constitute the mucus), and of being replaced from the tissues beneath. Now microscopical anatomists have learned that these epithelial scales or cells, which are so minute as to be undiscernible by the unaided eye, differ in appearance and arrangement in different parts of the body. Thus, those which line the gullet and the lower part of the throat are *tesselated*, or resemble the stones of a pavement; those that cover the root of the tongue are arranged in cylinders or tall cones, and are known as *columnar;* while those that line some of the viscera of the abdomen carry little waving hairs (*cilia*) at their tips, and are known as ciliated epithelium.

The result of the investigation left no doubt remaining that with that knife the *throat* of a *living human* being, which throat had been protected by some *cotton* fabric, had been cut. The accumulation of evidence was fatal to the prisoner, who without the microscopic testimony might have escaped.

But what was there in the dried brown stain that determined it to be blood? And, particularly, how was it proved not to be the blood of an ox, as the prisoner averred? To these points we will now give a moment's attention.

With this fine needle I make a minute prick through the skin of my hand. A drop of blood oozes out, with which I smear this slip of glass. The slip is now on the stage of the instrument, with a power of 600 diameters. You see an infinite number of small roundish bodies, of a clear yellowish colour, floating in a colourless fluid, but so numerous, that it is only here and there, as near the edges of the smear, that you can detect any interval in their continuity.

These bodies are what we frequently call the blood-globules, or, more correctly, blood-*disks;* since their form is not globular, but thin and flat, like a piece of money. The slightness of their colour is dependent on their extreme tenuity: when a larger number lie over each other the aggregated colour is very manifest, as it then becomes either a full dark red, or bright rich scarlet; for to these disks blood is entirely indebted for its well-known hue. All vertebrate blood is composed principally of these bodies, which when once seen are easily recognised again: the microscope then readily determines whether any given red fluid or dried stain is composed of blood.

The disks in the blood of Mammalia are circular, or nearly so, and slightly concave on both of the surfaces. On the other hand, in Birds, Fishes, and Reptiles their form is elliptical, and the surfaces are flat, or slightly convex. This distinction, then, will at once enable us to determine Mammalian blood.* But to determine the various tribes of this great class

* The Camels among Mammalia, and the Lampreys among Fishes, are exceptions to the above rule; the former having elliptical and convex blood-disks, the latter circular, and slightly concave.

among themselves, we must have recourse to another criterion,—that of dimensions.

The blood-disks of Man nearly agree in size with those of the Monkey tribe, of the Seals and Whales, of the Elephant, and of the Kangaroo. Most other quadrupeds have them smaller than in Man; the smallest of all being found in the ruminating animals. The little Musk-deer of Java has disks not more than one-fourth as large as the human, but these are remarkably minute; no other known animal approaches it in this respect: those of the Ox are about three-fourths, and those of the Sheep little more than half the human average.

Tables have been made out showing the comparative size of these corpuscles in various animals, and such tables are very useful; but we must bear in mind that the average dimensions only are to be looked for; since in any given quantity of blood, under examination, we shall not fail to see that some disks exceed, while others come short of, the dimensions of the majority.

Generally speaking, the blood-disks in Birds and in Fishes are about equal in size: their form is, however, that of a more elongated ellipse in Birds than in Fishes. They may be set down as averaging in breadth the diameter of the human disks, while their length is about half as much again, or a little more, in most Birds.

It is in Reptiles that we meet with the largest disks, and especially in those naked-skinned species, the Frogs and Newts. A large species inhabiting the American lakes—*Siren lacertina*—has disks of the extraordinary size of 1-400th of an inch long by 1-800th broad, or about eight times as large as those of Man. Our common Newts afford us the largest examples among British animals, but they do not reach above half the size just mentioned.

Taking this drop of blood from my finger as a standard of comparison, we find, on applying the micrometer, that the disks run from 1-2500th to 1-5000th of an inch; but that the great majority are about 1-3300th in diameter. On these slides are samples of other kinds. This is the blood of a Fish, —the common Blenny or Shanny (*Blennius pholis*). Here we see at once the oval form of the disks; their average is 1-2800th by 1-3300th of an inch. Here is the blood of a Frog (*Rana temporaria*); these are more than twice the size of the fish's; for they average 1-1250th by 1-1800th of an inch. And, finally, I can show you a drop of blood from this Smooth-newt (*Lissotriton punctatus*). The large size of the disks is

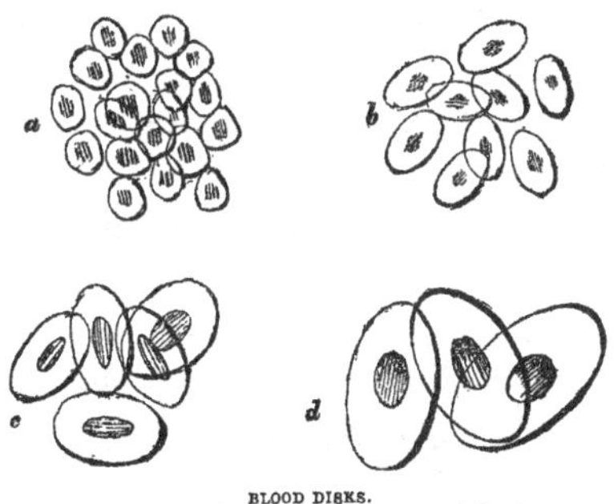

BLOOD DISKS.
a Man. *b* Blenny. *c* Frog. *d* Newt.

now conspicuous, and so indeed is the elegance of their form: in this case, as in the last, we see in each disk a distinct roundish nucleus. These run from 1-700th to 1-950th in length, by 1-1100th to 1-1600th in breadth; but the average are about 1-800th by 1-1300th of an inch.

It may interest you to see these blood-disks in their proper situation, and to observe the motion which they possess during the life of their owners. It is, indeed, one of the most instructive modes of using this wonder-working instrument to look through it at living structures, and watch the different processes of life as they are carried on under our eyes. Nor is this at all difficult to accomplish; for a large number of animals are so small that we can easily put them upon the stage of the microscope, and withal so transparent that their integuments and various tissues offer little or no impediment to our discerning the forms and movements of the contained viscera. And in cases where the entire animal is too large to be viewed microscopically as a whole, it sometimes happens that, by a little contrivance, we can so secure the creature as to look without interruption on certain parts of the body which afford the requisite minuteness and transparency.

I have here a living Frog. You perceive that the web which connects the toes is exceedingly thin and translucent, yet arteries and veins meander through its delicate tissues, which are then clothed on both surfaces with the common skin. But you ask how we can induce the Frog to be so polite as to hold his paw up and keep it steady for our scientific investigation. We will manage that without difficulty.

Most microscopes are furnished (among their accessory apparatus) with what is called a frog-plate, provided for this very demonstration. Here is mine. It is a thin plate of brass, two inches and a half broad and seven long, with a number of small holes pierced through it along the margins, and a large orifice near one end, which is covered with a plate of glass. This is to be Froggy's bed during the operation, for we must make him as comfortable as circumstances will admit.

Well, then, we take this strip of linen, damp it, and proceed to wrap up our unconscious subject. When we have passed two or three folds round him, we pass a tape round the whole, with just sufficient tightness to keep him from struggling. One hind-leg must project from the linen, and we now pass a needle of thread twice or thrice through the drapery and round the small of this free leg, so as to prevent him from retracting it.

Here then he lies, swathed like a mummy, with one little cold foot protruded. Lay him carefully on the brass plate, so that the webbed toes shall stretch across the glass. Now, then, we pass another tape through the marginal holes, and over the body, to bind it to the brass; of course taking care not to cut the animal, but only using just as much force as is needful to prevent his wrigglings. Now a bit of thread round each toe, with which we tie it to as many of the holes, so as to expand the web across the glass. A drop of cold water now upon the swathes to keep him cool, and a touch of the same with a feather upon the toes to prevent them from drying (which must be repeated at intervals during the examination),—and he is ready.

What a striking spectacle is now presented to us, as with a power of 300 diameters we gaze on the web of the foot! There is an area of clear colourless tissue filling the field, marked all over with delicate angular lines, something like scales; this is the tesselated epithelium of the surface. Our attention is caught by a number of black spots, often taking fantastic forms, but generally somewhat star-like: these are pigment cells, on which the colour of the animal's skin is dependent. But the most prominent feature is the blood. Wide rivers, with tortuous course, roll across the area, with many smaller streams meandering among them; some pursuing an independent course below the larger, and others branching out of them, or joining them at different angles. The larger rivers are of a deep orange-red hue, the smaller faintly tinged with reddish-yellow. In some of these channels the stream rolls with a majestic evenness; in others it shoots along with headlong impetuosity; and in some it is almost, or even quite, stagnant. By looking with a steady gaze, we see that in all cases the stream is made up of a multitude of thin reddish disks, of exactly the same dimensions and appearance as those we saw just now in the Frog's blood; only that here, being in motion, we see very distinctly, as they are rolled over each other, that they *are* disks, and not spherules; for they forcibly remind us of counters, such as are used for play, supposing they were made out of pale red glass.

It is charming to watch one of these streams, selecting one of medium size, where the density is not too great to see the individual disks, and fixing our eye on the point where a branch issues from one side of the channel, mark the disks shoot by one after another, some pursuing their main course, and others turning aside into the branch, perhaps so small as to allow of only a single disk to pass at once.

The streams do not pursue the same uniform direction. The larger ones do indeed; and their course is from the extremity of the toes towards the body: these are the veins; but the smaller streamlets flow in any direction, and frequently send out side-branches, which presently return into the stream from which they issued, or unite with others in a very irregular network. These are the capillaries, which feed the veins, and which are themselves fed by the arteries, whose course is in the opposite direction, viz. from the body. These, however, are with difficulty seen: they are more deeply seated in the tissues, and are less spread over the webs, being generally placed along the borders of the toes; they are, moreover, fewer and smaller than the veins; but the blood in them usually flows with more impetuous rapidity.

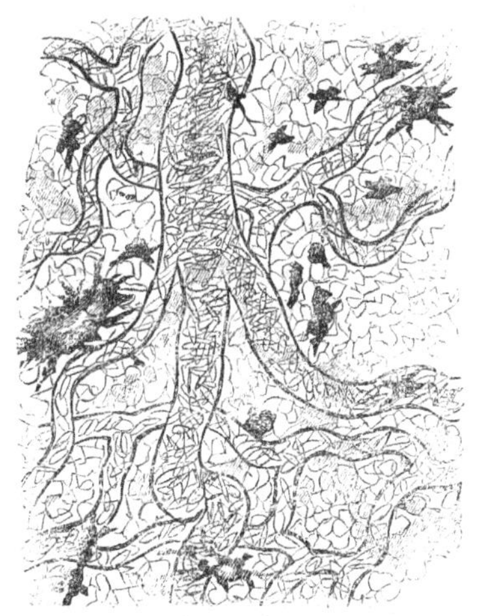

CIRCULATION IN FROG'S FOOT.

The variations in the impetus of the current which we observe in the same vessel are probably owing to the mental emotions of the animal; alarm at its unusual position, and at the confinement which it feels when it endeavours to move, may suspend the action of the heart, and thus cause an interruption in the flow; or analogous emotions may quicken the pulse. We will, however, now release our little prisoner, who, though glad to be at liberty, is, as you see, none the worse for his temporary imprisonment.

Let us now look at the circulation of the blood in one of the Invertebrate Animals. In this thin glass cell of sea-water is a small fragment of sea-weed, and attached to one of its slender filaments you may see three or four tiny knobs of jelly, clustered together like a bunch of grapes. These are animals; each endowed with a distinct life, but associated together by a common stalk, which maintains the mutual vital connexion of the whole. It is one of the Social Tunicata, and is named *Perophora Listeri*.

Though each globose knob is no larger than a small pin's head, it is full of organs which carry on the various functions of life; and, because the whole tissues are as transparent as crystal, they allow us to watch the processes with perfect ease. Take a peep at it.

It is a gelatinous sac, of a form intermediate between globular and cubical, flattened on two opposite sides, with a sort of wart at the summit and another at the side, each of which is pierced with a pursed orifice. The upper of these orifices admits water for respiration and food; the latter passes through a digestive system, and is discharged through the side orifice. The digestive organs lie on that flattened side, which is farthest from your eye, and are therefore dimly seen.

The globose body is inclosed in a coating of loose shapeless jelly, that passes off from one of the lower corners, and forms a short foot-stalk, which unites with similar foot-stalks from the sister-globules, and all together are attached to the sea-weed. Each foot-stalk has an organic core, into which a vessel passes from the body.

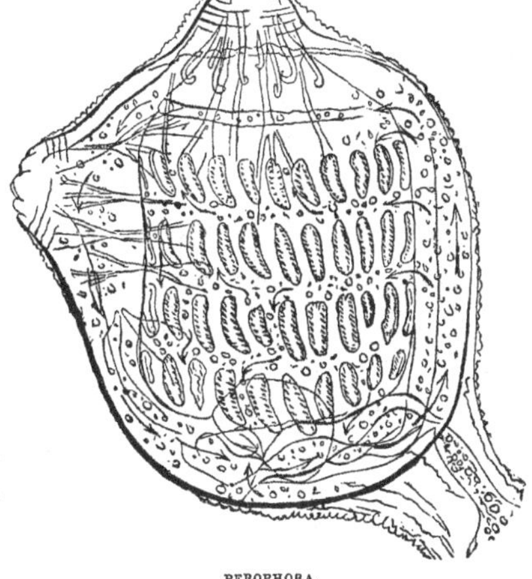

PEROPHORA.

Your attention is first arrested by the breathing sac, with its rows of oblong cells all in wheel-like motion. It is indeed a wonderful object; but for the present neglect this, as we will return to it presently, and direct your consideration to the course of the blood.

It is true the fluid which I so name is not red, like that of the Frog which you have just been gazing at, nor does it carry disks of the same elegantly regular form. But you have the advantage here of tracing, at one view, the whole course of the circulation, from its first rush out of the heart to its return into that organ again.

At the bottom of the interior, below the breathing sac, there is an oblong cavity, through whose centre there runs a long transparent vessel, formed of a delicate membrane, the appearance of which resembles that of a long bag, pointed (but not closed) at either end, and then twisted in some unintelligible manner so as to make three turns. This is the heart; and within it are seen many minute colourless globules, floating freely in a subtle fluid: this is the nutrient juice of the body, which we may, without much violence, designate the blood. Now see the circulation of this fluid. The membranous bag gives a spasmodic contraction at one end, and drives forward the globules contained there; the contraction in an instant passes onward along the three twists of the heart (the partbehind expanding immediately as the action passes on), and the globules are forcibly expelled through the narrow but open extremity. Meanwhile, globules from around the other end have rushed in as soon as that part resumed its usual width, which in turn are driven forward by a periodic repetition of the *systole* and *diastole*.

The globules thus periodically driven forth from the heart now let us watch, and see what becomes of them. They do not appear to pass into any defined system of vessels that we may call arteries, but to find their way through the interstices of the various organs in the general cavity of the body.

The greater number of globules pass immediately from the heart through a vessel into the short footstalk, where they accumulate in a large reservoir; but the rest pass up along the side of the body, which (in the aspect in which we are looking at it) is the right. As they proceed (by jerks, of course, impelled by the contractions of the heart), some find their way into the space between the breathing surfaces, through narrow slits along the edges of the sac, and wind along between the oval ciliary wheels, which we will presently consider. Besides these, however, other globules wind along between the outer surfaces of the sac and the inner surface of the body-walls.

But to return to the current which passes up the right side: arriving at the upper angle of the body, the stream turns off to the left abruptly, principally passing along a fold or groove in the exterior of the breathing-sac until it reaches the left side, down which it passes, and along the bottom, until it arrives at the entrance of the heart, and rushes in to fill the vacuum produced by the expansion of its walls after the periodic contraction. This is the perfect circle; but the minor streams that had forked off sideways in the course, as those within the sac for example, find their way to the entrance of the heart by shorter and more irregular courses.

One or two things connected with this circulatory system are worthy of special notice. The first is, that its direction is not constant but reversible. After we have watched this course followed with regularity for perhaps a hundred pulsations or so, all of a sudden the heart ceases to beat, and all the globules rest in their circling course, that we had supposed incessant. Strange to behold, after a pause of two or three seconds, the pulsation begins again, but at the opposite end of the heart, and proceeds with perfect regularity, just as before, but in the opposite direction. The globules, of course, obey the new impulse, enter at their former exit, and pass out at their former entrance, and perform their circulation in every respect the same as before, but in the reverse direction.

Those globules that pass through the vessel into the foot-stalk appear to accumulate there as in a reservoir, until the course is changed, when they crowd into the heart again and perform their grand tour. Yet there is a measure of circulation here, for even in the connecting vessel one stream ascends from the reservoir into the body as the other (and principal one) descends into it from the heart; and so, *vice versâ.*

I have spoken of these motions as being performed with regularity; but, if you look closely, you will see that this must be understood with some qualification. The pulsations are not quite uniform, being sometimes more languid, sometimes more vigorous; perhaps forty beats in a minute may be the average; but I have counted sixty, and presently after thirty; I have counted twenty beats in one half-minnte, and only fifteen in the next. The period during which one course continues is equally uncertain; but about two minutes may be the usual time. Sometimes the pulsation intermits for a second or so, and then goes on in the same direction; and sometimes there is a curious variation in the heart's action,—a faint and then a strong beat, a faint and a strong one, and so alternately for some time.

The phenomena of respiration are so closely connected with those of circulation that it is not at all *mal-apropos* to turn from the latter to the former; not to say that it would be high treason against scientific curiosity if I were to remove this object without explaining to you that marvellous play of wheels that occupies the largest part of the area that you behold. As you look on the globe, you observe, hanging down from the upper extremity, and reaching nearly to the bottom in one direction and almost from side to side in another, a transparent square veil, which is indeed a flat membranous bag, having its sides pretty close together, with small openings along its edges, and an orifice at the bottom leading into the stomach.

The mouth of this sac is in close connexion with the upper or principal orifice, and therefore receives the water, which is constantly flowing in, while that aperture is expanded. This fluid then bathes the whole interior of the sac, but a portion of it escapes by the lateral openings into the cavity of the body, between the sac and the mantle, and is discharged through the secondary, or side orifice.

The inner surface of this transparent sac is studded with rings of a long oval figure, set side by side in four rows. These rings appear to consist of a slight elevation of the general membranous surface, so as to make little shallow cells, the whole edges of which are fringed with cilia, whose movements make waves, that follow each other round the course in regular succession. In truth it is a beautiful sight to see forty or more of these oblong rings, all set round their interior with what look like the cogs on a watch-wheel, dark and distinct, running round and round with an even, moderately rapid, ceaseless motion. These black running figures, so like cogs and so well defined as they are, are merely an optical delusion; they do not represent the cilia, but merely the waves which the cilia make; the cilia themselves are exceedingly slender close-set hairs, as may be seen at the ends of the ovals, where a slight alteration of position prevents the waves from taking the tooth-like appearance. Sometimes one here and there of the ovals ceases to play, while the rest continue; and, now and then, the whole are suddenly arrested simultaneously as if by magic, and presently all start together again, which has a most charming effect. A still more singular circumstance is, that while in general the ciliary wave runs in the same direction in the different ovals, there will be one here and there in which the course is reversed; and I think that the animal has the power of choosing the direction of the waves, of setting them going and of stopping them, individually as well as collectively.

The object of these ciliary wheels is to keep up a constant current in the water. This fluid, as I have said, enters from without, through the upper orifice of the body, and is hurled over the whole surface of the breathing sac by means of the ciliary waves, parting with its oxygen, as it goes, to the blood, which streams, as we saw, everywhere between the rows of wheels. But the water has another function: it carries particles of organic matter with it, which are suitable for the nourishment of the creature;

these atoms are carried by the currents with the effete water to the bottom of the sac, and are poured into the stomach, where they are digested; the innutritive remains, together with the waste water, being discharged through the lateral orifice.

Thus we see how closely connected are the three cardinal processes of circulation, respiration, and digestion.

CHAPTER III.

MOLLUSCA: THEIR SHELLS, TONGUES, EYES, AND EARS.

ONE of the most interesting aspects of microscopic study is that in which it reveals the intimate structure of objects, which to the unassisted eye appear simple or nearly so, but which prove by the aid of magnifying power to be complex. Thus we are often introduced to very curious *contrivances* (if I may use such a word in reference to the works of God), by which difficulties are overcome, and substances, which would seem at first wholly unfit for certain duties, are in the most admirable manner adapted to fulfil them.

The combination of strength and lightness is always a difficult problem in human art; its successful solution always excites our admiration. In the Divine mechanics, too, it is very often required, and the variety of modes in which it is accomplished are in the highest degree novel and suggestive. We lately saw one of these in the structure of a feather, in the contrivance by which extreme lightness of material was made, by a most remarkable arrangement, to offer a firm resistance to opposing force. I have now another example to show you, in which a material, in itself heavy, is by its arrangement made very light, while it preserves its aggregate strength.

You have seen many times, when walking along the yellow sands kissed by the rippling waves, the shell, or bone as it is sometimes called, of the Cuttle-fish. You know that it consists of a shallow boat-shaped shell, the hollow of which is filled with a white substance, which can be scraped away even with the finger-nail, and which is sometimes used as *pounce*, to rub on paper from which writing has been erased. It is this substance of which I mean now to speak.

The possessor of this structure is a member of the numerous class MOLLUSCA, which are generally characterised by being inclosed in shells. Now shell, as we all know, is a solid, stony substance, much heavier than water; take into your hand that large *Cassis* on the mantel-piece, and observe its great weight and compactness. It is, in fact, real limestone; differing from that of the rocks only in this, that it has been deposited by the living organic cells of an animal, and arranged in a definite form. We will presently examine other examples. The "cuttle-bone" is a shell, not indeed inclosing the animal, but inclosed by it; being contained within a cavity in the substance of the fleshy mantle; cut open the mantle, and the shell instantly drops out.

The Cuttle is a rapid swimmer through the open sea. A shell so large as this, if solid and compact like that of the *Cassis*, would condemn it to grovel on the bottom, and frustrate all the instincts of its nature. On the other hand, it needs the strength and support of a solid column. Wonderful to tell, the calcareous shell is made not only to be no hindrance to its swimming, but to contribute greatly to its buoyancy: it is what the string of corks is to the bather who cannot swim, it is a *float*. Throw this entire cuttle-shell into water; it floats on the surface as buoyantly as if it were actually carved out of cork.

I cut with a keen knife a little cube out of the "pounce," and, fixing it on the end of the revolving stage-needle, apply a low power, say 70 diameters, using reflected light. We are looking now at the perpendicular section; is it not a beautiful object? you might fancy yourself looking at one of the noble icebergs that majestically navigate the polar seas, when it is rendered porous and laminated by the rains of spring. You see a number of thin horizontal tiers or stages, perfectly parallel and equidistant, about one-fortieth of an inch apart, rising above each other like the floors of an edifice. These are connected together by an infinite multitude of thin pillars of crystal, or rather leaves, some of which show their edges towards us, others their broader sides, and others are broken off at various distances, the fragments standing up from the floor, or depending from the roof, like stalactites and stalagmites in a cavern.

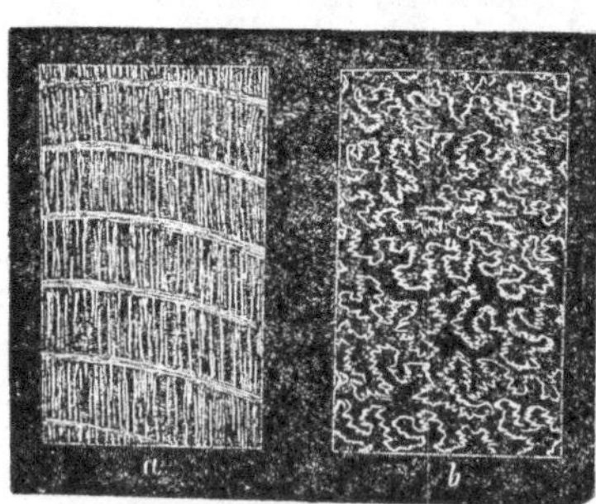

CUTTLE-SHELL.
a Perpendicular; *b* Horizontal section.

This whole series of crystal floors and supporting plates is formed of calcareous matter,—limestone, in short; but though the latter are set in such close array that the eye cannot penetrate to any appreciable distance between them, their extreme thinness renders the whole structure very light, the interstices being occupied by air.

But now if I give the stage-needle half a revolution, we shall have the horizontal section presented to the eye. In this aspect we acquire much more information as to the structure. The cut has been made very close to one of the horizontal floors, which we see marked all over with a great number of lines, each of which runs hither and thither, in a very sinuous pattern. The lines are made up of a brilliant sparkling substance; they are in fact the basal portions of what we saw in the other section as thin perpendicular plates; I have cut off the plates close to the bottom, and what we see is their insertion into the floor.

Thus we perceive that what we took for a multitude of plates, were but the various doublings and infoldings of a single plate of great length, running quite across the floor; an arrangement by which the strength of the material is greatly augmented. You have often seen the mode in which light walls are made of corrugated iron, especially at railway stations, and are doubtless aware that the corrugation, or bending in and out, imparts a strength to it which the mere sheet iron, if set up as a smooth, plane surface, would in no wise possess. The principle is exactly the same in the two cases; but the corrugation of the limestone plates in the cuttle-shell is far more perfect than that of the iron; added to which there is the other advantage, that the aggregate mass of material is made highly buoyant by the large bulk of empty space that intervenes between the sinuous folds of the crystal plates.

It may be interesting to compare with this the structure of the more solid shells of bivalves, which have been so elaborately studied by Dr. Carpenter. In general, these consist of two very distinct layers, well seen in the valve of the Pearl Oyster, and its allies. The Pinna, or Wing-shell, the largest of our native bivalves, affords us a good example, especially of the external layer, since here this layer projects beyond the inner one, in thin transparent edges, which give us an opportunity of examining their structure, without any artificial preparation. This fragment, taken from the edge of one of those leafy expansions, we will examine with a low magnifying power. Each of its surfaces has a sort of facetted, or honeycombed appearance, and the broken edges, which even to the naked eye appear fibrous, are seen to resemble a number of basaltic columns. "The shell is thus seen to be composed of a vast number of prisms, having a tolerably uniform size, and usually presenting an approach to the hexagonal shape. These are arranged perpendicularly, or nearly so, to the surface of the lamina of the shell; so that its thickness is formed by their length, and its two surfaces by their extremities." *

The inner layer of such shells is remarkable for

* Carpenter, The Microscope, p. 590.

possessing in different degrees the property of reflecting rainbow-like colours, often with great delicacy and splendour; and this is termed *nacre*, or familiarly "mother-of-pearl." This iridescent lustre depends, as Sir David Brewster has shown,* upon a multitude of grooves, or fine lines, which run in a very waved pattern, but nearly parallel to each other, across the surface of the nacre. "As these lines are not obliterated by any amount of polishing, it is

SECTION OF NACRE FROM PEARL OYSTER.

obvious that their presence depends upon something peculiar in the texture of this substance, and not upon any mere superficial arrangement. When a piece of nacre is carefully examined, it becomes evident that the lines are produced by the cropping out of laminæ of shell, situated more or less obliquely to the plane of the surface. The greater the dip of these laminæ, the closer will their edges be; whilst

* Phil. Trans. 1814.

the less the angle which they make with the surface, the wider will be the interval between the lines. When the section passes for any distance in the plane of a lamina, no lines will present themselves on that space. And thus the appearance of a section of nacre is such, as to have been aptly compared by Sir J. Herschel to the surface of a smoothed deal board, in which the woody layers are cut perpendicularly to their surface in one part, and nearly in their plane in another. Sir D. Brewster appears to suppose that nacre consists of a multitude of layers of carbonate of lime, alternating with animal membrane, and that the presence of the grooved lines on the most highly polished surface, is due to the wearing away of the edges of the animal laminæ, whilst those of the hard calcareous laminæ stand out. . . . . There is one shell, however, the well-known *Haliotis splendens*, which affords us the opportunity of examining the plaits without any disturbance of their arrangement, and thus presents a clear demonstration of the real structure of nacre. This shell is for the most part made up of a series of plates of animal matter, resembling tortoise-shell in its aspect, alternating with thin layers of nacre; and if a piece of it be submitted to the action of dilute acid, the calcareous portion of the nacreous layers being dissolved away, the plates of animal matter fall apart, each one carrying with it the membranous residuum of the layer of nacre that was applied to its inner surface. It will be found that the nacre-membrane covering some of these horny plates, will remain in an undisturbed condition; and their surfaces then exhibit their iridescent lustre, although all the calcareous matter has been removed from their structure. On looking at the surface with reflected light under a magnifying power of seventy-five diameters, it is seen to present a series of folds or plaits, more or less regular; and the iridescent hues which these exhibit, are often of the most gorgeous description. If the membrane be extended, however, with a pair of needles, these plaits are unfolded, and it covers a much larger surface than before; but its iridescence is then completely destroyed. This experiment, then, demonstrates that the peculiar lineation of the surface of nacre (on which its iridescence undoubtedly depends, as originally shown by Sir D. Brewster) is due, not to the outcropping of alternate layers of membranous and calcareous matter, but to the disposition of a single membranous layer in folds or plaits, which lie more or less obliquely to the general surface." *

Those beautiful objects,—so much prized for personal adornment,—pearls, are concretions accidentally formed within the shells of such mollusks, and are wholly composed of the inner layer. Drs. Kelaart and Möbius have recently published some highly interesting observations on the causes both of the iridescence and of the pearly lustre; and these I will cite from the abstract translation of them made by Mr. Dallas.

"The surface of pearls is not perfectly smooth, but covered with very fine microscopic elevations and depressions. These are more or less irregular in

* Carpenter. The Microscope, p. 594.

their altitude, but approach most nearly to equality in pearls of the finest water. In pearls which exhibit a certain iridescence, and which, when turned in different directions towards the eye, present even very faint bluish, greenish, and reddish tints, the surface is found to present delicate irregular curved furrows, which either run tolerably parallel to each other, or form small irregular closed curves. This is due to the mode of growth of the pearl, in which thin layers of nacre, of small dimensions, have been laid over each other. There is no continuous layer over the pearl, but a number of small portions which sometimes overlie the margins of the subjacent layers, and sometimes leave them uncovered. This structure is seen most distinctly in the pearl shell, where the conditions are rendered more simple by the layers being deposited on a flat, or but slightly curved, surface. The distance of the furrows from each other is not always the same; sometimes they may be recognised with the simple lens, whilst on other parts they approach within $\frac{1}{3000}$th of an inch of each other. That the iridescence of nacre, or the nacreous colour, as distinguished from pearly lustre, is caused by the interference of the light reflected from these furrows and the intervening edges of the strata, is proved by the circumstance, ascertained by Brewster, that impressions of mother-of-pearl taken in red or black sealing-wax exhibit the same phenomena of colour distinctly. In pearls, in consequence of their spherical form, the different masses of coloured light are so diffused that they unite to form white light; and this takes place with the greater perfection in proportion as the furrows are lost, and become converted into a surface of fine elevations and depressions.

"For their lustre, pearls are indebted to their being composed of fine layers, which allow light to pass through them, whilst the numerous layers lying one under the other, disperse and reflect the light in such a manner that it returns and mixes with that which is directly thrown back from the outer surface. It is the co-operation of light reflected from the surface, with light dispersed and reflected in the interior, that gives rise to lustre; for this reason the knots of window-glass exhibit pearly lustre, and the membranes of pearls deprived of their lime are almost as lustrous as solid pearls, except that their whiteness is destroyed. 'The two masses of light entering the eye act upon it from different distances. Now, as it adapts itself to the body seen through the transparent layer, it cannot distinctly see the light reflected from the surface, and the consciousness of this infinitely perceptible reflection produces the phenomena of lustre.'* The thinner and the more transparent the layers of which the pearl consists, the more beautiful is its lustre; and in this respect the sea-pearls excel those of our river-mollusks."†

We will pass now, by an easy transition, from the *shells* of the Mollusca to their *tongues*. Who that looks at the weather-worn cone of the Limpet, as he adheres sluggishly to the rock between tide-levels, would suspect that he carries coiled up in his throat a tongue twice as long as his shell? And that this tongue is

* Dove. Farbenlehre, 117. † Ann. & Mag. N. H.; Feb. 1858

armed with thousands of crystal teeth, all arranged with the most consummate art in a pattern of perfect regularity? It sounds almost like a fable to be told that the great Spotted Slug, which we sometimes find crawling in damp cellars, carries a tongue armed with 26,800 teeth! Yet there is no doubt of the fact.

You see on this slip of glass a very slender band about two inches in length. This is the tongue of the common Periwinkle. While in the living animal, its fore-part occupied the floor of the mouth, whence it passed down below the throat, and turning towards the right side, formed a close spire of many whorls, exactly like a coil of rope, which rested on the gullet. Here we have it extracted, uncoiled, cleansed, and affixed to a slip of glass for microscopical examination.

Only a small portion of the ribbon is visible at a time with such a power as is necessary to display the structure, but by means of the stage-movement we can bring the whole in succession under the eye, and discover that, with some modifications of form, the same essential plan of structure, and even the same elements, exist throughout. Concentrating our attention on a single transverse series of the numerous curved lines that at first sight bewilder the mind, we perceive by delicate focussing, that the object before us consists of a number of hooks projecting from the surface of the translucent ribbon, and arching downward. In this case a single row consists of seven such hooked plates or teeth; one in the centre and three on each side. Each hooked plate has its arching tip cut into five toothlets, of which the central one is the largest; and its base is united with the cartilaginous substance of the ribbon. Only the middle plate is symmetrical; the lateral ones bend inwards towards the central one, and are symmetrical only when considered in pairs, each associated with its opposite. The plates are perfectly transparent, but of a yellow horny colour; they are very hard, and as they are not dissolved by acids, it has been supposed that their substance is siliceous (having the nature of flint); but they are more probably chitinous, or formed of the substance of which the hard parts of insects are composed. The tongue before us has 600 rows such as these, each, as we see, closely following, and indeed overlapping, its predecessor; so that we can never look at a single row without at the same time seeing others which it overlaps, or by which it is overlapped.

The specimen which I will now show you is broader, but shorter. It is the tongue of *Trochus ziziphinus*, a large and handsome shell of regularly conical form, not uncommon on our rocky shores. It is perhaps a more interesting study than that of the Periwinkle. There are here, you observe, three constituent elements in the pattern. First, a delicate glassy central tooth, tapering to a fine point, and cut into minute saw-teeth along each edge. Then a series, of five on each side, of similar glassy pointed leaves, bending inward; and outside these, on either hand, are a great number of stout dark-coloured hooks, arching forward and inward, each notched with saw-teeth, and diminishing in thickness as they recede from the centre.

The manner of using this elaborate organ is no less

curious than is its structure. During life it is only the front portion—not more than one-third—of the ribbon that is in use; this is spread out on the floor

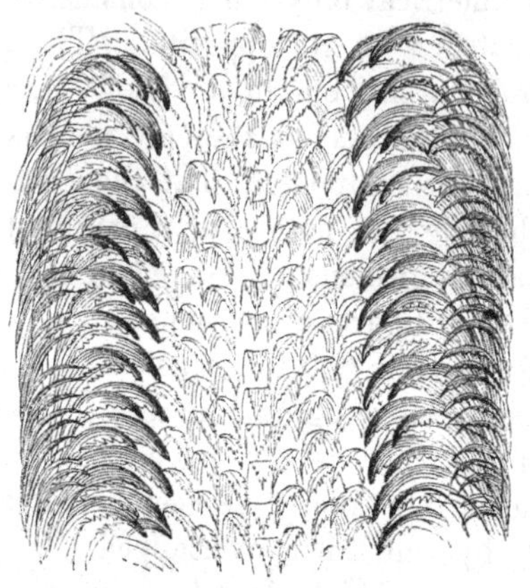

TONGUE OF TROCHUS.

of the mouth, with the teeth projecting and hooking backwards. The remainder has its edges rolled over towards each other, forming a tube closed at its extremity, which, as I have already observed, is coiled away (in the long-tongued kinds) among the viscera.

The mode in which the tongue is used may be readily seen by watching the actions of a Periwinkle in a marine, or a Pond-snail in a fresh-water aquarium. When the conferva has begun to form a thin green growth on the glass sides of the tank, the Mollusca are incessantly engaged in feeding on it, and rasping it away with this toothed ribbon. "The upper lip with its mandible is raised; the lower lip expands; the tongue is protruded, and applied to the surface for an instant, and then withdrawn; its teeth glitter like glass-paper, and in the Pond-snail it is so flexible that frequently it will catch against projecting points, and be drawn out of shape slightly as it vibrates over the surface."*

Perhaps every variety is accompanied by some variation in food or manner of feeding. With the *Trochus,* the proboscis, a tube with thick fleshy walls, is rapidly turned inside out to a certain extent, until a surface is brought into contact with the glass, having a silky lustre: this is the tongue; it is moved with a short sweep, and then the tubular proboscis infolds its walls again; the tongue disappearing, and every filament of conferva being carried up into the interior from the little area which had been swept. The next instant, the foot meanwhile having made a small advance, the proboscis unfolds again, the tongue makes another sweep, and again the whole is withdrawn; and this proceeds with great regularity. I can compare the action to nothing so well as to the manner in which the tongue of an ox licks up the grass of the field, or to the action of a mower cutting down swathe after swathe as he marches along. The latter comparison is more striking, for the marks of progress which each operator leaves behind him. Though the confervoid plants are swept off by the tongue of the Mollusk, it is not done so cleanly but that a mark is left where they grew; and the peculiar form and structure of the tongue, which I have above noticed, leave a series of successive curves all along the course which the Mollusk has

* Woodward's "Mollusca," 161.

followed, very closely like those which mark the individual swathes cut by the mower in his course through the field.

The Periwinkle's table-manners differ slightly from those of his relations. When he eats, he separates two little fleshy lips, and the glistening glass-like tongue is seen, or rather the rounded extremity of a bend of it, rapidly running round like an endless band in some piece of machinery; only that the tooth-points, as they run by, remind one rather of a watch-wheel. For an instant this appears, then the lips close again, and presently re-open, and the tongue again performs its rasping. It is wonderful to see;—perhaps not more wonderful than any other of God's great works, never more great than when minutely great; but the action and the instrument, the perfect way in which it works, and the effectiveness with which the vegetation is cleared away before it, all strike the mind with more than usual force, as exhibitions of creative skill.

As the Periwinkle moves along, mowing his sea-grass as he goes, he carries before him two soft and flexible horns, marked with zebra-like bands of black and white, which he constantly waves about. These are organs of some sense, probably of touch, and are therefore called *tentacles* (or tryers); but they bear on their outer sides, near the base, a pair of other organs, which are more closely analogous to what we ourselves possess. You see on each tentacle a little wart, which when you look at it with a lens you perceive to have a round black glossy extremity. This is the eye. By careful dissection under the microscope, we find it to contain a beautiful transparent crystalline lens, with a thick and glutinous vitreous humour adhering to it behind, bounded by a retina or curtain to receive the optic image, and an optic nerve.

But much more attractive you will find the eyes in this little Scallop. It is a half-grown individual of what is provincially known as the Squin (*Pecten opercularis*), much prized for its delicate sapidity. Belonging to the bivalve class of the Mollusca, the animal is inclosed within two shallow shelly plates, concave internally, and convex externally, which are united by a hinge, just as the works of a watch are protected by the case. When the little creature is at its ease, as when the water is pure and clear, it lies on one side, its valves being separated as we see them now, a quarter of an inch or so apart, allowing us to discern what is contained between them.

Well, we see first a number of slender white pointed threads, peeping out from each valve, and spreading on all sides, waving hither and thither, groping, now contracting, now expanding, with incessant but deliberate motion. These are tentacles. If we trace them to their origin, we find them attached to a fleshy sort of veil that lines each valve to near its edge, and then abruptly falls at an angle towards the opposite valve, where it meets a corresponding veil. These two veils form the mantle. It is from each of these that the tentacles spring; and we discover that there are four rows of these organs, one row set along the angle, and one along the edge of each veil.

But as we peer among these slender threads, our

attention is riveted by some tiny points that are seated near their bases, which glitter like brilliant gems. They are seen only in those rows of tentacles which spring from the angles of the veils, and not in those which fringe their edges. Even the unassisted sight can detect the gleam and glitter of these little specks; but it is only when we bring the lens to bear upon them that we see all their beauty. Then they look like diamonds or emeralds, each set in a broad ring of dark red substance, which greatly enhances their beauty. They are inserted into the mantle in the line of the tentacles, alternating with them, yet not with absolute regularity, for there are more tentacles than gem-points; they are about half as numerous again as the radiating ridges of the shell. Some are much larger and more prominent than others, but they have all the same structure and appearance.

These little organs are eyes. As its movements are far more extensive, and more fitful and rapid than is common in this class of animals, the little Pecten probably needs these brilliant organs of vision to guide its wayward rovings, as well as to guard it from hostile assaults. The animal is very sensitive, withdrawing its tentacles and mantle, and bringing the valves of its shell together, on any shock being given to the vessel in which it resides. It manifests, however, a wisely measured degree of caution, for it does not actually *close* the valves, unless it be repeatedly disturbed, or unless the shock be violent, contenting itself with narrowing the opening to the smallest space appreciable; yet even then the two rows of gem-like eyes are distinctly visible, peeping out from the almost closed shell, the two appearing like one undulating row from the closeness of their proximity.

If you are familiar with the pin-cushions which children often make with a narrow ribbon round the edges of these very Scallop-shells, you can scarcely fail to be struck with the resemblance borne by the living animal to its homely but useful substitute; and the beautiful eyes themselves might be readily mistaken for two rows of diamond-headed pins, carefully and regularly stuck along the two edges of the pin-cushion ribbon,—the ribbon itself representing the satiny and painted mantle. A friend of mine, to whom I once was showing this object, compared it not unaptly to a lady's ring set with diamonds.

You will not fail to remark, how the position of these beauteous organs is suited for their most extensive usefulness consistent with their safety. In the ordinary condition of the animal's expansion, and especially when it is about to make its sudden and vigorous leaps, the gemmeous points are so situated as just to project beyond the margin of the shell. So that when we view the creature perpendicularly as it lies, our eyes looking down on the convexity of the upper valve, the minute eyes are seen, all round its circumference, just, and but just, peeping from under its edge. It is clear that this arrangement secures to them the widest range of vision with the least possible exposure. As Divine contrivance has been often most deservedly recognised in the projection of the bony ridge over the human eye, which we call the brow, we surely cannot fail to recognise, and admire it also, in the position of these delicate organs, either beneath the margin of the solid shell, or, if projected, projected only in the smallest degree, and endowed with the power of retreating beneath its barrier with the rapidity of thought, on the least alarm.

There can be no doubt that these points, numerous as they are, are true eyes, endowed with the faculty of vision in a well-developed degree. For when their structure is carefully examined by the skilful anatomist, each is found to be covered with the proper sclerotic tunic which becomes a perfectly transparent cornea in front, and to possess a coloured iris,—perforated with a well-defined pupil, and connected with a layer of pigment which lines the sclerotic tunic,—a crystalline lens, and a vitreous humour for the due refraction of the rays of light, and a retina in their focus, formed by an expansion of the optic nerve, and fitted to receive the picture; the sensation of which is then conveyed by an optic nerve from each eye to the common nerve-trunk, which runs along the border of the mantle. Thus there exists in each of these lustrous points every element needful for the due performance of vision, though, probably, the impressions thus conveyed may be neither so powerful, nor so distinct as those which are conveyed by the eyes of vertebrate animals. They are, however, we may be sure, amply sufficient for the wants of the pretty Scallop, and are fresh proofs of the Divine wisdom and benevolence.

We have been accustomed, from childhood, to recognise as eyes the shining black extremities of the upper pair of "horns" in the Garden Snail. And though some naturalists have doubted, and even denied that the tentacle was anything more than a very delicate organ of touch, yet it has been abundantly proved by dissection, and is now incontrovertibly established, that its tip carries an eye, even more completely developed than those of the Pecten, which we have just been looking at. The eye is situated, not indeed on the very summit of the tentacle, but on one side of a movable bulb there placed. It is very minute, almost spherical, but slightly flattened in front. It is protected by a very thin transparent layer of the common skin, and is surrounded at the side and behind, by a perfectly black membrane called the *choröid*, or pigment-membrane. This black globule contains a transparent and semi-fluid substance, with which it is completely filled; towards the bottom it is of thinner consistence, and appears to contain many brilliant particles when the eye is dissected under the microscope; this may be considered as the vitreous humour. In the front part of the eye there is a crystalline lens, a small, circular, flattish, or rather lenticular body, perfectly clear and translucent, but a little more solid than the vitreous humour.

Now protection for these so delicate organs is provided in a way quite different from, yet equally effective with, that which we just now admired in the case of the Pecten. You know that if you touch, though ever so tenderly, the eye of the Snail, it is instantly drawn into the horn by a most curious process of inversion. This action is performed by means of a long muscular ribbon, which originates from the great muscle that retracts the head within the shell, and which is inserted into the extremity of the hollow tentacle. When this ribbon contracts at the will of the animal, and still more forcibly, when it is aided by the contraction of the great head-muscle, the tip of the tentacle with its eye is drawn within the surrounding parts, just like the finger of a glove. When the animal would again protrude its eye, the fibres which surround the tentacle, like so many rings throughout its whole length, successively contract, and thus gradually squeeze out, as it were, the inverted part, until it is turned back to its original position.

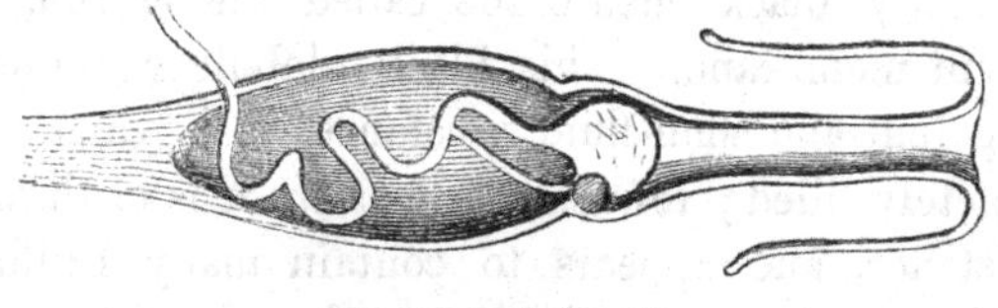

STRUCTURE OF EYE IN SNAIL.

But the ears of this homely "creeping thing" are, perhaps, even more curious than its eyes; though far less elaborate in their structure. You will imagine now, that I refer to the other pair of tentacles, as you are accustomed to associate the idea of ears with projecting organs situated on the head. No, you must not look there for them. Here, in this young Garden Slug, which is so small as to be conveniently examined on the stage of the microscope, and so devoid of colour that we can readily look through its tissues,—we shall easily find its ears, though they are not quite so prominent as those of an ass.

I subject the animal to a gentle pressure by means of the compressorium, just sufficient to flatten its soft body a little, without injuring it. And now, with this low power, you may see what Siebold, a learned zoologist and comparative anatomist, familiar with the curious phenomena of life, truly calls "a wonderful spectacle." In the neck of the little animal you discern, deep-seated in the soft flesh, a pair of perfectly transparent globules, or bladders, without any opening, but filled with a clear fluid, in which there are some minute bodies performing the most extraordinary evolutions. They constantly keep up a series of swinging or balancing movements, sometimes rotating, sometimes forcibly driven in a certain direction, then in the opposite, yet no single one ever by any accident touching the walls of the capsule in which they are contained. If the capsule be ruptured, the motions instantly cease. These little bodies are of a calcareous nature; and they are called *otolithes*, that is, ear-stones. The most that we know of these curious capsules, which are indubitably ascertained to be organs of hearing, we owe to the observations of the eminent zoologist just named, and you may perhaps like to know a little more about them.

Siebold says that a concentric depression is evident in these otolithes, and that there may be seen in the centre of the greater number of them a shaded spot, or rather a minute aperture, which penetrates through the concretion from the one flattened surface to the other. Subjected to a strong pressure, the otolithes crack in radiating lines, separating often into four pyramidal pieces. This separation also ensues, after a longer time, when the otolithes are immersed in diluted nitric acid; and, if we touch them with the concentrated acid, they suddenly dissolve with the disengagement of a gas, whence Siebold concludes them to be composed of carbonate of lime. The size of the otolithes is not equal, and in the same capsule there are always some which are smaller than others. Within the capsule they have, during life, a very remarkable, and in some respects peculiar, lively, oscillatory movement, being driven about as particles of any light insoluble powder might be in boiling water. The otolithes in the centre have the appearance of being pressed together so as to form a sort of solid nucleus, and towards this centre the otolithes towards the circumference seem even to be violently urged, their centripetal rush being invariably repulsed, and as often driven again into a centrifugal direction. Removed from the capsule, the motions of the otolithes instantly cease. The cause of these curious oscillations remained undiscovered. Siebold could detect no vibratile *cilia* on the surfaces of the capsule, and the cessation of the motion when the otolithes are removed, proves them to be unciliated themselves, and, at the same time, distinguishes the motion from that of inorganic molecules.

It has been more recently ascertained that the movements of the otolithes are due to very minute cilia with which the interior surface of the capsule is covered. This had been long suspected, and some eminent physiologists, as Wagner and Kölliker, have distinctly seen the cilia themselves.

If you ask what can be the use of ears to a class of animals which are invariably dumb, I answer that though this is true with respect to the great majority, yet it may be only that our senses are too dull to perceive the delicate sounds which they utter, and which may be sufficiently audible to their more sensitive organs; and besides, some Mollusca can certainly emit sounds audible by us. Two very elegant species of Sea-slug, viz. *Eolis punctata*, and *Tritonia arborescens*,* certainly produce audible sounds. Professor Grant, who first observed the interesting fact in some specimens of the latter which he was keeping in an aquarium, says of the sounds, that "they resemble very much the clink of a steel wire on the side of the jar, one stroke only being given at a time, and repeated at intervals of a minute or two; when placed in a large basin of water the sound is much obscured, and is like that of a watch, one stroke being repeated, as before, at intervals. The sound is longest and oftenest repeated when the Tritoniæ are lively and moving about, and is not heard when they are cold and without any motion; in the dark I have not observed any light emitted at the time of the stroke; no globule of air escapes to the surface of the water, nor is any ripple produced on the surface at the instant of the stroke; the sound, when in a glass vessel, is mellow and distinct." The Professor has kept these Tritoniæ alive in his room for a month, and, during the whole period of their confinement, they have continued to produce the sounds, with very little diminution of their original intensity. In a small apartment they are audible at the distance of twelve feet. "The sounds obviously proceed from the mouth of the animal; and, at the

* Now called *Dendronotus arborescens*.

instant of the stroke, we observe the lips suddenly separate, as if to allow the water to rush into a small vacuum formed within. As these animals are hermaphrodites, requiring mutual impregnation, the sounds may possibly be a means of communication between them; or, if they be of an electric nature, they may be the means of defending from foreign enemies one of the most delicate, defenceless, and beautiful Gasteropods that inhabit the deep."*

* Edinb. Phil. Journ. xiv. 186.

CHAPTER IV.

SEA-MATS AND SHELLY CORALLINES.

When we were at the sea-side last summer we bought, you may remember, of a poor widow whom we met on the beach, a little basket of dried sea-weeds. Fetch it: it is on the chimney-piece up-stairs.

Now all of these objects are not sea-weeds. I mean they are not all plants; some of them are animals, and these I want to bring under your notice this evening for our microscopical entertainment. Here are exquisitely delicate crimson leaves, as thin or thinner than the thinnest tissue-paper, with solid ribs and sinuous edges. Here is a tall and elegant dark red feather, quite regularly pinnated. Here is a tuft of purple filaments as "fine as silkworm's thread." And here is a broad irregular expanse of the richest emerald-green, crumpled and folded, yet as glossy as if varnished.

Well, all of these are plants, certainly: they are veritable *Algæ*, or sea-weeds. But here are other plant-like objects of a pale brown, drab, or snowy-white hue. Let us take this flattened brown leaf, divided into irregular broad lobes; it looks almost like a thickish paper, and is about as flexible. But pass your finger over it, and you feel that its surface is evenly roughened; and on close and careful scrutiny you discern, even by the naked eye, that its surface is covered with a delicate network of minute shallow cells.

"Broad Hornwrack," and "Leafy Sea-mat," are the names which the old collectors gave to this object; and modern naturalists have given it the scientific appellation of *Flustra foliacea*, and arrange it in the class *Polyzoa*, a group of animate beings, which have much of the form of Polypes, and much of the structure of Mollusks.

We cut off a little piece from the end of one of the lobes, and put this upon the stage of the microscope. We now see that the cells are disposed in nearly parallel rows; but so that those of one row alternate with those of the next, *quincunx* fashion, the middle of one cell being opposite the end of its right and left neighbours;—or like the meshes of a net. The cells extend over the whole leaf, and are spread over both its surfaces in this case; the united depth of two cells constituting the thickness of the leaf-like structure. There are other species, more delicate, which have but a single series of cells, all opening on the same side of the leaf.

Each individual cell is shaped like a child's cradle; and if you will imagine 20,000 wicker cradles stuck together side by side in one plane, after the quincunx pattern I have just mentioned; and then the whole broad array turned over, and 20,000 more glued on to these, bottom to bottom,—you will have an idea of the framework of this pale-brown leaf;—dimensions, of course, being out of the consideration. The number may appear somewhat immense, yet it is no larger than the ordinary average, as I will soon show you. I measure off a square half-inch of this leaf, which I carefully cut out with scissors; now with the micrometer count the cells in the square piece.—You find 60 longitudinal rows, each containing 28 cells, or thereabouts. Very well; a simple arithmetical process shows that there are 1,680 cells in this square half-inch; or 6,720 in a square inch. Now this very specimen, before I mutilated it, contained an area of about three square inches; which would give 20,160 cells. This is the number on one surface; the other contains an equal number; and thus you see that I have not exaggerated the population of this tiny marine city. This, however, is by no means a specimen of unusual size.

These cells, which I compare to cradles, are of shallow depth, but the head-part rises to a much greater height than the foot. All round this elevated portion the margin is armed with short blunt spines, two on each side, which stand obliquely erect, projecting outwards over the middle of the next cell, which thus, in concert with the spines of the cell on the opposite side, they protect.

If you search carefully over the aggregation of cells with this pocket-lens, you will perceive that on some of them are seated minute white globules, which look like tiny pearls. These are not placed in any regular order, two being sometimes found on contiguous cells, but generally they are scattered at more or less remote intervals. If we now apply the microscope to these appendages, each globule is seen to be flat on that perpendicular side which faces the foot of the cradle; and this flat side is a movable door, with a hinge along its lower edge. The door is of a yellow hue; the globule itself being, as I said, of a pearly white hue.

This is all that we can see in this dried specimen; but if we had been fortunate enough to have examined it when first it was torn from its attachment to an old shell at the bottom of the sea, you would have seen much more. And what would then have appeared I will describe to you.

Suppose, then, that a coverlid of transparent skin were stretched over each cradle, from a little within the margin all round, leaving a transverse opening just in the right place, viz. over the pillow, and you would have exactly what exists here. There is a crescent-form slit in the membrane of the upper part of the cell, from which the semicircular edge, or lip, can recede if pushed from within.

Suppose, yet again, that in every cradle there lies a baby, with its little knees bent up to its chin, in that zig-zag fashion that children, little and big, often like to lie in. But stay, here is a child moving! Softly! He slowly pushes open the semicircular slit in the coverlid, and we see him gradually protruding his head and shoulders in an erect position, straightening his knees at the same time. He is raised half out of bed, when lo! his head falls open, and becomes a bell of tentacles! The baby is the tenant-polype!

"This is a very amusing romance," you say. Nay, it is no romance at all. If you will excuse the homeliness of the comparisons, I venture to affirm that a personal examination of the creature itself would justify their correctness, and you would acknowledge that they could scarcely be more apt.

Moreover, the globular chambers show signs of life; their front doors suddenly open, gape widely, and then shut with a snap; and presently this opening and shutting is repeated. The meaning of this action you will better understand when we see analogous organs in another form of the same class of animals. Meanwhile, I will just point out a beautiful though minute proof of design in a point of the structure of the cells connected with these pearly chambers. If you look closely, you will see that the spines of the margin are not found on those cells that carry the pearls; and moreover, that they are also wanting on the approximate edges of the two cells that lie behind every such pearl-bearing cell. Now the reason of this omission is obvious. The spines projecting obliquely would

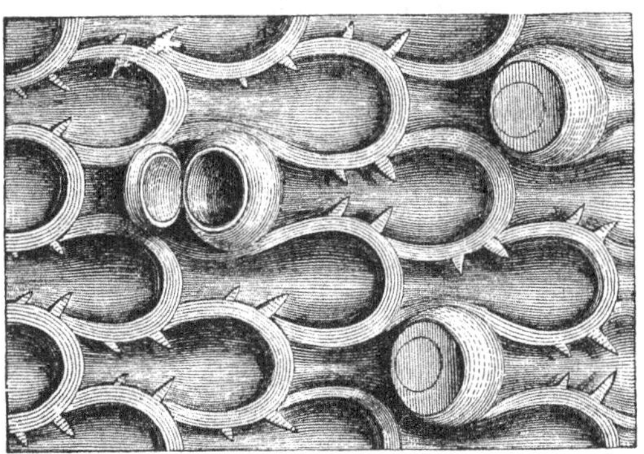

LEAFY SEA-MAT.
(A portion magnified.)

interfere with the gaping of the door; and hence they are invariably absent there.

I happen to have in my aquarium a living individual of another species belonging to the same class, and agreeing with this in all essential particulars of structure, though widely different in form. The difference, however, is mainly dependent on a rather unimportant point of arrangement; for the cells, instead of being set side by side and end to end in quincunx fashion, to an indefinite extent, on two surfaces of a plane, are disposed on one single surface, and in longitudinal rows of two or three cells abreast; thus narrow ribbon-like branches are formed, which now and then divide into two, then these into two more, and so on. These branches thus become fan-shaped, which, by being slightly curved, become segments of funnels; and the peculiar elegance of this coralline consists in the mode in which these branches are set on the stem, viz. in an ascending spiral curve, so that the effect is that of several imperfect funnels set one within another, but which yet you perceive, by turning the whole gradually round, to compose a single corkscrew band of successive fans. This whole structure stands upright in its natural state, like a little compact shrub growing from a root; and as a good many are commonly associated together, they form a sort of mimic grove, fringing the sides of dark rocky sea-pools.

The species is called the Corkscrew Coralline, or sometimes, the Bird's-head Coralline, the latter name being assigned to it for a reason which you will presently perceive. The appellation by which it is known to naturalists is *Bugula avicularia.*

We drop our specimen into a very narrow cell,

composed of parallel walls of thin glass, a very minute flattened tank, in fact, such as can be put on the stage of the microscope. Here, bathed in its native sea-water, as clear as crystal, we shall see it opening and expanding its numerous polypides with the utmost activity and evident enjoyment.

You gaze; but you know not what you see. The presence of many lines representing transparent vessels of strange and dissimilar shapes, overlying each other; and the swaying to and fro of curious objects, which strike now and then forcibly across the field of view, are quite bewildering. I must act the showman, and tell you what to see.

The cells are oblong, shaped much like a sack of corn, with a spine ascending from each of the upper corners. Each stands on the summit of its predecessor in the same row, and side by side with those of its fellow-rows, in such an order that the top of one cell comes opposite the middle of the one beside it. The top of the sack is rounded, and appears closed, but we shall presently find an opening there. The broad side that faces inwardly has a large elliptical transparent space occupying nearly its whole surface; this is covered with a very thin and elastic membrane, and answers a peculiar end. Just below one of the spines that crowns the summit of the cell on one of the edges, is situated a little lump, to which is attached, by a very free joint, an object which you will perceive to bear a remarkable resemblance to the head of a bird of prey. It has a beak strongly hooked, with two well-formed mandibles, of which the lower is movable, shutting into the cavity of the upper; you observe it deliberately opening, like that of a bird, only stretching to an enormous width of gape, and then closing with a strong and sudden snap. Now and then the whole head sways backward and forward on its joints; and these movements, combined with the fitful and apparently spiteful snappings, performed by many birds' heads scattered about the branch, are highly curious and amusing.

The birds' heads, however, are not the living inhabitants of the cells; they are not integral parts of them. The cells have their proper inhabitants, each dwelling in its own; and each essentially formed on the same plan as the "baby with the tucked-up knees," that makes the Sea-mat for his cradle-house.

In order to get a good view of the tenant here, you must move the stage about till you find that the branch is presented to your eye sidewise. Directing your attention then to the lateral edge of a single inhabited cell, its summit is seen to protrude diagonally towards the inner side (*i. e.* towards the axis of the spire), a tubular mouth, which is membranous and contractile. When the animal wishes to emerge, this tubular orifice is pushed out by evolution of the integument, and the tentacles are exposed to view, closely pressed into a parallel bundle; the evolution of the integument, that is attached at their base, goes on till the whole is straightened, when the tentacles diverge and assume the form of a funnel, or rather that of a wide-mouthed bell, the tips being slightly everted. They are furnished with a double row of short cilia in the usual order, one set working upward, the other downward. Their base surrounds a muscular thick ring, the entrance to a funnel-shaped sac, the substance of which is granular, and evidently muscular, for its contractions and expansions are very vigorous, and yet delicate. Into this first stomach passes, with a sort of gulp, any animalcule whirled to the bottom of the funnel by the ciliary vortex, and from thence it is delivered through a contracted, but still rather wide gullet, into an oblong stomach, the lower portion of which is obtuse. An extremely attenuated duct connects this, which is probably the true stomach, with a globular, rather small, intestine, which is again connected by a lengthened thread with the base of the cell. By an arrangement common to the ascidian type of the digestive function, the food is returned from the intestine into the true stomach, whence the effete parts are discharged through a wide and thick tube that issues from it close behind the point where the gullet enters. This rectal tube passes upwards parallel to the gullet, and terminates by an orifice outside and behind the base of the tentacles. All these viscera are beautifully distinct and easily identified, owing to the perfect transparency of the walls of the cell, the simplicity of the parts, and their density and dark yellow colour. All of them are manifestly granular in texture, except the slender corrugated tube which connects the stomach with the globose intestine: this is thin and membranous, and is doubtless, if I may judge from analogy, capable of wide expansion for the passage of the food-pellet.

The sudden contraction of the polypide into its cell upon disturbance or alarm, and its slow and gradual emergence again, afford excellent opportunities for studying the forms, proportions, and relative positions of the internal organs. In contraction, the globular intestine remains nearly where it was, but the stomach slides down into the cell behind it, as far as the flexible duct will allow, and the thick gullet bows out in front, showing more clearly the separation between it and the rectum, and the insertion of both into the stomach. This retraction is, in part, effected by a pair of longitudinal muscular bands, which are inserted at the back of the bottom part of the cell, and into the skin of the neck below the tentacles. The contraction of these bands draws in the integument, like the drawing of a stocking within itself, and forces down the viscera into the cavity of the cell, which is probably filled with the vital juices.

Besides the hind bands, there is one, or a pair of similar muscular bands attached on each side of the front part of the base of the cell, and inserted similarly into the neck. By watching the contraction of these, you will be enabled to determine the use of the membrane-covered aperture up the front of the cell. At the moment of the retraction of the viscera into the cell, a large angular membrane is forced outward from the front side, which is protruded more or less in proportion to the degree of withdrawal of the polypide; and as the latter emerges again, the membrane falls back to its place. It is evident, then, that this is a provision for enlarging the cavity; the walls are horny, and probably almost inelastic; but when the stomach forces the intestine forward, and the thick gullet is bent outward by the withdrawal of the neck and tentacles, the needful room is provided by the bulging out of this elastic membrane, which recovers its place by the pressure of the surrounding water, when the pressure of the fluids within is removed.

Now, after watching these movements of the polypides, and the various structures whose forms and limits those movements reveal, it will become manifest to you that there is no visible organic connexion between the animal distinctively so called and the curious bird's head. This latter has a muscular system of its own, by means of which its energetic motions are performed; but it appears quite isolated on the outside of the calcareous cell, and wholly cut off from the interior by the knob on which it works, and by the thickness of the cell-wall. Both knob and wall appear quite imperforate; and yet we cannot but presume that some connexion exists, perhaps through the medium of an excessively delicate and subtile, but living tissue, which may be presumed not only to *line*, but also to *cover* the strong cell; just as the strong envelope and spines of a Sea-urchin are covered with a thin film of living flesh.

The functions and use of these singular processes are as obscure as their connexions with the animal. Yet that they play some important part we may almost certainly infer, from the general prevalence of similar or analogous appendages among the various forms of this class. The globular pearls which you lately saw on the sea-mat, is but another form of bird's head; and the falling-door answers to the opening and shutting mandible. The forms, indeed, of these organs are very diverse, and sometimes they are greatly disguised. But what about their function?

More than one observer has noticed the seizure of small roving animals by these pincer-like beaks; and hence the conclusion is pretty general, that they are in some way connected with the procuring of food. But it seems to have been forgotten, not only that these organs have no power of passing the prey thus seized to the mouth, but also that this latter is situated at the bottom of a funnel of ciliated tentacles, and is calculated to receive only such minute prey as is drawn within the ciliary vortex. I have ventured to suggest a new explanation. The seizure of a passing animal, and the holding of it in a tenacious grasp until it dies, may be a means of attracting the proper prey to the vicinity of the mouth. The presence of decomposing animal substance in water invariably attracts crowds of infusory animalcules, which then breed with amazing rapidity, so as to form a cloud of living atoms around the decaying body, quite visible in the aggregate to the unassisted eye; and these remain in the vicinity, playing round and round until the organic matter is quite consumed. Now a tiny Annelid or other animal caught by the bird's head of a Polyzoan, and tightly held, would presently die; and though in its own substance it would not yield any nutriment to the capturer, yet by becoming the centre of a crowd of busy infusoria, multitudes of which would constantly be drawn into the tentacular vortex, and swallowed, it would be ancillary to its support, and the organ in question would thus play no unimportant part in the economy of the animal.

CHAPTER V.

INSECTS: WINGS AND THEIR APPENDAGES.

I PROPOSE now to reveal to you some of the microscopic marvels of the insect world; a race vastly more populous than all of the other animate tribes put together; for the most part so minute as to be peculiarly suitable subjects for our present investigations, and so furnished with elaborate contrivances and exquisite pieces of mechanism, as to elevate our thoughts at every turn to the majesty of the Divine wisdom displayed in the most minute of His creatures. Let us begin with their *wings*.

The most perfect fliers in existence are Insects. The swallow and the humming-bird are powerful on the wing, and rapid; but neither these nor any other "winged fowl" can be compared with many of the filmy-winged Insects. The common House-fly, for example, will remain for hours together floating in the air beneath the ceilings of our dwelling-rooms, hovering and dancing from side to side, without effort and without fatigue. It has been calculated that in its ordinary flight the House-fly makes about 600 strokes with its wings every second, and that it is carried through the air a distance of five feet during that brief period. But, if alarmed, the velocity can be increased six or seven-fold, as every one must have observed, so as to carry the insect thirty or five and thirty feet in the second. In the same space of time, observes Mr. Kirby, a race-horse could clear only ninety feet, which is at the rate of more than a mile in a minute. Our little fly, in her swiftest flight, will in the same space of time go more than the third of a mile. Now compare the infinite difference of the size of the two animals (ten millions of the fly would hardly counterpoise one racer), and how wonderful will the velocity of this minute creature appear! Did the fly equal the race-horse in size, and retain its present powers in the ratio of its magnitude, it would traverse the globe with the rapidity of lightning.*

Bees, again, are accomplished masters of aërial motion. The Humble-bees, notwithstanding their heavy bodies, are the most powerful fliers of this class. The same excellent entomologist tells us that they "traverse the air in segments of a circle, the arc of which is alternately to right and left. The rapidity of their flight is so great that, could it be calculated, it would be found, the size of the creature considered, far to exceed that of any bird, as has been proved by the observations of a traveller in a railway carriage proceeding at the rate of twenty miles an hour, which was accompanied, though the wind was against them, for a considerable distance by a Humble-bee (*Bombus subinterruptus*), not merely with the same rapidity, but even greater, as it not unfrequently flew to and fro about the carriage, or described zig-zag lines in its flight. The aërial movements of the Hive-bee are more distinct and leisurely." †

* Intr. to Entom. Lett. xxii.

† Ibid.

You have doubtless often admired the noble Dragon-fly, with its four ample and wide-spread wings of gauze, hawking in a green lane, or over a pool in the noon of summer. It sails, or rather shoots with arrowy fleetness hither and thither, now forwards, now backwards, now to the right, now to the left, without turning its body, but simply by the action of its powerful and elegant wings. Leeuwenhoek once saw an insect of this tribe chased by a swallow in a menagerie a hundred feet long. The Dragon-fly shot along with such astonishing power of wing, to the right, to the left, and in all directions, that this bird of rapid flight and ready evolution was unable to overtake and capture it, the insect eluding every attempt, and being in general fully six feet in advance of the bird. A Dragon-fly has been known to fly on board a ship at sea, the nearest land being the coast of Africa, five hundred miles distant, a fact highly illustrative of its power of wing.

It is a point of interest to know the structure of the organs by which such results are accomplished, and therefore we will devote an hour to the microscopical examination of the wings of one or two Insects. Let us begin with the common Fly, one of which, a fine blue-bottle, is somewhat noisily buzzing in the window—

> "The blue-fly sung i' the pane,"—

as if to invite our attention to him. Well, we will borrow one of his wings for the lesson: and putting it into the stage-forceps, we shall be able to turn it in any direction for observation beneath the microscope.

At first it seems a very thin transparent membrane, of a shape between triangular and oval, with a few fine black lines running through it, and along one edge. But on bringing a greater magnifying power to bear on it, we see that the clear surface is covered with minute short stiff hairs, each of which has an expanded base. And still further, by delicate focussing, we find that there are two sets of these hairs, which come into view alternately, those of one row projecting upward towards our eye, those of the other downward. They are placed on both the upper and under-surface, and are in fact appendages of two distinct membranes, applied to each other. There is some reason to believe that these hairs are delicate organs of touch communicating impressions through the skin to a sensitive layer beneath; at least such seems their function on the body, and we may judge from analogy that it is not different here.

The black lines are elastic, horny tubes, over which the membranes are spread and stretched, like the silk of an umbrella by its ribs. The upper membrane is firmly attached to the tubes (which are called nervures); the lower has but a slight adhesion, and is easily stripped from them. The nervures originate in the body, and diverge like a fan to various points of the tip, and to the upper and lower edges; some of them, however, terminate in the substance of the wing without reaching the edge, and some send off cross branches by which two are connected together. They generally maintain the same thickness throughout, but there are enlargements where the branches join the main trunks. These nervures are hollow, and are during life filled with a subtle fluid, which is supplied from the vessels of the body. They contain also ramifications of the exquisite spiral air-vessels, which we shall presently consider, so that both air and blood circulate in them.

In this wing of the Bee all of these structures may be seen to greater advantage. The membrane appears perfectly homogeneous by transmitted light, even with so high a magnifying power as 600 diameters, at least on a cursory examination; though, by careful manipulation, we may discern faint traces of angular lines which divide the whole surface into irregular areas. But by using reflected light at an oblique angle, this areolation, which indicates the primary cells of the structure, is much plainer, and each area is perceived to carry a single hair in its centre.

The hairs themselves here take the character of curved spines, not unlike those of a rose-tree. Along the front edge of the wing they are straight, stout, densely crowded, and overlapping in an inclined position; but the most interesting modification of these organs is seen at the front edge of the posterior pair. Unlike the Fly, which has but a single pair of wings, the Bee has two pairs, of which the fore pair is the larger and more horny, the hinder pair seeming to be, as it were, cut out of the hinder and inner side of the fore ones. The two edges—the hinder edge of the fore pair and the front edge of the hind pair—then correspond, but it is necessary that, during flight, when the wings are expanded, the two wings on each side should *maintain* this relative position, neither overlapping the other, but together presenting one broad surface, wherewith to beat the air. There must be, therefore, some contrivance for locking together the two edges in question, which yet shall be capable of being unlocked at the pleasure of the animal; for the wings during repose slide over one another. This contrivance is furnished by a series of hairs or spines running along the front edge of the hind-wing; they are bent up into strong semicircular hooks, arching outwards, looking, under a high power, like the hooks on a butcher's stall. On the other hand, the margin of the fore-wing is strengthened, and is turned over with a shallow doubling, so as to make a groove into which the hooks catch; and thus, while the fore-wings are expanded, the hooks of the other pair are firmly locked in their doubled edge, while, as soon as flight ceases, and the wings are relaxed, there is no hindrance to the sliding of the front over the hind pair.

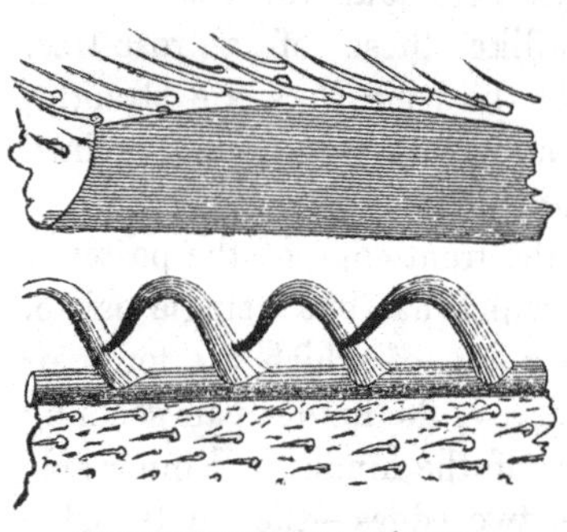

DOUBLING AND HOOKS IN A BEE'S WING.

The wings of many Insects are interesting on account of the organs with which they are clothed. A familiar example is furnished by the common Gnat, a wing of which is on the slide now before me. There is the same general structure as before,—two clear elastic membranes stretched over slender horny tubular nervures, and studded on both surfaces with short spine-like hairs, which in this case, however, are excessively numerous and minute. But along the nervures, and along other lines which run (generally) parallel with the front margin, and also along the whole margin, there are set long leaf-like scales of very curious appearance and structure.

Confining our attention to one of these lines, suppose one of the nervures, we see that its course is marked on the upper membrane by five rows of minute elevated warts, arranged obliquely with one another. From each of these warts springs a slender stem which gradually dilates into a thin leaf-shaped plate of transparent substance, having from four to eight or ten longitudinal ribs. They project in a radiating manner, all inclined towards the tip of the wing. The same line on the under-surface carries the like number of leaf-like plates, corresponding in arrangement, structure, form and direction with those on the upper side. The margins of the wing all round are furnished with similar organs, with this difference, that whereas the plates along the lines are as it were cut off abruptly at their greatest diameter, the marginal ones converge again with a gracefully curved outline, to a fine point: a form which is seen to the greatest advantage along the hind edge of the wing, while those of the front margin are thicker, and more densely crowded.

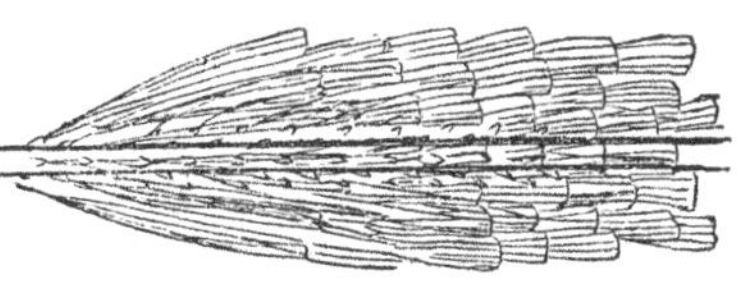

SCALES ON A GNAT'S WING.

There are, however, other Insects which display these or similar appendages in far greater profusion, and in much variety of form and appearance. In the fissures of cliffs that border the sea-shore may often be found some wingless but active Insects, which are endowed with the power of leaping in great perfection. From their hinder extremity being furnished with long projecting bristles, they are sometimes called Bristle-tails, but naturalists designate the genus *Machilis*. If you can get one sufficiently still to examine it, you will be delighted with the lustre of its clothing, which appears dusted all over with a metallic powder of rich colours,—red, brown, orange and yellow, foiled by dull lead-grey in places.

BRISTLE-TAIL.
(Slightly enlarged.)

If you touch one of these nimble leapers, though ever so lightly, you will see the result on your finger-ends; for they will be found covered with a thin stratum of the finest dust, which displays the coloured metallic reflections seen on the insect. By touching one with a plate of glass instead of your finger, you will get the same dust to adhere to this transparent medium by applying which to the microscope you may at once discern the marvellous nature of the raiment with which the little creature is bedecked.

The dust is now seen to be composed of myriads of thin scales, mostly regular and symmetrical in their forms, though varying exceedingly among themselves in this respect. Some are heart-shaped, some shovel-shaped, some round, oval, elliptical, half round, half elliptical, long and narrow, sometimes irregular and unequal, and of various other undescribable outlines. Perhaps the most common forms are ovate, heart-shaped, and that of the pan of a fire-shovel. Each thin scale has a minute foot-stalk, which is not connected with it at either extremity, but at a point of one surface a little way from the smaller end, whence it projects at an oblique angle; so that when the stalk was inserted in its proper cell in the skin of the insect, the scale lay horizontally, covering the insertion. This is a peculiarity not found in some other scales that I shall presently introduce to you.

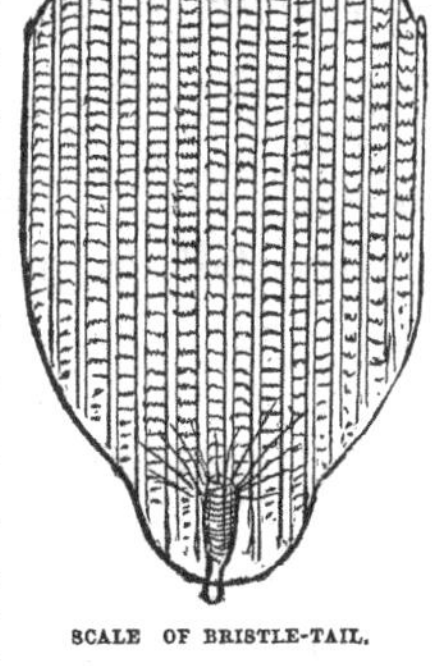
SCALE OF BRISTLE-TAIL.

The whole body of the scale is traversed by a series of fine close-set parallel lines, running longitudinally from end to end. At least this is the ordinary arrangement; but occasionally you see scales, in which there are two series of parallel lines, arranged on either side of an imaginary central line, towards which they respectively converge, but never, I think, diverge from it. These lines appear to form thickened ribs, but seem to be made by elevations of the membrane both above and below. Between the ribs, on the larger scales, we see a number of very delicate cross lines, which are probably regular wrinklings of the depressed surface.

There is another little Insect of the same family, commonly found in cupboards, and in chinks of old damp houses, and called the Sugar-louse (*Lepisma*); very much like this, but of a silvery lustre, and carrying the three bristles of the tail diverging widely. This also is covered with scales, some of which are preserved on this glass slide. Here, while the general appearance and structure agree with those of the scales which we have just been considering, there is considerable diversity in details. The form is usually ovate or shovel-like; the foot-stalk, projecting at a similar angle, is not set on the inferior surface, but in the bottom of a deep narrow incision; and the ribs are invariably divergent like those of a fan. In these, however, there is a peculiarity of arrangement, which I have never seen noticed, but which is obvious enough in the specimens before us. The ribs on the two surfaces diverge at a different angle, those of the upper surface being the more divergent, divaricating from the foot-stalk, while those of the lower membrane are coarser, and much more nearly parallel, their bases ranging along the hind edge of the scale. The effect of the intersection of the sets of lines at so acute an angle, is to convey the optical impression that the scale is covered with short irregular dashes.

Such is the arrangement on these scales, which I prepared myself from the common *Lepisma*; but I have a slide marked "Lepisma," from one of the dealers in microscopic objects, in which the ribs on the two sides concur; but, on one side, there are obliquely divergent lines visible only near the margin, which appear to be produced by wrinkles of the membrane analogous to the transverse dashes on those of *Machilis*.

Scales much more delicate than either of these sorts are found on the *Podura*, a minute insect of which there are several species; which leap, jerking out the bristles of the tail, that are ordinarily carried under the body, like a coiled spring. They are common in cellars, in hot-bed frames, on dunghills, on the surface of water in road-ruts, &c. On the slide before you are some of the smaller scales from one of these insects; they are exceedingly delicate, and the clearness with which you discern the character of their markings, is a proof of the good definition of my microscope; this is what is called a "test-object." At first sight they seem covered with ribs like those of the *Machilis* and *Lepisma*; but, by the use of a magnifying power of 600 diameters (as I have now put on), you perceive that, in the first place, the lines are not straight nor parallel, but curve irregularly, and are often branched; and, in the second place, that they are not uninterrupted, but made up of a series of successive wedge-shaped warts, which lie nearly flat, but project a little at the larger end, where each overlaps the next. The scale we are looking at measures ·0014 inch in length, and ·0009 in width; here the marks are well defined; here are smaller scales ·0008 in length by ·00035 in width, but these are more dim and difficult to resolve.

The beautiful and extensive order called Lepidoptera or Scale-winged, *par excellence*, including the gay tribes of Butterflies and Moths, present us with many exceedingly interesting varieties in these singular coverings. The study of these might be almost as wide as the immensity of species; I can only show you a few examples.

Here are specimens from the pretty little white Five-plume Moth (*Pterophorus*), so common in meadows in summer. The general shape of the scales from the body and wings is that of a willow-leaf, some singly pointed, but more cut at the tip into two, three, or four notches. Those from the legs are longer and slenderer in proportion; and among the others from the wings, there are some which take the form of hairs, which send forth one or more branches from one side, that form a very acute angle with the main stem. The scales proper are all marked with longitudinal lines, very minute and close, but they mostly bear a central band, and sometimes a marginal one on each side, of spots set in sinuous lines like the bands on a mackerel's back; these are probably composed of pigment-granules.

These from the pretty Six-spot Burnet Hawk-moth, are nearly opaque, especially those from the red parts of the wings, which have a rich ruddy glow by transmitted light. They are narrow in shape, tapering gradually forward from the foot-stalk, and terminate mostly in two blunt points. The ribs are coarse for the size of the scales, and the depressed spaces are marked with irregular pigment-grains.

The hairs with which the bodies of Moths are

invested are essentially of the same character as the scales which clothe their wings. Here are examples from the glowing sides of the abdomen of that richly coloured insect, the Cream-spot Tiger-moth (*Arctia villica*). You see they are simple scales, drawn out to an inordinate length and great tenuity; each has its quill-like foot-stalk, and we may trace on some of them the ribs and transverse dotting, while here we see all intermediate stages between the slenderest hair and the broadly ovate, bluntly-pointed scales from the wings.

You are familiar of course with the brilliant little Blue Butterfly (*Polyommatus Alexis*), which dances and glitters in the sunshine on waste places in June. Among the scales of ordinary form which clothe the lovely little wings will occur one here and there of a different shape from the rest. Here you may see one; it is much smaller than the average; the foot-stalk is very long, and the shape of the entire scale is that of a battledoor. The ribs are rather few and coarse, and they have this peculiarity, that each rib swells at intervals into rounded dilatations, each of which has a minute black point in its centre. In some of these battledoor scales there is, near the lower part of the expansion, a crescent of minute pigment-grains.

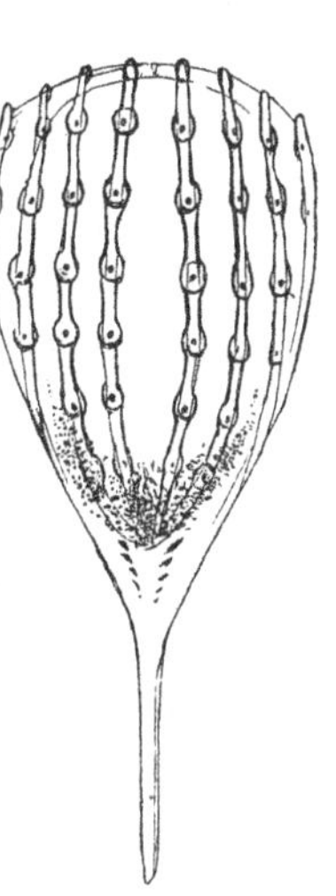

BATTLEDOOR SCALE OF POLYOMMATUS ALEXIS.

The silvery grey surface of the front wings of a common moth, known as the Buff-tip (*Pygæra bucephala*), is composed of scales of unusual magnitude, and of a remarkable form, their shape being that of an expanded fan, and being quite distinguishable by the unassisted eye. The ribs are very fine and numerous, and there are diverging lines of pigment-grains running through the scale.

Those of the Emperor Moth (*Saturnia pavonia-minor*), are likewise triangular in outline, and are remarkable for being deeply notched at the end; so deeply as to leave projecting points (from two to five) as long as, or even longer than, the integral portions of the scale.

In some species we find scales the tips of which are furnished with a curious sort of fringe. This slide presents several such in the midst of many of a more ordinary shape and appearance. The scales in question are straight, and parallel-sided, rather narrow, with the basal end rounded, and the terminal extremity tapered abruptly to a point; it is on each slope of this point that the fringe is arranged. The surface does not appear to be elevated in ribs, but smooth; while the whole interior, except a crescent around the foot-stalk, is filled with pigment-grains, imparting a mottled appearance. It is remarkable that all the scales of this form have the foot-stalk turned in under the expanse. The

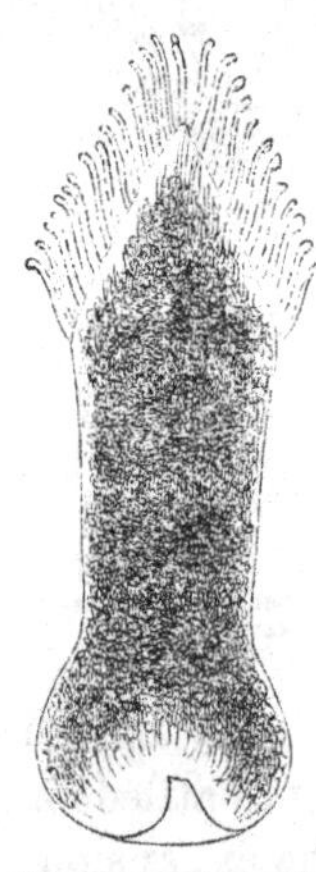
FRINGED SCALE OF PIERIS.

example which we are considering is from the white portion of the wing of *Pieris Glaucippe*, a fine butterfly from China; but a similar structure is found in our own Garden Whites, and Meadow Browns, (*Pieridæ* and *Satyridæ*).

Scales taken from the brilliant changeable blue-green patch in the hind-wing of *Papilio Paris*, a fine Indian butterfly, have an interesting appearance. They are simply pear-shaped in outline, with few longitudinal ribs set far apart, and numerous strongly-marked corrugations running across between them. That these are really elevations of the surface, is well seen in some scales, even with transmitted light, and a high power; for the slopes of the wrinkles that face the light display the lustrous emerald reflection proper to the wing, while the transmitted colour of the whole scale is a rich transparent red.

The dimensions of the scales do not bear any certain proportion to the size of the insect which is clothed with them; those from the broad wings of the noble *Saturnia Atlas*, for example, eight or nine inches in expanse, being exceeded in size by some from those of our little native Muslin Moth, an inch wide.

You will say that what I am about to show you is a lovely object; but for its right display I must use a low magnifying power,—not higher than a hundred diameters,—with the Lieberkuhn to reflect the light of the mirror full upon the surface. It is a small fragment cut from the wing of *Papilio Paris*, showing several rows of the scales in their natural arrangement. The gemmeous radiance of the glittering green scales on the black ones, by which they are environed, glares out with a splendid effect; and what is more interesting, you can trace the manner in which they are set,—those of each row slightly overlapping the bases of another row, like slates on a roof,—and also the mode in which they are inserted. The clear horn-coloured membrane of the wing is seen raised in shallow transverse *steps* (if I may use such a term), so that if it were divided longitudinally, the edge would appear cut into saw-like teeth. Along the margins of these ridges are set minute sockets, which are very distinctly seen, where the scales have been displaced; in these the tiny footstalks of the scales are inserted.

The little Beetles which we are familiar with under the name of Weevils, characterised by their long slender snouts, at the end of which they carry curiously folding antennæ, and which constitute the family *Curculionidæ*, are in many cases clothed with scales, to which they owe their colours and patterns. Several of our native species display a green or silvery lustre, which under the microscope is seen to be produced by oval scales. But these are eclipsed by the splendour of many tropical species; especially that well-known one from South America, which is called the Diamond Beetle, and scientifically *Entimus imperialis*, from its unparalleled magnificence.

A piece of one of the wing-cases of this beetle is gummed to the slide now upon the stage. We look at it by reflected light with a magnifying power of 130 diameters. We see a black ground, on which are strewn a profusion of what look like precious stones blazing in the most gorgeous lustre. Topazes, sapphires, amethysts, rubies, emeralds seem here sown broadcast; and yet not wholly without regularity, for there are broad bands of the deep black surface, where there are no gems, and, though at considerable diversity of angle, they do all point with more or less precision in one direction, viz. that of the bands.

SCALES OF DIAMOND-BEETLE.

These gems are flat transparent scales, very regularly oval in form, for one end is rather more pointed than the other; there is no appearance of a footstalk, and by what means they adhere, I know not; they are evidently attached in some manner by the smaller extremity to the velvety black surface of the wing-case. The gorgeous colours seem dependent in some measure on the reflection of light from their polished surface, and to vary according to the angle at which it is reflected. Green, yellow, and orange hues predominate; crimson, violet, and blue are rare, except upon the long and narrow scales that border the suture of the wing-cases, where these colours are the chief reflected. Yet there appears to be some positive colour in their substance; for in these latter scales, which projecting beyond the edge of the wing-case can be examined as transparent objects, and that with a high power, the transmitted light is richly coloured with the same tints as the same scales displayed under the Lieberkuhn.

We may derive pleasant instruction from continuing our observations on a few other wings of insects. If you have ever thought on the subject, you have probably taken for granted that the various sounds produced by insects are voices uttered by their mouths. But it is not so. No insect has anything approaching to a voice. Vocal sounds are produced by the emission of air from the lungs, variously modified by the organs of the mouth. But no insect breathes through its mouth; no air is expelled thence in a single species; it is a biting, or piercing, or sucking organ; an organ for the taking of food, or an organ for offence or defence; but never an organ of sound.

The wings are in most cases the immediate causes of insect sounds. On this subject you will read with pleasure some very interesting remarks by the learned Mr. Kirby, inquiring, "by what means these sounds are produced."

"Ordinarily, except perhaps in the case of the gnat, they seem perfectly independent of the will of the animal; and, in almost every instance, the sole instruments that cause the noise of flying insects are their wings, or some parts near to them, which, by their friction against the trunk, occasion a vibration—as the fingers upon the strings of a guitar—yielding a sound more or less acute in proportion to the rapidity of their flight, the action of the air perhaps upon these organs giving it some modifications.

Whether, in the beetles that fly with noise, the elytra [or wing-sheaths] contribute more or less to produce it, seems not to have been clearly ascertained; yet since they fly with force as well as velocity, the action of the air may cause some motion in them, enough to occasion friction. With respect to *Diptera*, Latreille contends that the noise of flies on the wing cannot be the result of friction, because their wings are then expanded; but though to us flies seem to sail through the air without moving these organs, yet they are doubtless all the while in motion, though too rapid for the eye to perceive it. When the aphidivorous flies are hovering, the vertical play of their wings, though very rapid, is easily seen; but when they fly off it is no longer visible. Repeated experiments have been tried to ascertain the cause of sound in this tribe, but it should seem with different results. De Geer, whose observations were made upon one of the flies just mentioned, appears to have proved that, in the insect he examined, the sounds were produced by the friction of the root or base of the wings against the sides of the cavity in which they are inserted. To be convinced of this, he affirms, the observer has nothing to do but to hold each wing with the finger and thumb, and, stretching them out, taking care not to hurt the animal, in opposite directions, thus to prevent their motion—and immediately all sound will cease. For further satisfaction he made the following experiment. He first cut off the wings of one of these flies very near the base; but finding that it still continued to buzz as before, he thought that the winglets and poisers, which he remarked were in a constant vibration, might occasion the sound. Upon this, cutting both off, he examined the mutilated fly with a microscope, and found that the remaining fragments of the wings were in constant motion all the time that the buzzing continued; but that by pulling them up by the roots, all sound ceased. Shelver's experiments go to prove, with respect to the insects that he examined, that the winglets are more particularly concerned with the buzzing. Upon cutting off the wings of a fly—but he does not state that he pulled them up by the roots—he found the sound continued. He next cut off the poisers—the buzzing went on. This experiment was repeated eighteen times with the same result. Lastly, when he took off the winglets, either wholly or partially, the buzzing ceased. This, however, if correct, can only be a cause of this noise in the insects that have winglets. Numbers have them not. He next, therefore, cut off the poisers of a crane-fly (*Tipula crocata*), and found that it buzzed when it moved the wing. He cut off half the latter, yet still the sound continued; but when he had cut off the whole of these organs the sound entirely ceased." *

There is a pretty little beetle (*Clytus*), not uncommon in summer in gardens, remarkable for the brilliant gamboge-yellow lines across its dark wing-cases, which makes a curious squeaking sound when you take it in hand. You think it is crying; but if you carefully examine it with a lens while the noise is uttered, you will perceive that the cause is the

* Introd. to Entom. Lett. xxiv.

grating of the thorax against the front part of the two wing-cases. Several other beetles produce similar sounds when alarmed, by rubbing the other end of the wing-sheaths with the tip of the abdomen. Many of those genera which feed on ordure and carrion do this.

But the noisiest of all insects are those of the classes *Orthoptera* and *Homoptera*, the Crickets and Grasshoppers, and the Treehoppers. And these shall bring us back to our microscope, to which we shall return with the more zest, after this little interval of repose for our strained eyes.

Listen! we hear coming up the kitchen-stairs, the stridulous chirping of the House-cricket (*Acheta domestica*).

"The cricket chirrups on the hearth."

The cook shall catch us one for investigation. "Please, sir, here's the crickets: here's half a dozen on 'em. I don't like 'em, I don't; nasty noisy varmint!" Thank you, cook; we'll try and turn them to some useful purpose to-day, at least.

Now, you see, each of the upper wings or wing-cases has a clear space near the centre, of a triangular form, crossed by one or two slender nervures. This space has received the name of the tympanum or drum. It is bounded externally by a broad dark nervure, which with a low power we see is scored with three or four longitudinal furrows, of course separated by as many horny ridges. In front of the clear drum, and forming a curved base to the triangle, there passes across a horny ridge, tapering outwards, which is roughened throughout its length by close-set teeth exactly like a file. When the insect chooses to be musical, it partially opens and closes its wing-sheaths, causing the two files to rub across each other; and this gives rise to the peculiar ringing vibration, the intensity of which is heightened by the tense "drum" acting as a sounding-board.

So at least some say; but M. Goureau, who has published some elaborate observations on the chirping of insects,* asserts that the sound is chiefly owing to the action of the "file" (which he calls the "bow") on the longitudinally-ridged nervure, which he calls the "treble-string."

We see in this individual, that is so obliging as to produce what cook calls its "nasty noise" before us, that he elevates the wing-sheaths so as to form an acute angle with the body, and then rubs them together with a very brisk horizontal motion; but which of the nervures it is that actually produces the sounds, it would require a very careful and elaborate series of experimental researches to determine. It has been asserted that the legs play a part in the music by being rubbed against the bows; this, however, seems improbable from their relative position.

In the Southern United States, I have had opportunities of seeing and of hearing a very noisy performer of the Gryllus tribe, called the Katedid (*Pterophylla concava*), which sings through the night in the foliage of the trees. The sounds, reiterated on every side, resemble a score or two of quarrelsome people with shrill voices, divided into pairs, the individuals of each pair squabbling with each other; "I did!"

* Ann. Soc. Ent. de France.

"You didn't!" "I did!" "You didn't!" the objurgation maintained with the most amusing pertinacity, and without a moment's intermission. Here the wing-sheaths, which are large and as it were inflated, are certainly the organs of sound. A portion of each is turned, at right-angles to the remainder, over the back, so that the one partly overlaps the other. The musical organ consists of a hard glassy ridge in front, behind which is a transparent membrane, which appears tightly stretched over a semicircular rim, like the parchment of a drum, answering in structure and in function to the part so compared in the cricket.

This Gryllus I found would crink freely, when held in my fingers, provided I held it by the head or thorax, so as not to interfere with the freedom of the wing-cases; though these needed only to be partially opened, the bases being merely slightly separated without affecting the general contiguity. The two glassy ridges were rubbed across each other, making the sharp crink. Ordinarily this was done thrice, three distinct but rapid crossings making the sound represented by the word "Katedid;" but occasionally the insect gave but a single impulse, uttering as it were but one syllable of the word.

The Locusts and Grasshoppers, however, do, it appears, make use of their hind legs in producing their crink. If you look at this Grasshopper's leg, you will see that the thigh is marked with a number of transverse overlapping angular plates, and that the shank carries a series of short horny points along each side. The insect when it crinks, brings the shank up to its thigh, and rubs both to and fro against the wing-sheaths, doing this by turns with the right and left legs, which causes the regular breaks in the sound. The drum, on which this rubbing vibrates, has been described by the anatomist, De Geer:— "On each side of the first segment of the abdomen," says he, "immediately above the origin of the posterior thighs, there is a considerable and deep aperture of rather an oval form, which is partly closed by an irregular flat plate or operculum of a hard substance, but covered by a wrinkled flexible membrane. The opening left by this operculum is semilunar, and at the bottom of the cavity is a white pellicle of considerable tension, and shining like a little mirror. On that side of the aperture which is towards the head there is a little oval hole, into which the point of a pin may be introduced without resistance. When the pellicle is removed, a large cavity appears. In my opinion this aperture, cavity, and above all the membrane in tension, contribute much to produce and augment the sound emitted by the grasshopper."*

In this case we may without hesitation conclude that the friction of the thigh-plates and shank-points on the rough edges of the wing-cases, produces the musical vibration on the tense membrane, as rubbing a wet glass with the finger will yield a loud musical note.

The most elaborate contrivance for the production of sounds among the Insect races, however, is found among the Cicadæ, celebrated in classical poetry as the very impersonations of song and eloquence. I regret I cannot show you this apparatus; for though we

* De Geer, iii. 471.

have a British species,—lately discovered in the New Forest,—it is very rare. Should you travel, however, either in the old or new world, you will have abundant opportunities of using your microscope to verify the following description by our prince of entomologists, Mr. Kirby.

"If you look at the under side of the body of a male, the first thing that will strike you is a pair of large plates of an irregular form—in some semi-oval, in others triangular, in others again a segment of a circle of greater or less diameter—covering part of the belly, and fixed to the trunk between the abdomen and the hind legs. These are the drum-covers or opercula, from beneath which the sound issues. At the base of the posterior legs, just above each operculum, there is a small pointed triangular process, the object of which, as Réaumur supposes, is to prevent them from being too much elevated. When an operculum is removed, beneath it you will find on the exterior side a hollow cavity, with a mouth somewhat linear, which seems to open into the interior of the abdomen: next to this, on the inner side, is another large cavity of an irregular shape, the bottom of which is divided into three portions; of these the posterior is lined obliquely with a beautiful membrane, which is very tense—in some species semi-opaque, and in others transparent—and reflects all the colours of the rainbow. This mirror is not the real organ of sound, but is supposed to modulate it. The middle portion is occupied by a plate of a horny substance, placed horizontally, and forming the bottom of the cavity. On its inner side this plate terminates in a *carina* or elevated ridge, common to both drums. Between the plate and the after-breast (*postpectus*) another membrane, folded transversely, fills an oblique, oblong, or semilunar cavity. In some species I have seen this membrane in tension; probably the insect can stretch or relax it at its pleasure. But even all this apparatus is insufficient to produce the sound of these animals; one still more important and curious yet remains to be described. This organ can only be discovered by dissection. A portion of the first and second segments being removed from that side of the back of the abdomen which answers to the drums, two bundles of muscles meeting each other in an acute angle, attached to a place opposite to the point of the *mucro* of the first ventral segment of the abdomen, will appear. In Réaumur's specimens, these bundles of muscles seem to have been cylindrical; but in one I dissected (*Cicada Capensis*) they were tubiform, the end to which the true drum is attached being dilated. These bundles consist of a prodigious number of muscular fibres applied to each other, but easily separable. Whilst Réaumur was examining one of these, pulling it from its place with a pin, he let it go again, and immediately, though the animal had been long dead, the usual sound was emitted. On each side of the drum-cavities, when the opercula are removed, another cavity of a lunulate shape, opening into the interior of the abdomen, is observable. In this is the true drum, the principal organ of sound, and its aperture is to the *Cicada* what our larynx is to us. If these creatures are unable themselves to modulate their sounds, here are parts enough to do it for them:

for the mirrors, the membranes, and the central portions, with their cavities, all assist in it. In the cavity last described, if you remove the lateral part of the first dorsal segment of the abdomen, you will discover a semi-opaque and nearly semicircular concavo-convex membrane with transverse folds; this is the drum. Each bundle of muscles, before mentioned, is terminated by a tendinous plate nearly circular, from which issue several little tendons that, forming a thread, pass through an aperture in the horny piece that supports the drum and are attached to its under or concave surface. Thus the bundle of muscles being alternately and briskly relaxed and contracted, will by its play draw in and let out the drum: so that its convex surface being thus rendered concave when pulled in, when let out a sound will be produced by the effort to recover its convexity; which, striking upon the mirror and other membranes before it escapes from under the operculum, will be modulated and augmented by them. I should imagine that the muscular bundles are extended and contracted by the alternate approach and recession of the trunk and abdomen to and from each other.

"And now, my friend," adds the excellent author, "what adorable wisdom, what consummate art and skill are displayed in the admirable contrivance and complex structure of this wonderful, this unparalleled apparatus! The great Creator has placed in these insects an organ for producing and emitting sounds, which in the intricacy of its construction seems to resemble that which He has given to man and the larger animals for receiving them. Here is a *cochlea*, a *meatus*, and, as it should seem, more than one *tympanum!*"

In some instances the sounds of insects more nearly approach the character of true voices; at least so far as they are produced by the emission of air from the breathing organs. One of the most eminent of continental entomologists, Dr. Burmeister, tells us so. Finding that the buzz of a large fly (*Eristalis tenax*) still continued after the winglets, the poisers, and even the wings, had been quite cut off except their stumps (only in this last case the sound was somewhat weaker and higher), he conceived that the spiracles lying between the *meso-* and *meta-thorax* must be the instruments of the sound; which, accordingly, he found to cease entirely when they were stopped with gum, though while the wings were in vibration. Pursuing his researches, he extracted one of these spiracles, and opening it carefully, found its posterior and inner lip, which is directed towards the commencement of the *trachea*, to be expanded into a small, flat, crescent-haped plate, upon which are nine parallel, very delicate, horny laminæ, the central one being the largest, while those on each side become gradually smaller and lower; so it is, he is persuaded, in consequence of the air being forcibly driven out of the *trachea* and touching these laminæ that they are made to vibrate and sound, precisely in the same way with the *glottis* of the *larynx*. Dr. Burmeister (who remarks that Chabrier, in his *Essai sur le Vol des Insectes*, p. 45, &c., has also explained the hum of insects as produced by the air streaming from the thorax during flight, and also speaks of laminæ which lie at the aperture of the spiracles), in order to be certain that the laminæ in question in the posterior spiracles of the thorax are alone concerned in producing sound, also inspected the anterior ones, but without finding in them any trace of these laminæ. He explains the weaker and sharper tones produced when the wings, all but the very roots, are cut off, as resulting from the weaker vibrations of the contracting muscles, and consequently less forcible expulsion of the air when the vibratory organs are removed; and he thinks with Chabrier that some air may escape through the open *tracheæ* of the wings which are cut off. Though he regards these laminæ as the cause of humming in bees and flies, he does not decide that other causes may not produce the buzz of cockchafers, &c., in the thoracic spiracles of which he could not discern them.*

* Man. of Entom. 468.

CHAPTER VI.

INSECTS: THEIR BREATHING ORGANS.

In order to understand the passage last quoted from Burmeister, you ought to know something of the manner in which breathing is performed among insects. Essentially, breathing is the same function, wherever it occurs; and it does occur, doubtless, in all animals under some form or other. It is the absorption of oxygen from without to the fluids within, to repair the waste constantly produced by vital energy. But it may be obtained from different sources, and imbibed in various modes.

All insects in the perfect state are air-breathers; that is, they procure their oxygen from the air as we do; and most of them are so in their earlier stages. Even in exceptional cases, viz. such larvæ or pupæ as are provided with what represent gills, and appear to be dependent on the water for their respiration, the exception is rather apparent than real, for the function is performed in air-vessels still. Now these air-vessels shall afford us some interesting microscopical observations.

This brown fly, which is buzzing and hovering on invisible wings over the flowers in the garden, you perhaps take for a bee. No; it has but two wings; for I have caught it, and you may ascertain the fact for yourself; it belongs to the genus *Syrphus*. Having caught it, I deprive it of life by means of the very organs I am going to examine, for I turn a tumbler over it and insert under the edge a lighted lucifer-match. In a few seconds it is dead,—suffocated; for phosphoric and sulphuric acids introduced into the breathing tubes quickly destroy life. I presently take it out, and putting it into a dissecting-trough under a lens, cut up the abdomen with a pair of fine-pointed scissors. Then I pin open the divided abdomen to the bottom of the trough, which is coated with wax for the purpose; and, looking at it with the lens—but you shall look for yourself.

Well, you see little else but the polished brown walls of the body and a number of fine white threads. It is those threads that we want. With a small camel's hair pencil I move them to and fro in the water, and soon perceive that they are like little trees with comparatively thick trunks, sending off many branches, and gradually becoming excessively slender. Here and there short thick branches break out on two opposite sides, and on each side are connected with the wall of the abdomen. Here then with the fine scissors I snip them across, and lift up a portion with the hair pencil into a drop of water which I have already put into the live-box. The cover now flattens the drop, spreads the white threads,—and the object is ready for our eye.

We have before us a considerable portion of the tracheal system of the fly. And though, owing to the involution of the parts and the injury our rude anatomy has done, we cannot trace the beautiful regularity which exists in life, we may see the principle on which they are arranged, and much of the perfection with which they are constructed.

Here then is a system of pipes,—some large, some small; the smaller branching forth from the large, and themselves sending off yet smaller branches, which in their turn divide and subdivide until the final ramifications are excessively attenuated. Besides these, we see here and there ovate or barrel-shaped reservoirs, having the same appearance and intimate structure as the pipes, but of much larger calibre and connected with them by a branch.

This, I say, is the breathing system, or a large portion of it. These pipes receive the air from without through trap-doors, which we will examine presently, and convey it to the most distant parts of the body. In ourselves the air is inhaled into a great central reservoir—the lungs, and the blood dispersed through every part is brought to this reservoir to be oxygenated. In insects it is the blood that is collected into a great central reservoir, and the air is distributed by a minutely divided system of vessels over the blood-reservoir.

The *tracheæ* or air-pipes have a silvery white appearance by reflected light; but if we use transmitted light and put on a high power, we discern a wonderful structure, which I will describe in the eloquent language of Professor Rymer Jones, and you shall estimate its truth as you examine the object:—

"There is one elegant arrangement connected with the breathing-tubes of an insect specially worthy of admiration; and perhaps in the whole range of animal mechanics it would be difficult to point out an example of more exquisite mechanism, whether we consider the object of the contrivance or the remarkable beauty of the structure employed. The air-tubes themselves are necessarily extremely thin and delicate; so that on the slightest pressure their sides would inevitably collapse and thus completely put a stop to the passage

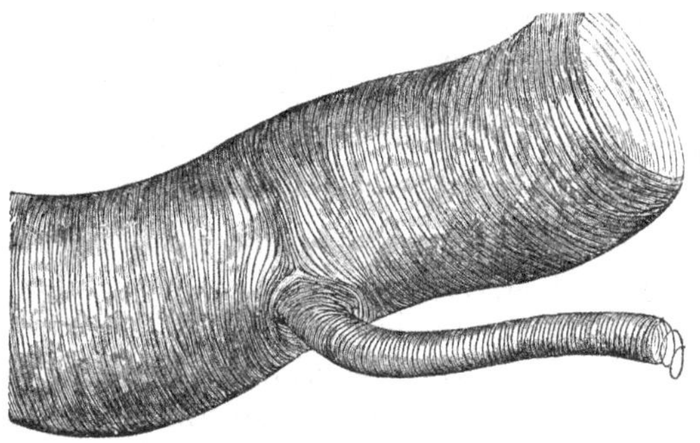

AIR-PIPE OF FLY.

of air through them, producing, of course, the speedy suffocation of the insect, had not some means been adopted to keep them always permeable; and yet to do so, and at the same time to preserve their softness and perfect flexibility, might seem a problem not easily solved. The plan adopted, however, fully combines both these requisites. Between the two thin layers of membrane which form the walls of every air-tube a delicate elastic thread (a wire of exquisite tenuity) has been interposed, which, winding round and round in close spirals, forms by its revolutions a cylindrical pipe of sufficient firmness to preserve the air-vessels in a permeable condition, whilst at the same time it does not at all interfere with its flexibility; this fine coil is continued through every division of the *tracheæ*, even to their most minute ramifications, a character whereby these vessels are readily distinguishable when examined under the microscope."*

Man has imitated this exquisite contrivance in the spiral wire-spring which lines flexible gas-pipes; but his wire does not pass between two coats of membrane. One of the most interesting points of the contrivance is the way in which the branches are (so to speak) inserted in the trunk, the two wires uniting without leaving a blank. It is difficult to describe how this is done; but by tracing home one of the ramifications you may see that it is performed most accurately,—the circumvolutions of the trunk-wire being crowded and bent round above and below the insertion (like the grain of timber around a knot), and the lowest turns of the branch-wire being suitably dilated to fill up the hiatus.

You must not suppose, however, that the whole of one tube is formed out of a single wire. Just as in a piece of human wire-work the structure is made out of a certain number of pieces of limited length, and joinings or interlacings occur where new lengths are introduced, so, strange to say, it seems to be here. It is strange, I say, that it should be so, when there can be no limit to the resources, either of material, or skill to use it; but so it is, as you may see in this specimen, which has been dissected out of the body of a silkworm. The spiral is much looser here than in the air-tube of the fly, the turns of the wire being wider

* Nat. Hist. of Anim. i. 6.

apart; and hence its structure is much more easily traced. Here you see in many places the introduction of a new wire, always commencing with the most fine-drawn point, but presently taking its place with the rest so as to be undistinguishable from them. In some cases certainly (perhaps this may be the explanation of the phenomenon in all) the wire so introduced may be found to terminate with the like attenuation before it has made a single volution, and seems to be inserted when the permanent curvature of the pipe would leave the wires on the outer side of the curve too far apart, half a turn, or even much less, then being inserted of supernumerary wire.

I told you that the air enters these tubes through certain "trap-doors." This is not the term which the physiologist employs, certainly: he calls them spiracles. In our own bodies the air enters only at one spiracle, a curiously defended orifice opening just in front of the gullet, at the back of the mouth. But in the class of animals we are now considering there are a good many such breathing orifices. You may see them to great advantage in any large caterpillar, the silkworm for example, where all along the sides of the pearl-grey body you perceive a row of dots, which with a lens you discover to be little oval disks sunken into little pits, of a black hue with a white centre, through which is a very slender slit. There are nine of these organs on each side, a pair to each segment or division of the body, with the exception of the first, which is the head, and of the third and fourth, which are destined to bear the wings; these are destitute of spiracles.

Essentially, these organs, under whatever modifications of form and position they may appear, have the same structure. They are narrow orifices, with two lips capable of being opened at the will of the animal, or accurately closed; and in many soft-skinned insects, such as the silkworm, and most larvæ, they are set in a horny ring, by which means they are prevented from collapsing, through the unresisting character of the general integument. The opening and shutting of them is performed by an internal apparatus of muscles, which is sometimes strengthened by being attached to two horny plates, which project inwardly.

But the most curious thing to be noted in the structure of these spiracles is the contrivance which induced me to call them trap-doors. Small as are their openings, they are still large enough to admit many floating particles of dust, soot, and other extraneous matters, which would tend to clog up the delicate air-passages, and to impede the right performance of their important functions. Hence they need to be guarded with some sort of sieve, or filter, which, while admitting the air, shall exclude the dust.

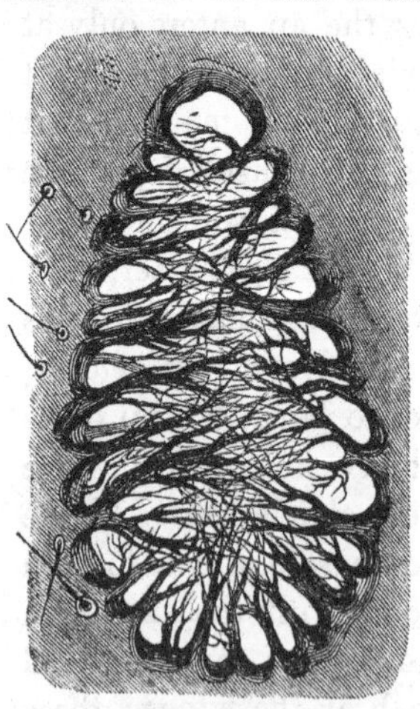

SPIRACLE OF FLY.

Various and beautiful are the modes in which

this common purpose is effected, but I can show you only two or three. This is one of the breathing orifices of the common House-fly, in which, as you see, minute processes grow from the margin all round, which extend partly across the open area, branching and ramifying again and again, and spreading and interlacing with those of the opposite side, so as to form a perfect sieve, which the finest atoms of dust cannot penetrate.

The same end is attained, in another way, in the dirty cylindrical grub, which is found so abundantly at the roots of grass in pasture lands, and which country folk call, from the toughness of its skin, "leather-coat." It is the larva of the Crane-fly (*Tipula oleracea*), so familiar to us under the *soubriquet* of Daddy Long-legs. I can easily procure one of these, for, unfortunately, they are but too ubiquitous. Here is one, who shall have the honour of being martyred for the benefit of science. Before we assassinate him, however, just look here, at the hinder extremity of his body, where there is an area, surrounded and protected by several points, and in this area, two black spots.

With the dissecting-scissors I have carefully cut out one of these specks, and now I put it under the Lieberkuhn, for illumination on the stage of the microscope. There is, first of all, a dark horny ring of an oval figure, a little way within which there is an opaque, dark plate of the same figure, but smaller, occupying the central portion of the area. The space between the margin of the plate and the bounding ring is occupied by a series of slender filaments, placed side by side, proceeding from one to the other, through the interstices of which the air is filtered. The central plate seems to be quite imperforate.

The fat, thick-bodied grubs of those beetles called chafers, exhibit, in their spiracles, a modification of this structure, rendered still more elaborate. In the case of the larva of the common Cockchafer (*Melolontha vulgaris*), for example, the central plate is a projection from one side of the margin of the spiracle, — to use a geographical simile, we may say that, instead of being an island in the midst of a lake, it is a promontory. Thus, the breathing space is a crescent-shaped band, which is crossed in every part by bars passing from the margin to the projecting plate. But, as if the interstices left by these bars would be too coarse for the purpose, they are further sublimated by a membrane, which is stretched across them, and which is perforated with a number of excessively minute round holes, through which alone the air is admitted.

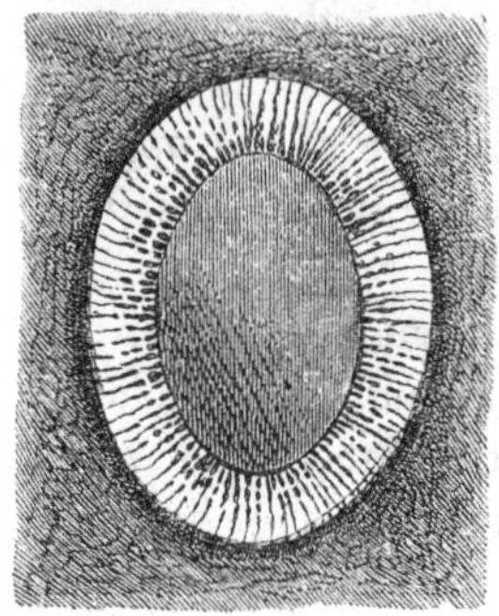

SPIRACLE OF LEATHER-COAT.

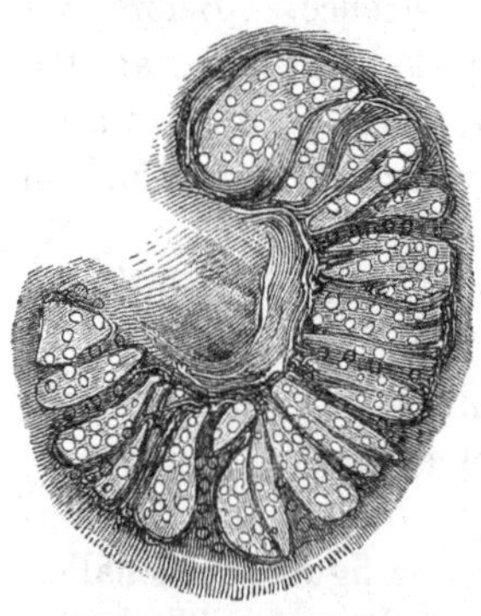

SPIRACLE OF COCKCHAFER-GRUB.

In many of the two-winged flies, which inhabit the water in their earlier stages, there are some interesting contrivances and modifications connected with the organs of respiration. It is necessary that the orifices of the air-tubes should be brought at intervals to the surface of the water, in order to come into contact with the external air; while, at the same time, it is important that as small a portion as possible of the animal's body be exposed to danger, by being protruded from its sheltering element. An example in point you may see in this vase.

Here is a slender worm, an inch and a half in length, thickest a little behind the head, and tapering gradually to a lengthened tail, the twelve segments of the body being very conspicuous. It swims up and down or to and fro in the clear water with a not very rapid, wriggling movement, throwing its body alternately from side to side in the form of the letter S.

This is the maggot of a handsome dipterous fly, sometimes called the Chameleon-fly (*Stratiomys chamæleon*). There is much about it to reward observation and careful examination with a low magnifying power, especially the head, with its pointed snout, and its pair of foot-like palpi. These are situated one on each side of the head, are three-jointed, the last joint being studded with short stiff spines, and the second having a thumb-like projection. With these organs, the grub roots and burrows among the decaying vegetable matter at the bottom for its food; and when not so engaged, they are often rapidly vibrated in a singular manner, the sight of which might induce a feeling of fear, as if they were threatening weapons of offence,—a pair of poisonous stings, for instance; they have, however, no such function, the poor grub being perfectly harmless.

What I wish you chiefly to observe, however, is the tail, with its curious organization. With the naked eye, you can perceive that the last joint is much slenderer and more lengthened than the rest, and that it is tipped with a beautiful crown of feathers, like the diadem of some semi-savage prince. This is best seen when the animal comes to the surface, which it always does tail uppermost, for as soon as the tip reaches the air, the plumes instantly open, and form an exquisite cone or funnel, from which every drop of moisture is excluded, though the water stands around at the level of the brim. A few seconds it remains motionless thus, the whole body hanging downwards, suspended from the caudal coronet, then suddenly the tips of the plumes curve inward toward each other, inclosing a globule of air, and the animal wriggles away into the depths, carrying its burden, like a pearl, or a glittering bubble of quicksilver, behind it.

This you may observe with the unassisted sight, and you may mark, also, how, from time to time, a portion, more or less, of the bubble of gleaming air is inhaled or expired by the animal, causing a diminution or increase of its volume; and this of itself would convince you that it is the spiracles of the animal which are thus protected.

The application of a low magnifying power, say

from thirty-five to fifty diameters, for we can hardly use a higher magnification than this to the animal while alive, will reveal a few more of the details.

We see, then, that the extremity of the last segment forms a circular disk, hollowed in the centre, where it is perforated with the two orifices of the air-pipes. The margin of this disk carries about thirty stiff but slender spines or bristles, some of which are branched in a forked manner. Each bristle bears, on its two opposite sides—viz. on those aspects which face the next bristle on either hand,—two series of not very close-set branchlets, set like the plumes of a feather, or the pinnæ of a fern-leaf, which give it the elegant plumose appearance which the unassisted eye recognises. The bristles have a granulose surface near the extremity, and terminate in fine points.

The curious faculty of repelling water, which the interior surface of this plumy coronal possesses, is of the highest value in the economy of the insect; for, on the one hand, it permits the breathing orifice to be brought into contact with the air, even when nearly a quarter of an inch below the surface; and on the other hand, it allows the volume of air inclosed within the funnel to be perfectly isolated and carried securely away, as a reservoir for the wants of the animal, when engaged in its avocations of necessity or pleasure, in the recesses of its sub-aquatic groves. It is remarkable that so complete is this repellent power, that when the tail is at the surface, the animal may make a very perceptible descent without breaking the continuity of the air, the surface presenting the curious phenomenon of a deep funnel-shaped dimple leading down to the tail of the animal.

The chameleon-fly is not, however, so abundant and so universally distributed as that you may always calculate upon being able to repeat these observations when you will. I shall, therefore, show you an analogous example, much more easily obtained. Both are inhabitants of our fresh waters: the chameleon-grub lives in ponds, crawling among the stems of aquatic plants, and occasionally visiting the surface in the manner you have seen; but it is precarious—in some seasons not uncommon, in others, scarcely to be met with by the most persevering search. For my next specimen, I have but to go with a basin to the water-butt in the yard, and take a dip of the surface-water at random: I shall be pretty sure of a score at least.

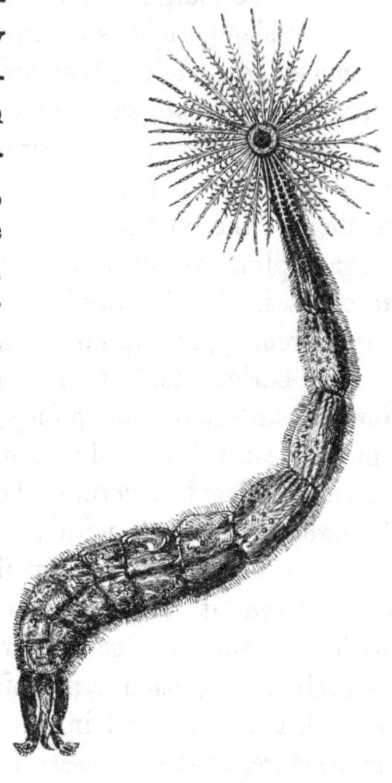

GRUB OF CHAMELEON-FLY.

Here they are swarming, as I told you. What, those things? why, they are gnat-grubs. Well, don't despise them, you will find them worth looking at. I dare say you have never submitted them to half-an-hour's microscopical examination. I have caught one with a spoon, and put it into this narrow glass trough of water that it may rest conveniently on the stage.

We will take a cursory glance at its entire person. Here is a flat, roundish head, a great globose, swollen thorax, and a long, slender, many-jointed body, ending in a curious fork. But all is curious:—the head, with its horny transparency; its pair of rod-like antennæ, covered with minute points; its two black eye-patches; and its jaws, beset with strong, curved hairs, set in radiating rows, and, ever and anon, working to and fro with the most rapid vibrations:—the thorax,—so transparent, with its amber-like clearness, that you can discern the dorsal vessel, which contains the blood, ever dilating and collapsing with the most beautiful regularity; and, beneath this, the gullet, through which, now and then, descends a dark pellet of food, to join the mass already lodged in the stomach farther down,—a result, by the way, that explains that incessant vibration and pumping motion of the mouth-organs, which thus evidently are engaged in collecting food from the water; though, even with this power, we can see no solid matter taken in, till we discern it agglomerated in the swallowed pellets:—the body, or abdomen, with its ten joints, all (with a slight exception) the counterparts of each other; and each carrying its own dilatation of the dorsal vessel, and its own portion of the long and well-filled intestinal canal:—all these, I say, are very interesting and curious to observe; especially when we select, as I have done, a young individual for examination; since the tissues then possess a translucency which is essential to our seeing with distinctness anything of the internal organization, but which soon gives place to opacity, as the insect advances in age.

Very curious, too, are the hairs with which the whole surface of the animal is furnished at certain determinate points. But these are seen to more advantage in an older specimen; for, in this one of tender hours, they are nearly simple; whereas, in an opaque, nearly full-grown individual, every hair is seen to be studded with secondary points, that project from its surface throughout its length. These hairs are arranged in beautiful radiating pencils or tufts, and scattered, as I have said, at definite points over the whole body;—there is a tuft on each antenna; one on the forehead; one in front of each eye-spot; several circles of them set round the thorax; one circle of scanty pencils set round each segment of the body, and a few smaller tufts scattered about besides; all of them springing from minute round warts.

The extremity of the abdomen deserves, however, a separate investigation, and we will now direct our attention to the tail-end of our tiny grub. There are ten segments to the abdomen; at the eighth it seems to divide into two branches, one longer than the other. This appearance, however, is due to the circumstance that the respiratory tube is sent forth from the eighth segment, and that the ninth and tenth segments are bent away at an angle from the general line of the body.

The ninth segment is very small: the tenth is squarish, with rounded corners, and is brought to a thin edge. Around the margin there is the most exquisite array of hairs possible; at one corner there are three pencils; while round the opposite, and down the corresponding side, run in two rows twelve pencils, set very close to each other, and each containing a large number of very slender hairs. The extreme end of the segment is ornamented with four diverging organs of taper conical form and crystalline clearness, through the midst of each of which passes a very fine branch of the air-tube system, which gives off still more attenuated branchlets in its course.

We have not yet, however, examined the origin of this air-breathing system. There is but one entrance to the air, or rather two placed close together, at the end of that round column, which is sent off from the eighth segment of the abdomen. This column, which is roughened all over with minute points, and fringed with rows of hairs, ends in a horny, conical point, which seems entire while under water, but no sooner does it come to the surface, than it is seen to split into five triangular pieces, which open widely, and expose a hollow, at the bottom of which are the two spiracles.

From these the two main air-pipes are seen to commence and to proceed along the centre of the column, thence into the abdomen, which they traverse one along each side, sending off slender branchlets all along, and becoming more and more attenuated themselves; till, at length, we trace them into the thorax, and thence through the slender neck into the head itself, until they terminate in fine points close to the back of the mouth. It needs, it is true, a very transparent specimen to follow the tracheal tubes thus through their entire course; but in such it can be done without difficulty. And it is very instructive to do so, inasmuch as one such personal examination of an insect under a good microscope will make you far more familiar with the peculiarities of its physiology, than the clearest book-descriptions, or even the best and most elaborate plates, alone.

Perhaps you may think I have kept you too long over these gnat-grubs, but my reason for being more minute in the examination of this creature is, that its extreme abundance in every place, and through the greatest part of the year, puts it in the power of everyone to procure a specimen alive and healthy, almost whenever he chooses, and, therefore, it is peculiarly available for microscopic study; while the transparency of its tissues, and its generally simple organization, make it a more than usually suitable object for investigation: besides which, there are the beautiful and interesting points in the details of its structure which I have been endeavouring to bring before you.

Not less interesting and remarkable is the change in the position of the spiracles, which takes place as soon as this grub arrives at the pupa or chrysalis state. The skin of the active, fish-like larva splits down the back, and out presses an equally active little monster; which, if you did not know it, you would never think of connecting with the grub from which it has proceeded; so totally different is it in form, in structure, and in motions.

We shall easily find some in our basin that have passed into this stage. Yes, here is one, which will please to take its place in the glass trough with its younger brothers. How strange the transformation! It reminds us of a lobster, though, of course, the resemblance is only cursory. With the naked eye we see that the thorax is greatly enlarged, not only actually, but proportionally; that it forms an oval mass, occupying some five-sixths, at least, of the entire animal; the rest apparently being taken up by a slender, many-jointed abdomen, which curves round the great thorax, and, bending under it, ends in an excessively delicate, transparent, swimming-plate. It is this curving abdomen, with its terminal swimmer, and its backward strokes in swimming, that constitute the resemblance to a prawn or lobster.

If we now bring a low power with the reflected light of the Lieberkuhn to bear on it, we shall see the progress the animal has made in this its change of raiment. The thorax shows on its sides the future wings, crumpled and folded down, the nervures of which we can discern distinctly. The elegant little head, too, can be well made out; its eyes now perfectly marked with the numerous hexagonal facets that belong to the matured organs of vision in these creatures; its antennæ, like slender rods, folded down side by side along the inferior edge of the thorax; the short palpi lying outside these; and within, both the lancets and piercers that are destined to subserve the blood-sucking propensities of our sanguinary little subject, when it attains its winged condition;—all encased in the transparent pupa-skin, that lies like a loose wrapper around everything.

The extremity of the abdomen has now nothing to do with respiration, and hence it is never brought to the surface of the water, as it was constantly before. The little animal still habitually lives in contact with the air, coming up to it with rapid, impatient jerks, whenever it has descended; but it is invariably the summit of the thorax that is uppermost, and when the creature rests, it is this part that touches the surface.

Why is this? you ask. Look, and you will see why. From the summit of the thorax project two little horns, which, under the microscope, are seen to be clear trumpet-shaped tubes with open mouths, cut as it were obliquely off. These enter the thorax close to the bases of the wings; and when we confine the animal in a glass cell, exercising a gentle pressure upon the thorax, we see bubbles of air alternately projected from the trumpet mouths of the tubes and sucked in again. These, then, are the spiracles, the orifices of the air-tubes, where the vital fluid enters the body, and whence it is carried to every part of the system.

There is something curiously beautiful about the structure of these spiracular tubes, of which I cannot attempt to explain the object. With a high magnifying power, their whole exterior surface is seen to be covered with regular rounded scales, overlapping each other, and very closely resembling those of a fish.

CHAPTER VII.

INSECTS: THEIR FEET.

I HAVE here inclosed a small window-fly in the live-box of the microscope, that you may examine the structure of its feet as it presses them against the glass cover; and thus not only get a glimpse of an exquisitely-formed structure, but acquire some correct ideas on the question of how a fly is enabled to defy all the laws of physics, and to walk jauntily about on the under surface of polished bodies, such as glass, without falling, or apparently the fear of falling. And a personal examination is the more desirable because of the hasty and erroneous notions that have been promulgated on the matter, and that are constantly disseminated by a herd of popular compilers, who profess to teach science by gathering up and retailing the opinions of others, often without the slightest knowledge whether what they are reporting is true or false.

The customary explanation has been that given by Derham in his "Physico-theology;" that "divers flies and other insects, besides their sharp-hooked nails, have also skinny palms to their feet, to enable them to stick to glass, and other smooth bodies, by means of the pressure of the atmosphere, after the manner as I have seen boys carry heavy stones, with only a wet piece of leather clapped on the top of a stone." Bingley, citing this opinion, adds that they are able easily to overcome the pressure of the air "in warm weather, when they are brisk and alert; but towards the end of the year this resistance becomes too mighty for their diminished strength; and we see flies labouring along, and lugging their feet on windows as if they stuck fast to the glass: and it is with the utmost difficulty they can draw one foot after another, and disengage their hollow cups from the slippery surface."*

But long ago another solution was proposed: for Hooke, one of the earliest of microscopic observers, described the two palms, pattens, or soles (as he calls the *pulvilli*), as "beset underneath with small bristles or tenters, like the wire teeth of a card for working wool, which, having a contrary direction to the claws, and both pulling different ways, if there be any irregularity or yielding in the surface of a body, enable the fly to suspend itself very firmly." He supposed that the most perfectly polished glass presented such irregularities, and that it was moreover always covered with a "smoky tarnish," into which the hairs of the foot penetrated.

The "smoky tarnish" is altogether gratuitous; and Mr. Blackwall has exploded the idea of atmospheric pressure, for he found that flies could walk up the interior of the exhausted receiver of an air-pump. He had explained their ability to climb up vertical polished bodies by the mechanical action of the minute hairs of the inferior surface of the palms; but further experiments having showed him that flies

* Anim. Biogr.

cannot walk up glass which is made moist by breathing on it, or which is thinly coated with oil or flour, he was led to the conclusion that these hairs are in fact tubular, and excrete a viscid fluid, by means of which they adhere to dry polished surfaces; and on close inspection with an adequate magnifying power, he was always able to discover traces of this adhesive material on the track on glass both of flies and various other insects furnished with *pulvilli*, and of those spiders which possess a similar faculty.*

In the earlier editions of Kirby and Spence's "Introduction to Entomology," Mr. Kirby had adopted the suctorial hypothesis. But in a late one he made an allusion to Mr. Blackwall's opinion, and added the following interesting note:—

"On repeating Mr. Blackwall's experiments, I found, just as he states, that when a pane of glass of a window was slightly moistened by breathing on it, or dusted with flour, bluebottle-flies, the common house-flies, and the common bee-fly (*Eristalis tenax*) all slipped down again the instant they attempted to walk up these portions of the glass; and I moreover remarked that each time after thus slipping down, they immediately began to rub first the two fore tarsi, and then the two hind tarsi, together, as flies are so often seen to do, and continued this operation for some moments before they attempted again to walk. This last fact struck me very forcibly, as appearing to give an importance to these habitual procedures of flies that has not hitherto, as far as I am aware, been attached to them. These movements I had

* Linn. Trans. xvi. 490, 768.

always regarded as meant to remove any particle of dust from the legs, but simply as an affair of instinctive cleanliness, like that of the cat when she licks herself, and not as serving any more important object; and such entomological friends as I have had an opportunity of consulting tell me that their view of the matter was precisely the same; nor does Mr. Blackwall appear to have seen it in a different light, since, though so strongly bearing on his explanation of the way in which flies mount smooth vertical surfaces, he never at all refers to it. Yet, from the absolute necessity which the flies on which I experimented appeared to feel of cleaning their *pulvilli* immediately after being wetted or clogged with flour, however frequently this occurred, there certainly seems ground for supposing that their usual and frequent operation for effecting this by rubbing their tarsi together is by no means one of mere cleanliness or amusement, but a very important part of their economy, essentially necessary, for keeping their *pulvilli* in a fit state for climbing up smooth vertical substances by constantly removing from them all moisture, and still more all dust, which they are perpetually liable to collect. In this operation the two fore and two hind tarsi are respectively rubbed together for their whole length, whence it might be inferred that the intention is to remove impurities from the entire tarsi; but this I am persuaded is not usually the object, which is simply that of cleaning the under side of the *pulvilli* by rubbing them backward and forward along the whole surface of the hairs with which the tarsi are clothed, and which seem intended to serve as a brush for this particular purpose. Sometimes, indeed, when the hairs of the tarsi are filled with dust throughout, the operation of rubbing them together is intended to cleanse these hairs; because, without these brushes were themselves clean, they could not act upon the hairs of the under side of the *pulvilli*. Of this I witnessed an interesting instance in an *Eristalis tenax*, which by walking on a surface dusted with flour had the hairs of the whole length of the tarsi, as well as the *pulvilli*, thus clogged with it. After slipping down from the painted surface of the window-frame which she in vain attempted to climb, she seemed sensible that before the *pulvilli* could be brushed it was requisite that the brushes themselves should be clean, and full two minutes were employed to make them so by stretching out her trunk, and passing them repeatedly along its sides, apparently for the sake of moistening the flour and causing its grains to adhere; for after this operation, on rubbing her tarsi together, which she next proceeded to do, I saw distinct little pellets of flour fall down. A process almost exactly similar I have always seen used by bluebottle-flies and common house-flies which had their tarsi clogged with flour by walking over it, or by having it dusted over them; but these manœuvres are required for an especial purpose, and on ordinary occasions, as before observed, the object in rubbing the tarsi together is not to clean *them*, but the *pulvilli*, for which they serve as brushes. Besides rubbing the tarsi together, flies are often seen, while thus employed, to pass the two fore tarsi and tibiæ with sudden jerks over the back of the head and eyes, and the two hind tarsi and tibiæ over and under the wings, and especially over their outer margins, and occasionally also over the back of the abdomen. That one object of these operations is often to clean these parts from dust, I have no doubt, as on powdering the flies with flour they thus employ themselves, sometimes for ten minutes, in detaching every part of it from their eyes, wings, and abdomen; but I am also inclined to believe that, in general, when this passing of the legs over the back of the head and outer margin of the wings takes place in connexion with the ordinary rubbing of the tarsi together, as it usually does, that the object is rather for the purpose of completing the entire cleansing of the tarsal brushes (for which the row of strong hairs visible under a lens on the exterior margin of the wings seems well adapted), so that they may act more perfectly on the *pulvilli.* Here, too, it should be noticed, in proof of the importance of all the *pulvilli* being kept clean, that as the tarsi of the two middle legs cannot be applied to each other, flies are constantly in the habit of rubbing one of these tarsi and its *pulvillus*, sometimes between the two fore tarsi, and at other times between the two hind ones. . . . .

"Though the above observations, hastily made on the spur of the occasion since beginning this note, seem to prove that it is necessary the *pulvilli* of flies and of some other insects should be kept free from moisture and dust to enable them to ascend vertical polished surfaces, they cannot be considered as wholly settling the question as to the precise way in which these *pulvilli*, and those of insects generally, act in effecting a similar mode of progression; and my main reason for here giving these slight hints is the hope of directing the attention of entomological and microscopical observers to a field evidently, as yet, so imperfectly explored.

"After writing the above, intended as the conclusion of this note, I witnessed to-day (July 11, 1842), a fact which I cannot forbear adding to it. Observing a house-fly on the window, whose motions seemed very strange, I approached it, and found that it was making violent contortions, as though every leg were affected with St. Vitus' dance, in order to pull its *pulvilli* from the surface of the glass, to which they adhered so strongly that though it could drag them a little way, or sometimes by a violent effort get first one and then another detached, yet the moment they were placed on the glass again, they adhered as if their under side were smeared with bird-lime. Once it succeeded in dragging off its two fore legs, when it immediately began to rub the *pulvilli* against the tarsal brushes; but on replacing them on the glass they adhered as closely as before, and it was only by efforts almost convulsive, and which seemed to threaten to pull off its limbs from its body, that it could succeed in moving a quarter of an inch at a time. After watching it with much interest for five minutes, it at last by its continued exertions got its feet released and flew away, and alighted on a curtain, on which it walked quite briskly, but soon again flew back to the window, where it had precisely the same difficulty in pulling its *pulvilli* from the glass as before; but after observing it some time, and at last trying to catch it, that I might examine its feet with a lens, it seemed by a vigorous effort to regain its powers, and ran quite actively on the glass, and then flying away I lost sight of it. I am unable to give any satisfactory solution of this singular fact. The season, and the fly's final activity, preclude the idea of its arising from cold or debility, to which Mr. White attributes the dragging of flies' legs at the close of autumn. The *pulvilli* certainly had much more the appearance of adhering to the glass by a viscid material than by any pressure of the atmosphere, and it is so far in favour of Mr. Blackwall's hypothesis, on which one might conjecture that from some cause (perhaps of disease) the hairs of the *pulvilli* had poured out a greater quantity of this viscid material than usual, and more than the muscular strength of the fly was able to cope with."*

In the foot of the fly under our own observation you may see how well the joints of the tarsus are covered with hairs, or rather stiff pointed spines, of various dimensions and distances apart, and hence how suitable these are for acting the part of combs to cleanse the palms. But these last are the organs that most claim and deserve our examination. In the specimen of the little Musca that I have imprisoned, the last tarsal joint is terminated by two strong divergent hooks which are themselves well clothed with spines, and by two membranous flaps or palms beneath them. These are nearly oval in outline, though in some species they are nearly square, or triangular, and in some of a very irregular shape. They are thin,

* Intr. to Entom., 7th Ed., p. 458.

membranous, and transparent, and when a strong light is reflected through them by means of the achromatic condenser, we see their structure under this power of 600 diameters very distinctly.

The inferior surface of the palm, on which we are now looking, is divided into a vast number of lozenge-shaped areas, which appear to be scales overlapping each other, or they may be divided merely by depressed lines. From the centre of each area proceeds a very slender, soft, and flexible pellucid filament, which reaches downwards to the surface on which the fly is walking, and is there slightly hooked and enlarged into a minute fleshy bulb. Those from the areas near and at the margins of the palms more and more arch outwards, so that the space covered by the bulbs of the filaments is considerably greater than that of the palm itself.

Now it is evident that the bulbous extremities of these soft filaments are the organs of adhesion. We notice how they drag and hold, as the fly draws its foot from its place, and it seems almost certain that the adhesion is effected by means of a glutinous secretion poured out in minute quantities from these fleshy tips. When the foot is suddenly removed, we may often see a number of tiny particles of fluid left on the glass where the filaments had been in contact with it: but I do not build conclusively on this appearance, because the fly, having been confined for some quarter of an hour in this nearly tight glass cell, has doubtless exhaled some moisture, which has condensed on the glass; and the specks we see may possibly be due to the filaments of the palms having become wet by repeatedly brushing the moist surface.

Mr. Hepworth, however, asserts that a fluid is poured out from these filaments, and is deposited on the glass, when the fly is vigorous, with great regularity. He says that "when in a partially dormant state, the insect does not appear to be able to give out this secretion, though it can still attach itself: indeed, this fluid is not essential for that purpose."* It is asserted that the speckled pattern of fluid left on the glass by the fly's footsteps remains (if breathed on) when the moisture is evaporated; and hence it is presumed to be of an oily nature.

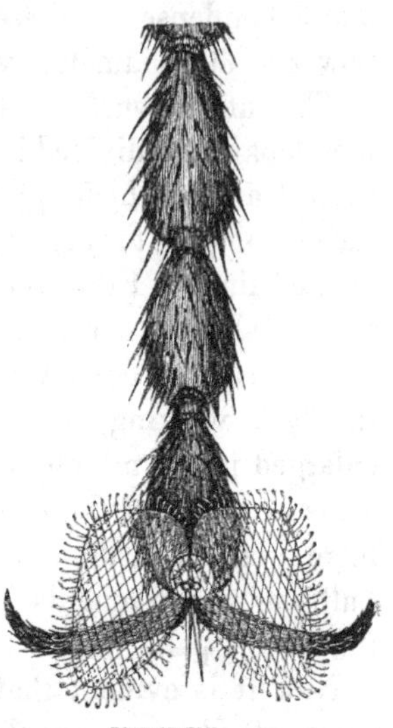

FOOT OF FLY.

In some Beetles the joints of the foot are furnished with similar appendages. I shall now show you the fore-foot of a well-known insect, called by children the Bloody-nose Beetle (*Timarcha tenebricosa*), a heavy-bodied fellow, of a blue-black colour, abundant in spring and summer on hedge-banks. You have doubtless often observed it, and have been amused, perhaps, at seeing the drop of clear scarlet fluid which exudes from its mouth when touched.

The feet in this species are broad and well developed. You may see with the naked eye, on turning it up, that its dilated joints are covered on the under surface with a velvety cushion of a rusty-brown

* Microsc. Journal, for April, 1854.

colour; and here, under a low power of the microscope with the Lieberkuhn, you can resolve the nature of the velvet

The foot, or *tarsus* as it is technically called, is composed of four very distinct pieces; of which the first is semicircular, the second crescent-shaped, the third heart-shaped, and the fourth nearly oval. The last is rounded on all sides, has no cushioned sole, and carries two stout hooks. The first three are flat or even, hollowed beneath into soles, something like the hoof of a horse, and the whole interior bristles with close-set minute points, the tips of which terminate at the same level and form a velvety surface. Now these points are the whitish bulbous extremities exactly answerable to those on the palms of the fly, and doubtless they answer the very same purpose. Only here they are set in far closer array and are a hundred times more numerous; whence we may reasonably presume a higher power of adhesion to be possessed by the beetle. The structure is best seen in the male, which may be distinguished by its smaller dimensions, and by its broader feet.

A still better example of a sucking foot is this of the *Dyticus marginalis*. It is the great flat oval beetle, which is fond of coming up to the surface of ponds, and hanging there by the tail with its pair of hind legs stuck out on each side at right angles; the redoubtable monster which little boys who bathe hold in such salutary awe under the name of Toe-biter. We have turned the tables upon the warrior, and have bitten *his* toe—off, and here it is. This is the tarsus of one of the fore limbs.

The peculiarity that first strikes us is that the first three joints are as it were fused into one, and dilated so as to make a large roundish plate. The under surface of this broad plate is covered with a remarkable array of sucking disks, of which one is very large, occupying about a fourth part of the whole area. It is circular, and its face is strongly marked with

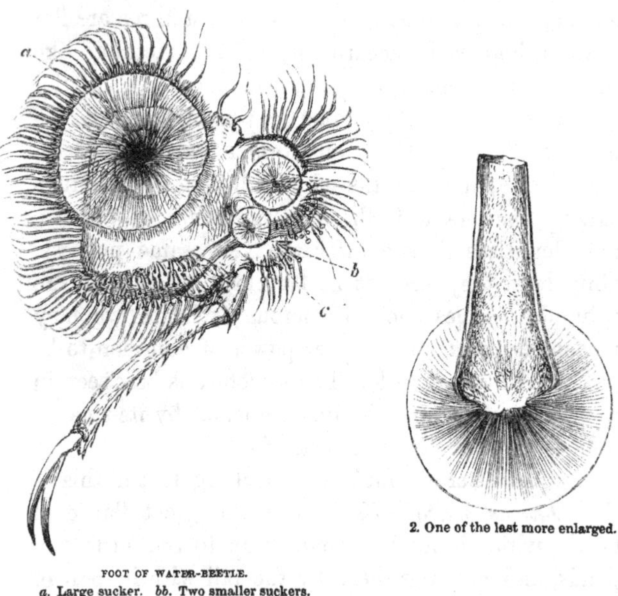

FOOT OF WATER-BEETLE.
a. Large sucker. *bb.* Two smaller suckers.
c. Small crowded suckers.

2. One of the last more enlarged.

numerous fibres radiating from the centre. Near this you perceive two others of similar form and structure, but not more than one tenth part of its size; one of these, moreover, is smaller than the other. Indeed, the size and number of these organs differ in different individuals of the same species.

The greater number of the suckers are comparatively minute; but they are proportionally multitudinous and crowded. Each consists of a club-shaped shaft, with a circular disk of radiating fibres attached to its end. The whole apparatus constitutes a very effective instrument of adhesion.

There is a somewhat similar dilatation of the first joints of the tarsus, but for a very different object, in the Honey-bee; and it is particularly worthy to be observed, not only for the interesting part which it plays in the economy of the insect, but for the example it affords us of the adaptation of one and the same organ to widely different uses, by a slight modification of its structure.

It is the hind foot of the Bee that we are now to examine. The first joint is, as you see, enlarged into a wide, long, and somewhat ovate form, constituting a flattish plate, slightly convex on both surfaces. The upper face presents nothing remarkable, but the under side is set with about nine stiff combs, the teeth of which are horny straight spines, set in close array, and arranged in transverse rows across the joint, nearly on a level with its plane, but a little projecting, and so ordered that the tips of one comb slightly overlap the bases of the next. We see them in this example very distinct, because their colour, a clear reddish-brown, contrasts with a multitude of tiny globules of a pale yellow hue, like minute eggs, which are entangled in the combs.

Now these globules serve to illustrate the object of

this apparatus. They are grains of pollen; the dust that is discharged from the anthers of flowers, which being kneaded up with honey forms the food of the infant bees, and is, therefore, collected with great perseverance by those industrious insects; and the way in which they collect it is, by raking or combing it from the anthers, by means of these effective instruments on their hind feet.

You see that in this specimen the combs are loaded with the grains, which lie thickly in the furrows between one comb and another. But how do they discharge their gatherings? Do they return to the hive, as soon as they have accumulated a quantity such as this, which one would suppose they could gather in two or three scrapes of the foot? No; they carry a pair of panniers, or collecting baskets, which they gradually fill from the combs, and then return to deposit the results of their collecting.

One of these baskets I can show you; and, indeed, we should be unpardonable to overlook it, for it is the companion structure to the former. I make the stage forceps to revolve on its axis, and thus bring into focus the joint (*tibia*) immediately above that of the combs, and so that we shall look at its opposite surface; that is, the outer. We notice at once two or three peculiarities, which distinguish the joint in this instance from other parts of the same limb, and from the corresponding part in the same limb of other insects.

First, the surface is decidedly concave, whereas it is ordinarily convex. Secondly, this concave surface is smooth and polished (except that it is covered with a minute network of crossed lines), not a single hair, even the most minute, can be discerned in any part; whereas the corresponding surface of the next joints, both above and below, is studded with fine hairs, as is the exterior of insects generally. Thirdly, the edges of this hollowed basin are beset with long, slender, acute spines, which pursue the same curve as the bottom and sides, expanding widely, and arching upward.

Here, then, we have a capital collecting-basket. Its concavity of course fits it to contain the pollen. Then its freedom from hairs is important: hairs would be out of place in the concavity. Thirdly, the marginal spines greatly increase the capacity of the vessel to receive the load, on the principle of the sloping stakes which the farmer plants along the sides of his waggon when he is going to carry a load of hay or corn.

But, you ask, how can the Bee manage to transfer the pollen from the combs to the basket? Can she bend up the tarsus to the tibia? or, if she can, surely she could only reach the inner, not the outer surface of the latter. How is this managed?

A very shrewd question. Truth to say, the basket you have been looking at never received a single grain from the combs of the joint below it. But the Bee has *a pair* of baskets and *a pair* of comb-joints. It is the *right* set of combs that fills the *left* basket, and *vice versâ*. She can easily cross her hind-legs, and thus bring the tarsus of one into contact with the tibia of the other; and if you will pay a moment's more attention to the matter, you will discover some further points of interest in this beautiful series of contrivances still. If you look at this living Bee, you notice that, from the position of the joints, when the insect would bring one hind-foot across to the other, the under surface of the tarsus would naturally scrape the edge of the opposite tibia in a direction from the bases of the combs towards their tips; and, further, that the edge of the tibia so scraped would be the *hinder* edge, as the leg is ordinarily carried in the act of walking.

Now, if you take another glance at the basket-joint in the forceps of the microscope, you will see—what, perhaps, you have already noticed—that the marginal spines have not exactly the same curvature on the two opposite edges, but that those of the one edge are nearly straight, or at most but slightly bowed, whereas those of the opposite edge are strongly curved, the arc in many of them reaching even to a semicircle, so that their points, after performing the outward arch, return to a position perpendicularly over the medial line of the basket.

It is the outer or hinder edge of the joint that carries the comparatively straight spines. These receive the grains from the combs, which, then falling into the basket, are received into the wide concavity formed partly by its bottom and sides, but principally *by the arching spines of the opposite edge.* Their curving form would have been less suitable than the straighter one to pass through the interstices of the combs, because it would be much more difficult to get at their points; while, on the other hand, the straight lines of these would have been far less effective as a receiver for the burden. The thickness of the spines is just that which enables them to pass freely through the interstices of the comb-teeth, and *no more.*

On the whole, this combination of contrivances reads us as instructive a lesson on the wisdom of God displayed in creation as any that we have had brought under our observation.

The end to be attained by all this apparatus is worthy of the wondrous skill displayed in its contrivance; for it is connected with the feeding of the stock, and whatever diminishes the labour of the individual bees enables a larger number to be supported. But valuable as is the Honey-bee to man, there are other important purposes to be accomplished, which are more or less dependent, collaterally, on this series of contrivances.

"In many instances it is only by the bees travelling from flower to flower that the pollen and farina is carried from the male to the female flowers, without which they could not fructify. One species of bee would not be sufficient to fructify all the various sorts of flowers, were the bees of that species ever so numerous; for it requires species of different sizes and different constructions. M. Sprengel found that not only are insects indispensable in fructifying different species of *Iris*, but some of them, as *I. Xiphium*, require the agency of the larger humble-bees, which alone are strong enough to force their way beneath the style-flag; and hence, as these insects are not so common as many others, this *Iris* is often barren, or bears imperfect seeds." *

* Penny Cyclop., *art.* BEE.

The legs and feet of Caterpillars are constructed on a very different plan from those of perfect insects, as you may see in this living Silkworm. The first three segments of the body, reckoning from the head, are furnished each with a pair of short curved limbs set close together on the under side. These represent the true legs of the future moth, and show, notwithstanding their shortness, four distinct joints, of which the last is a little pointed horny claw. The whole limb resembles a short stout hook. Then two segments occur which are quite smooth beneath, and destitute of limbs; and then on the sixth we begin to find another series, which goes on regularly, a pair on each segment, to the eleventh and final one, with the single exception of the tenth segment, which is again deprived of limbs.

But these organs are of a very peculiar character. They have no representatives in the mature insect, but disappear with the larva state, and they are not considered limbs-proper at all, but mere accessory developments of the skin to serve a special purpose. They are sometimes called claspers, sometimes false-legs, but more commonly pro-legs.

Each consists of a fleshy wart, which is capable to some extent of being turned inside out, like the finger of a glove. Partly around the blunt and truncate extremity are set two rows of minute hooks, occupying the side next the middle line of the caterpillar in a semicircle along the margin. These hooks arch outward as regards the axis of the pro-leg, though the majority of them point towards the medial line of the body. The double row is somewhat interrupted at its middle point; and just there, in each pro-leg, a clear vesicle or fleshy bladder protrudes from the sole, which may perhaps serve as a very delicate organ of touch, or may exude a viscid secretion helpful to progress on smooth bodies. The hooks seem adapted to catch and hold the fine threads of silk, which most caterpillars spin as a carpet for their steps.

In some cases the circle of hooks is complete, as in this example, which I find in one of the slides of my drawer, marked "Pro-leg of a Caterpillar." It is some large species, probably a Sphinx, for the hooks are very large, of a clear orange-brown hue, and set in a long oval ring—single as to their bases, but double as to their points—completely around the extremity of the foot. These hooks are simply cutaneous, as may be well seen in this prepared specimen,—doubtless mounted in Canada balsam;—for their origins are mere blunt points, set most superficially in the thin skin without any enlargement or apparent bulb.

CHAPTER VIII.

INSECTS: STINGS AND OVIPOSITORS.

PROBABLY at some period of your life you have been stung by a bee or wasp. I shall take it for granted that you have, and that having tested the potency of these warlike insects' weapons with one sense, you have a curiosity to examine them with another. The microscope shall aid your vision to investigate the morbific implement.

This is the sting of the Honey-bee, which I have but this moment extracted. It consists of a dark brown horny sheath, bulbous at the base, but suddenly diminishing, and then tapering to a fine point. This sheath is split entirely along the inferior edge, and by pressure with a needle I have been enabled to project the two lancets, which commonly lie within the sheath. These are two slender filaments of the like brown horny substance, of which the centre is tubular, and carries a fluid, in which bubbles are visible. The extremity of each displays a beautiful mechanism, for it is thinned away into two thin blade-edges, of which one remains keen and knife-like, while the opposite edge is cut into several saw-teeth pointing backwards.

The lancets do not appear to be united with the sheath in any part, but simply to lie in its groove; their basal portions pass out into the body behind the sheath, where you see a number of muscle-bands crowded around them: these, acting in various directions, and being inserted into the lancets at various points, exercise a complete control over their movements, projecting or retracting them at their will. But each lancet has a singular projection from its back, which appears to act in some way as a guide to its motion, probably preventing it from slipping aside when darted forth, for the bulbous part of the sheath, in which these projections work, seems formed expressly to receive them.

Thus we see an apparatus beautifully contrived to enter the flesh of an enemy: the two spears finely pointed, sharp-edged, and saw-toothed, adapted for piercing, cutting, and tearing; the reversed direction of the teeth gives the weapon a hold in the flesh, and prevents it from being readily drawn out. Here is an elaborate store of power for the jactation of the javelins, in the numerous muscle-bands; here is a provision made for the precision of the impulse; and finally, here is a polished sheath for the reception of the weapons and their preservation when not in actual use. All this is perfect; but something still was wanting to render the weapons effective, and that something your experience has proved to be supplied.

The mere intromission of these points, incomparably finer and sharper than the finest needle that was ever polished in a Sheffield workshop, would produce no result appreciable to our feelings; and most surely would not be followed by the distressing agony attendant on the sting of a bee. We must look for something more than we have seen.

We need not be long in finding it. For here, at the base of the sheath, into which it enters by a narrow neck, lies a transparent pear-shaped bag, its surface covered all over, but especially towards the neck, with small glands set transversely. It is rounded behind, where it is entered by a very long and slender membranous tube, which, after many turns and windings, gradually thickening and becoming more evidently glandular, terminates in a blind end.

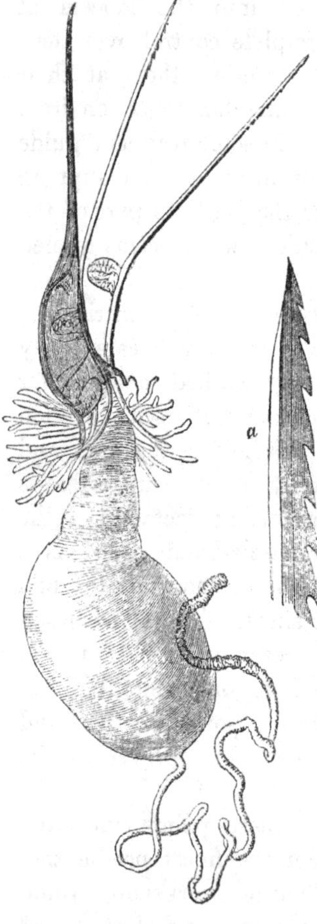

STING OF BEE.
a. Tip of Lancet, more enlarged.

This is the apparatus for preparing and ejecting a powerful poison. The glandular end of the slender tube is the secreting organ: here the venom is prepared; the remainder of the tube is a duct for conveying it to the bag, a reservoir in which it is stored for the moment of use. By means of the neck it is thrown into the groove at the moment the sting is projected, the same muscles, probably, that dart forward the weapon compressing the poison-bag and causing it to pour forth its contents into the groove, whence it passes on between the two spears into the wound which they have made.

A modification of this apparatus is found throughout a very extensive order of insects,—the *Hymenoptera;* but in the majority of cases it is not connected with purposes of warfare. Wherever it occurs it is always confined to the female sex, or (as in the case of some social insects) to the neuters, which are undeveloped females. When it is not accompanied by a poison-reservoir it is ancillary to the deposition of the eggs, and is hence called an *ovipositor*, though in many cases it performs a part much more extensive than the mere placing of the *ova*.

In the large tribe of Cuckoo-flies (*Ichneumonidæ*), which spend their egg and larva states in the living bodies of other insects, this ovipositor is often of great length; even many times longer than the rest of their bodies; for the larvæ which have to be pierced by it require to be reached at the bottom of deep holes and other recesses in which the providence of the parent had placed them for security. The structure of the organ may be seen in this little species, not more than one-sixth of an inch in entire length, of which the ovipositor projects about a line. Under the microscope you see that this projection consists of two black fleshy filaments, rounded without and flattened on their inner faces, which are placed together,—and of the true implement for boring, in the form of a perfectly straight awl, of a clear amber hue, very slender and brought to an abrupt oblique point, where there are a few exceedingly fine reverted teeth. It is probably double, though it refuses to open under the pressure which I bring to bear upon it. At the base are seen within the semi-pellucid abdomen the slender horns, on which the muscles act in projecting the borer.

You are doubtless aware that the little berries which look like bunches of green currants often seen growing on the oak, are not the proper fruit of the tree, but diseased developments produced by a tiny insect, for the protection and support of her young But perhaps you have never paid any special attention to the living atom whose workmanship they are, and are not familiar with the singular mechanism by which she works. I have not had an opportunity of seeing it myself, and therefore cannot show it to you; but as Gall-flies are by no means rare, and you may easily rear a brood of flies from the galls, you may have a chance of meeting with it. I will therefore quote to you what Rennie says about it.

"There can be no doubt, that the mother gall-fly makes a hole in the plant for the purpose of depositing her eggs. She is furnished with an admirable

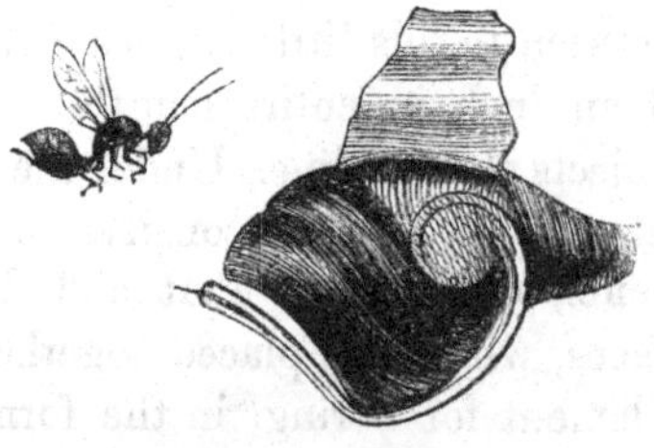

GALL-FLY AND MECHANISM OF OVIPOSITOR.

ovipositor for that express purpose, and Swammerdam actually saw a gall-fly thus depositing her eggs, and we have recently witnessed the same in several instances. In some of these insects the ovipositor is conspicuously long, even when the insect is at rest; but in others, not above a line or two of it is visible, till the belly of the insect be gently pressed. When this is done to the fly that produces the currant-gall of the oak, the ovipositor may be seen issuing from a sheath in form of a small curved needle, of a chesnut-brown colour, and of a horny substance, and three times as long as it at first appeared.

"What is most remarkable in this ovipositor is, that it is much longer than the whole body of the insect, in whose belly it is lodged in a sheath, and, from its horny nature, it cannot be either shortened or lengthened. It is on this account that it is bent into the same curve as the body of the insect. The mechanism by which this is effected is similar to that of the tongue of the woodpeckers (*Picidæ*), which, though rather short, can be darted out far beyond the beak by means of a forked bone at the root of the tongue, which is thin and rolled up like the spring of a watch. The base of the ovipositor of the gall-fly is, in a similar way, placed near the anus, runs along the curvature of the back, makes a turn at the breast, and then, following the curve of the belly, appears again near where it originates.

"With this instrument the mother gall-fly pierces the part of a plant which she selects, and, according to our older naturalists, 'ejects into the cavity a drop of her corroding liquor, and immediately lays an egg or more there; the circulation of the sap being thus interrupted, and thrown, by the poison, into a fermentation that burns the contiguous parts and changes the natural colour. The sap, turned from its proper channel, extravasates and flows round the eggs, while its surface is dried by the external air, and hardens into a vaulted form.' Kirby and Spence tell us, that the parent-fly introduces her egg 'into a puncture made by her curious spiral sting, and in a few hours it becomes surrounded with a fleshy chamber.' M. Virey says, the gall-tubercle is produced by irritation, in the same way as an inflamed tumour in an animal body, by the swelling of the cellular tissue, and the flow of liquid matter, which changes the organization, and alters the natural external form."*

Perhaps a still more charming example of animal mechanics is that furnished to us by the Saw-flies (*Teuthredinidæ*). These are very common four-winged insects of rather small size, many species of which are found in gardens and along hedges in summer, produced from grubs which are often mistaken for true caterpillars, as they strip our gooseberry and rose bushes of their leaves; but may be distinguished from them by the number of their pro-legs, and by their singular postures; for they possess from eight to fourteen pairs of the former organs, and have the habit of coiling up the hinder part of their body in a spiral ring, while they hang on to the leaf by their six true feet.

These saw-fly caterpillars are produced from eggs which are deposited in grooves, made by the parent-fly in the bark of the tree or shrub whose future leaves are destined to constitute their food; and it is for the construction of these grooves and the deposition of the eggs in them, that the curious mechanism is contrived which I am now bringing under your notice.

* Ins. Arch. 371.

Almost all our acquaintance with this instrument and the manner of its employment, we owe to the eminent French naturalist Réaumur, and to his Italian contemporary Valisnieri. Their details I shall first cite, as they have been put into an English dress by Rennie, and then show you a specimen dissected out by myself, and point out some agreements and some discrepancies between it and them.

"In order to see the ovipositor, a female saw-fly must be taken, and her belly gently pressed, when a narrow slit will be observed to open at some distance from the anus, and a short, pointed, and somewhat curved body, of a brown colour and horny substance, will be protruded. The curved plates which form the sides of the slit, are the termination of the sheath, in which the instrument lies concealed till it is wanted by the insect.

"The instrument thus brought into view, is a very finely contrived saw, made of horn, and adapted for penetrating branches and other parts of plants where the eggs are to be deposited. The ovipositor-saw of the insect is much more complicated than any of those employed by our carpenters. The teeth of our saws are formed in a line, but in such a manner as to cut in two lines parallel to and at a small distance from each other. This is effected by slightly bending the points of the alternate teeth right and left, so that one half of the whole teeth stand a little to the right, and the other half a little to the left. The distance of the two parallel lines thus formed is called the course of the saw, and it is only the portion of wood which lies in the course that is cut into sawdust by the action of the instrument. It will follow, that in proportion to the thinness of a saw there will be the less destruction of wood which may be sawed. When cabinet-makers have to divide valuable wood into very thin leaves, they accordingly employ saws with a narrow course; while sawyers who cut planks, use one with a broad course. The ovipositor-saw being extremely fine, does not require the teeth to diverge much, but from the manner in which they operate, it is requisite that they should not stand like those of our saws in a straight line. The greater portion of the edge of the instrument, on the contrary, is towards the point somewhat concave, similar to a scythe, while towards the base it becomes a little convex, the whole edge being nearly the shape of an italic *ſ*.

"The ovipositor-saw of the fly is put in motion in the same way as a carpenter's hand-saw, supposing the tendons attached to its base to form the handle, and the muscles which put it in motion to be the hand of the carpenter. But the carpenter can only work one saw at a time, whereas each of these flies is furnished with two, equal and similar, which it works at the same time—one being advanced and the other retracted alternately. The secret, indeed, of working more saws than one at once is not unknown to our mechanics; for two or three are sometimes fixed in the same frame. These, however, not only all move upwards and downwards simultaneously, but cut the wood in different places; while the two saws of the ovipositor work in the same cut, and, consequently, though the teeth are extremely fine, the effect is similar to [that of] a saw with a wide set.

"It is important, seeing that the ovipositor-saws are so fine, that they be not bent or separated while in operation—and this, also, nature has provided for by lodging the backs of the saws in a groove, formed by two membranous plates, similar to the structure of a clasp-knife. These plates are thickest at the base, becoming gradually thinner as they approach the point which the form of the saws requires. According to Valisnieri, it is not the only use of this apparatus to form a back for the saws, he having discovered between the component membranes two canals, which he supposes are employed to conduct the eggs of the insect into the grooves which she has hollowed out for them.

"The teeth of a carpenter's saw, it may be remarked, are simple, whereas the teeth of the ovipositor-saw are themselves denticulated with fine teeth. The latter, also, combines at the same time the properties of a saw and of a rasp or file. So far as we are aware, these two properties have never been combined in any of the tools of our carpenters. The rasping part of the ovipositor, however, is not constructed like our rasps, with short teeth thickly studded together, but has teeth almost as long as those of the saw, and placed contiguous to them on the back of the instrument, resembling in their form and setting the teeth of a comb." *

Now look at this object which I have just extracted from the abdomen of a rather large female Saw-fly, of a bright green hue, spotted with black. The first portion of the apparatus that protruded on pressure,

* Insect Architecture, 153.

was this pair of saws of an ∫-like figure. These agree in general with those described; here is, in each, the doubly-curved blade, the strengthened back, the rasp-like jagging of the lateral surfaces, the teeth along the edge, and the secondary toothlets of the latter. All these essential elements we see, but there is much discrepancy in the detail, and many points not noticed;—in part, doubtless, owing to its being another species which was under observation, and partly to the inferiority of the microscopes employed a hundred and fifty years ago, to those we are using.

In the first place the curve of the ∫ is different, the convexity of the edge being towards the point and the concavity nearest the base. Then the strengthening does not appear to me a groove in which the saw plays, but a thickening of the substance of the back. Each main tooth of the saw in this case is the central point in the edge of a square plate, which appears to be slightly concave on its two surfaces, being thickened at its two sides, at each of which, where it is united to the following plate, it rises and forms with it a prominent ridge running transverse to the course of the saw. Each of these ridges then forms a second tooth, as stout as the main edge-tooth, which, with the rest of the same series, forms a row of teeth on the oblique side of the saw, in a very peculiar manner, difficult to express by words. It is singular that this side of the saw should be studded with minute hairs, since these would seem to interfere with the action of the saw, or at least be liable to be themselves rubbed down and destroyed in its action.

But their existence is indubitable; there they are, pointing at a very acute angle towards the top of the saw. The back edge of the implement bristles with many close-set hairs or spines, forming a sort of brush, but pointing in the opposite direction.

Each main tooth of the edge-series is cut into one or two minute toothlets on its posterior side (next the base of the saw) and about half-a-dozen on its opposite side (next the tip). The texture is clear and colourless where thin; but in the thickened parts, as the main teeth, the transverse ridges, and the back, it is a clear amber-yellow; the strengthening back-piece deepening to a rich translucent brown.

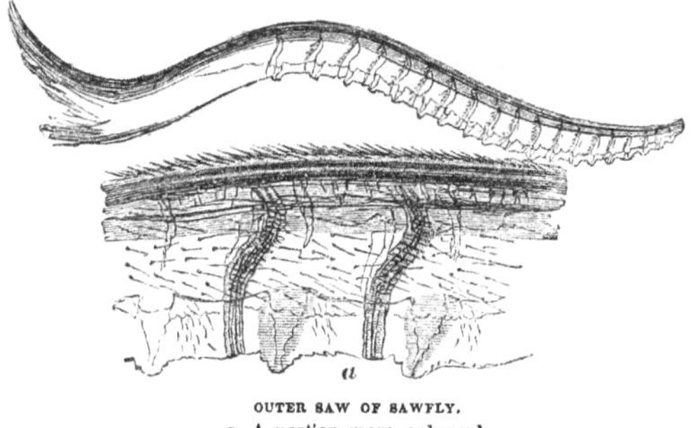

OUTER SAW OF SAWFLY.
a. A portion more enlarged.

There is, however, in this species of mine a second set of implements of which the French naturalist, observant as he was, takes not the slightest notice; and his English commentator appears to have as little suspected its presence. This pair of saws that we have been looking at is but the sheath of a still finer pair of lancets or saws, which you may see here. These are much slenderer than the former, and are peculiar in their construction. Their extreme tip only bears saw-teeth, and these are directed backwards, but one side of the entire length presents a succession of cutting edges, as if a number of short pieces of knife-blades had been cemented on a rod, in such a manner as that the cutting edges should be directed backwards, and overlap each other. The other lateral surface is plain, and both are convex in their general aspect. The appearance of these implements is very beautiful; for the texture is of a clear pale amber, but the structure is strengthened by a band which runs along each edge, and by transverse bands crossing at regular intervals, of a denser tissue; and these are of a rich golden translucent brown.

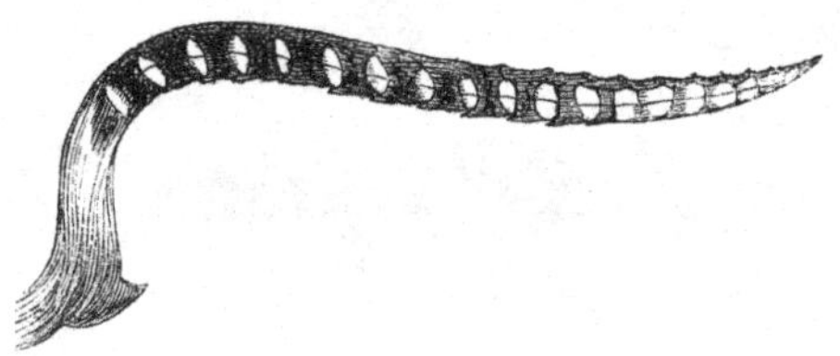

INNER SAW OF SAW-FLY.

From the construction of this implement I should infer that its force is exerted in pulling and not in pushing; the direction of the teeth and of the cutting plates shows this. The sharp horny point is probably thrust a little way into the solid wood or bark, and then a backward pull brings the teeth and cutting plates to act upon the material, and so successively And probably these points are the first parts of the whole apparatus that come into operation; the blunter saws of the sheath serving mainly to widen and deepen the course, after the finer points have pioneered the way.

You may like to hear what Réaumur has to say about the manner in which the fly works, especially as I have nothing of my own on the subject, which yet is a most interesting one :—

"When a female Saw-fly has selected the branch of a rose-tree, or any other, in which to deposit her eggs, she may be seen bending the end of her belly inwards, in form of a crescent, and protruding her saw, at the same time, to penetrate the bark or wood. She maintains this recurved position so long as she works in deepening the groove; but when she has attained the depth required, she unbends her body into a straight line, and in this position works upon the place lengthways, by applying the saw more horizontally. When she has rendered the groove as large as she wishes, the motion of the tendon ceases, and an egg is placed in the cavity. The saw is then withdrawn into the sheath for about two-thirds of its length, and at the same moment, a sort of frothy liquid, similar to a lather made with soap, is dropped over the egg, either for the purpose of gluing it in its place, or sheathing it from the action of the juices of the tree. She proceeds in the same manner in sawing out a second groove, and so on in succession, till she has deposited all her eggs, sometimes to the number of twenty-four. The grooves are usually placed in a line, at a small distance from one another, on the same branch; but sometimes the mother-fly shifts to another, or to a different part of the branch, when she is either scared or finds it unsuitable. She commonly, also, takes more than one day to the work, notwithstanding the superiority of her tools. Réaumur has seen a Saw-fly make six grooves in succession, which occupied her about ten hours and a-half.

"The grooves, when finished, have externally little elevation above the level of the bark, appearing like the puncture of a lancet in the human skin; but in the course of a day or two the part becomes first brown and then black, while it also becomes more and more elevated. This increased elevation is not owing to the growth of the bark, the fibres of which, indeed, have been destroyed by the ovipositor-saw, but to the actual growth of the egg; for, when a new-laid egg of the Saw-fly is compared with one which has been several days inclosed in the groove, the latter will be found to be very considerably larger. This growth of the egg is contrary to the analogy observable in the eggs of birds, and even of most other insects; but it has its advantages. As it continues to increase, it raises the bark more and more, and consequently widens, at the same time, the slit at the entrance; so that, when the grub is hatched, it finds a passage ready for its exit. The mother-fly seems to be aware of this growth of her eggs, for she takes care to deposit them at such distances as may prevent their disturbing one another by their development." *

The merry little jumping insects called Frog-hoppers (*Tettigonia*), one of which in its larva state emits the little mass of froth so common on shrubs, and

* Insect Architecture, 155.

called cuckoo-spit, are furnished with a set of tools for their own private carpentry, which, though less elaborate than those of the saw-flies, are worthy of a moment's glance. If we catch one of these vaulters and gently press the abdomen, we shall see proceeding from its hinder and lower part, a thickish piece, large compared with the size of the insect, which it is then easy to extract with a pair of fine pointed pliers. I have just done this, and here is the result on a slip of glass.

First there is a pair of brown protecting pieces, oblong in form, and studded with hairs like the rest of the exterior of the body. From between them projects what resembles a lancet, of the usual translucent amber-coloured horn, appropriated to these instruments (which is to them what steel is to us); and which we shall presently discover to be composed of two blades exactly alike, convex without and concave within, applied face to face. One edge of this pair of implements is quite smooth, but the other is cut into the most beautifully regular and most minute teeth.

This, however, is but the sheath. Within the two spoon-shaped faces there lie two other lancets, blade to blade, still finer and more delicate. Both edges of these blades are of the most perfect keenness, without a flaw; but their sides appear roughened with rows of very minute horny knobs, like a rasp.

I shall illustrate this demonstration by another extract from Réaumur, premising, however, that his observations refer to the large species of true *Cicadæ* from warmer latitudes, whose machinery seems to differ from that of our little friends in some particulars. For example, the two inner lancets seem to be united in one, in Réaumur's species, or else, which I think more probable, he did not succeed in separating them.

He describes the two curved spoon-shaped pieces, as finely indented on both sides with teeth; which are strong, nine in number, arranged with great symmetry, increasing in fineness towards the point. This instrument he describes as composed of three pieces, the two exterior, which he calls the *files*, and another pointed, which he compares to a lancet, which is not toothed. "The files are capable of being moved forward and backward, while the centre one remains stationary; and as this motion is effected by pressing a pin or the blade of a knife over the muscles on either side at the origin of the ovipositor, it may be presumed that those muscles are destined for producing similar movements when the insect requires them. By means of a finely-pointed pin carefully introduced between the pieces, and pushed very gently downwards, they may be, with no great difficulty, separated in their whole extent.

"The contrivances by which those three pieces are held united, while at the same time the two files can be easily put in motion, are similar to some of our own mechanical inventions, with this difference, that no human workman could construct an instrument of this description so small, fine, exquisitely polished, and fitting so exactly. We should have been apt to form the grooves in the central piece, whereas they are scooped out in the handles of the files, and play upon two projecting ridges in the central piece, by which means this is rendered stronger. M. Réaumur discovered that the best manner of showing the play of this extraordinary instrument is to cut it off with a pair of scissors near its origin, and then, taking it between the thumb and the finger at the point of section, work it gently to put the files in motion.

"Beside the muscles necessary for the movement of the files, the handle of each is terminated by a curve of the same hard horny substance as itself, which not only furnishes the muscles with a sort of lever, but serves to press, as with a spring, the two files close to the central piece." *

The use of these instruments is the same as I have already alluded to in the case of the saw-flies. The female Tree-hopper deposits her eggs in holes which she bores in dead twigs by means of these files and lancets. The branches chosen are said to be recognisable by being studded with little oblong elevations, caused by the partial raising of a splinter of wood at the orifice of the hole, to which it serves as a cover. These are arranged in a single line, the holes which they protect being only half-an-inch in length, and reaching to the pith, whose course they then follow. Not more than six or eight eggs are laid in each hole, but an idea of the labours of the industrious and provident mother will be formed from the fact that each lays six or seven hundred eggs in the course of the summer.

* Insect Architecture, 149.

CHAPTER IX.

INSECTS: THEIR MOUTHS.

THE parts of the mouth in different insects afford an almost endless store of delightful observations; and the more as, with all their variety, they are found to be in every case composed of the same essential elements. You would not think so, indeed; you would naturally suppose,—looking at the biting jaws of a Beetle, the piercing proboscis of a Bug, the long elegantly-coiled sucker of a Butterfly, the licking tongue of a Bee, the cutting lancets of a Horse-fly, and the stinging tube of a Gnat,—that each of these organs was composed on a plan of its own, and that no common structure could exist in instruments so diverse. But it is so, as we shall see.

We may consider the various organs of the mouth as most harmoniously and perfectly developed in the active carnivorous Beetles, the *Carabidæ*, or ground-beetles, for instance. Let us examine the head of this black *Scarites*, from the garden; and first from above.

In front of the polished headshield, and jointed to it by a broad transverse straight edge, is a four-sided piece, forming an oblong square, nearly twice as broad as long, a little convex, and marked with six little pits or sinkings of the surface, along its front edge. This is the upper lip; but, instead of being fleshy, as ours is, it is composed of a hard polished black shelly substance, of a peculiar nature, called chitine, the same substance as the hard parts of all Insects and Crustacea are made of.

From beneath the sides of this there project on each side two broad hooked pieces, which, as you see, I can with a needle force out laterally, so as to show their form better, for they hinge upon the sides of the face, beneath the head-shield. Each forms the half of a crescent, the curved points of which are turned towards each other, and can work upon each other, the points crossing, like shears. These are the proper biting jaws, or mandibles, and in many of the larger beetles they have great power of holding and crushing. Sometimes, their inner side is cut into strong teeth, but here this side forms a blunt cutting edge; the upper surface, however, is scored with ridges and furrows, like a file, and this structure is best seen in the left jaw, which, when the pair close, crosses over the right. This is an action of the jaws the reverse of ours, but it is characteristic of all the articulate classes of animals, in which the jaws, whenever present, always work horizontally, from right to left, and not vertically, up and down.

I will now, by making the forceps revolve, bring the under side of the head into view; for without separating the parts by dissection (which, however, is by no means difficult), it is impossible to see them all from one point of view. The part nearest our eye now is the chin, a wide horny piece, like the upper lip, jointed to the head by its straight hind edge, but, unlike it, having its front edge hollowed out with two deep notches, the central piece between them itself notched at its tip. Immediately above this notched central tooth (I speak of the relative position of the parts, supposing the insect to be crawling on the ground, without reference to the way in which we turn it about on the microscope), and

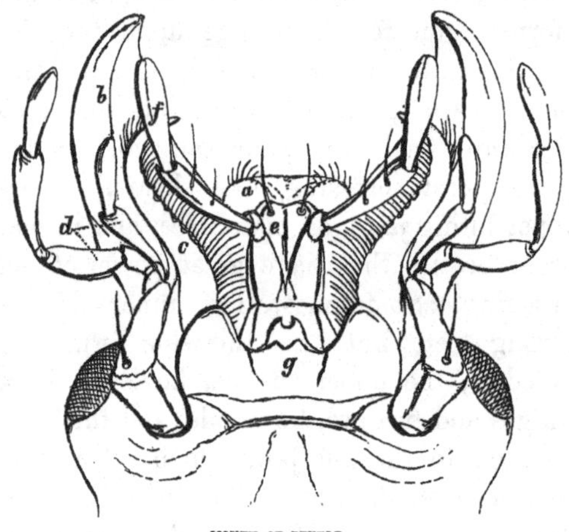

MOUTH OF BEETLE.
(*Seen from beneath.*)
a, upper lip; *b*, mandibles; *c*, maxillæ; *d*, maxillary palpi; *e*, tongue; *f*, labial palpi; *g*, chin.

united with it, there is a sort of solid square pedestal, on which stand a pair of jointed organs, and between them an oblong horny plate rounded at the tip, where it bears two bristles. This latter is the tongue; while the jointed organs on each side are called feelers,—*palpi;* though this is a begging of the question, for we do not really know the function of these organs. The chin, the tongue (*ligula*), and these palpi, constitute together the under lip.

Between the tongue and the biting jaws, or mandibles, we see a pair of organs similar to these latter, but smaller, less solid and more curved. These are the under or secondary jaws, *maxillæ*, the use of which is to hold the food, while the biting jaws work on it, and to convey it when masticated to the back of the mouth. Their whole inner edge is set with short stiff bristles, which towards the tips of the jaws become spines. Near the base of these jaws, on the outer edge, are jointed two pairs more of palpi, one pair to each jaw; of which the exterior is much stouter and longer than the interior. Thus this beetle has three pairs of these many-jointed organs, the *labial*, and the two pairs of *maxillary* palpi.

Now, in this form of mouth, which has been called a perfect or complete mouth,—that is, one in which all the constituent parts can be well made out, we find the following organs:—1. the upper lip (*labrum*); 2. the *mandibles;* 3. the *maxillæ;* with *a.* the *maxillary palpi;* 4. the lower lip (*labium*), comprising *β.* the *tongue*, *γ.* the *labial palpi*, *δ.* the chin (*mentum*).

I now exhibit to you the head of the Honey-bee. The front is occupied by an upper lip, and a pair of biting jaws (*mandibles*), which do not greatly differ from the same parts in a beetle. The jaws, however, are more hatchet-shaped, or rather like the hoof of a horse, supposing the soles to be the opposing surfaces. The other organs are greatly modified, so that you would scarcely recognise them. The under jaws (*maxillæ*) are greatly lengthened, and the two, when placed in contact, form a kind of imperfect tube, or sheath. Within these is the lower lip, divided into its constituent parts:—the thick opaque chin, at

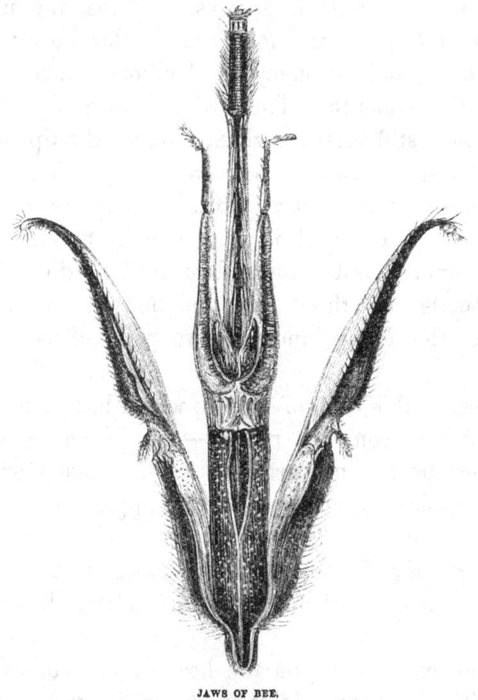

JAWS OF BEE.

its basal end; then the two labial palpi, each consisting of four joints, of which the two terminal ones are minute, while the two basal are large and greatly lengthened so as to resemble in appearance the *maxillæ*, whose function they imitate also; for the pair of palpi when closed form an inner sheath for the tongue (*ligula*). Finally you see this organ, which is the most curiously developed and modified of all; for it is drawn out to a long slender cylindrical tube, formed of a multitude of close-set rings, and covered with fine hairs. Some deny it to be tubular, and maintain that it is solid; but certainly it appears to me to have a distinct cavity throughout, with thickish walls.

Under a high power the structure of the investing hairs is very interesting; for they are seen to be flat filaments of the yellow chitine, very much dilated at their bases, and set side by side in regular whorls, the bottom edges of which form the rings of which the tongue is composed. The tip is probably a sensitive organ of taste, for it terminates in a minute globose pulpy body, whose surface is beset with tiny curved points. Thus I have pointed out to you all the parts which enter into the mouth of the beetle, except the *maxillary palpi;* and those, very small indeed, but quite distinct, you may see, on the outer edge of the *maxillæ*, just below the point where their outline begins to swell into its graceful curve.

The cylindrical tongue is capable of considerable extension and contraction at the will of the animal, being sometimes pushed far out of the mouth, and at others quite concealed within its sheath. "The manner," observes Mr. Newport, "in which the honey is obtained, when the organ is plunged into it at the bottom of a flower, is by lapping, or a constant succession of short and quick extensions and contractions of the organ, which occasion the fluid to accumulate upon it, and to ascend along its upper surface, until it reaches the orifice of the tube formed by the approximation of the *maxillæ* above, and of the labial palpi, and this part of the *ligula* below."

Well might Swammerdam, when describing this exquisite structure, humbly exclaim,—"I cannot refrain from confessing, to the glory of the Immense and Incomprehensible Architect, that I have but imperfectly described and represented this small organ; for, to represent it to the life in its full perfection, as truly most perfect it is, far exceeds the utmost efforts of human knowledge."

Here you may see the implement with which the Bug performs its much-dreaded operation of blood-sucking; for though this is not the head of the Bed-bug, but of one of the winged species that are found so abundantly on plants, and which I have just obtained by beating the hedge at the bottom of my garden,—yet the structure of the mouth is so exactly alike in all the members of this immense family, that one example will serve for all others.

From the front of the head, which, owing to the manner in which this part is carried, is the *lower* part, proceeds a fine thread, about four times as long as the head itself, which passes along between the fore legs, close to the body, beneath the breast. It is, however, at the pleasure of the animal, capable of being brought up so as to point directly forward, and even projected in front of the head, and in the same plane as the body; a fact which once came under my own observation. I found a Plant-bug

(*Pentatoma*) which had plunged this thread-like sucker of his into the body of a caterpillar, and was walking about with his prey, as if it were of no weight at all; carrying it at the end of his sucker, which was held straight out from the head and a little elevated. He fiercely refused to allow the poor victim to be taken away, being doubtless engaged in sucking its vital juices; just as the Bed-abomination victimises the unfortunates who have to sleep at some village inn.

Well, we put this head with its sucker between the plates of the compressorium, upon the microscope-stage. The thread is an organ composed of four lengthened slender joints, beset with scattered bristles, and terminating in a point on which are placed a number of excessively minute radiating warts,—probably the seat of some sensation,—perhaps taste. This jointed organ is the under-lip; it is slit all down one surface, so that it forms an imperfect tube, or furrow, within which lies the real weapon, a wire of far greater tenuity, which by pressure I can force out of its sheath. It is so slender that its average diameter is not more than $\frac{1}{1200}$th of an inch, and it ends in the most acute point; yet this is not a single body, but consists of four distinct wires, lying within one another, and representing the maxillæ and the mandibles. These can be separated by the insect, and will sometimes open when under examination; but no instrument that I can apply to them is sufficiently delicate to effect their separation at my pleasure. Just at the very tip, however, under this high power, we can see, by the semi-transparency of the amber-coloured chitine of which the organ is composed, that there is another tip a little shorter, and as it were contained within the other. This inner point is cut along its edges into saw-teeth pointing backward. Such exquisite mechanism is bestowed upon the structure, and such elaborate contrivance is displayed for the comfort of an obscure and obscene insect, by Him who has not disdained to exercise his skill and wisdom in its creation!

You know the stout flies which are denominated Horse-flies or Whame-flies (*Tabanus*), which are so numerous in the latter part of summer flying around horses, and men too, if we intrude upon their domains. They are continually alighting on the objects of their attentions, and though driven away, returning with annoying pertinacity to the attack. You may always recognise them by the brilliant metallic hues—reds, yellows, and greens,—with which their large eyes are painted, often in stripes or bands. These are voracious blood-suckers; and, as might be supposed from their propensities, they are well furnished with lancets for their surgery. Here you may see their case of instruments, which are so effective, that Réaumur tells us, that having compelled one to disgorge the blood it had swallowed, the quantity appeared to him greater than the whole body of the insect could have been supposed capable of containing.

All the parts here are formed of the common amber-coloured chitine, brilliantly clear and translucent. The upper lip forms a sort of straight sheath, in which all the other parts are lodged when not in use. The *mandibles* are narrow lancets; of which one edge near the tip is beset with reverted saw-teeth, and the opposite edge with excessively sharp points standing out at right angles, while the surface is roughened with lozenge-shaped knobs set in regular rows. Below these are the *maxillæ*, which are the principal cutting instruments; these are shaped like a carving-knife with a broad blade, strengthened at the basal part of the back by a thick ridge, but brought to a double edge near the tip. The back-edge is perfectly fine and smooth, so that the highest powers of the microscope can only just define its outline; while the other edge is notched into teeth so delicate, that twelve of them are cut in the length of a ten-thousandth part of an inch; and yet they are quite regular and symmetrical in length, height and form! I know of no structure of the kind which equals this. These teeth are continued throughout the inner edge of the blade from the tip to the base, and are about eight hundred in number; though the length of the entire blade is only such that upwards of a hundred and fifty of them, if laid end to end, would not reach to the extent of an inch!

The office of these wonderful instruments is doubtless to cut and enlarge the wound within, and thus promote the flow of blood. The whole apparatus is plunged into the flesh of the victim—horse or man; then the *maxillæ* expand, cutting as they go, and doubtless working to and fro as well as laterally, so as to saw the minuter blood-vessels. At the same time the *mandibles*, with their saw-teeth on one side, and pricking points on the other, work in like manner, but seem to have a wider range. Finally, there is an exceedingly delicate piece beneath all, which seems to represent the *labium* or under lip.

In the active and cunning little Flea, that makes his attacks upon us beneath the shelter of the blankets and under cover of night, the piercing and cutting blades are very minute, and have a peculiar armature. They remind me (only in miniature of course) of those formidable flat weapons which we often see in museums, the *rostrums* of the huge Saw-fishes (*Pristis*); a great plate of bone covered with grey skin, and set along each side with a row of serried teeth. Here the blades are similar in form, being long straight narrow laminæ of transparent chitine, set along each edge with a double row of glassy points, which project from the surface, and are then hooked backwards. These are the *mandibles*, and they closely fold together, inclosing another narrower blade, the upper lip, which has its two edges studded with similar points, but in a single row.

In general, as we have seen, the *maxillæ* are the specially armed weapons, the *mandibles* acting a secondary part, often serving as mere sheaths—in those insects which pierce other animals with the mouth. But in this case the *mandibles* are the favoured parts, the *maxillæ* being developed into broad leaf-shaped convex sheaths, inclosing the *mandibles*.

There are, however, two cutting blades besides,—the *labial palpi*, which have their upper edge thick, divided into four distinct joints, and set with bristles, —thus retaining the proper palpine character, while their under edge is thinned away to a fine keen blade, in which there is no sign of jointing. Then there are the *maxillary palpi*, of which the joints are furnished at their tips with tiny projecting warts, doubtless the seats of a delicate perception, and hollowed into a double series of chambers, which are filled with a dark-coloured fluid.

All this is very interesting to behold, and is calculated to exalt our ideas of the wonderful and inexhaustible resources of Omnipotence, as well as to humble us, when we reflect on how little we certainly understand even of what we see. But common as the Flea is, it is not a matter of course that you will be able to repeat these observations with the first specimen you put on the stage of your microscope. Several favourable conditions must combine in order to insure a successful examination. You should choose a female Flea, partly because of her greater size, and partly because the predatory weapons are better developed, in all these piercing and sucking insects, in the females,—true Amazons. Then you will find it needful to amputate the head, in order to get rid of the front legs, the thick thighs of which else impede your sight of the mouth, being projected on each side of it. And this is a delicate operation: it must be performed on a plate of a glass, under a lens, with one of those dissecting needles whose points are ground to a cutting-edge. Next, having severed the head, you must place it in a drop of water, between the plates of your compressorium, the graduated pressure of which, by means of the screw, will cause the organs of the mouth to open and expand separately. Finally, you must have a good instrument, and a high power: less than 600 diameters will not avail to bring out distinctly the toothing of the *mandibles* and *labrum;* and even then you will need delicate manipulation and a practised eye. But the object is worthy of the care bestowed upon it.

Once more. Let us submit to examination the complex case of instruments wherewith the Gnat performs her unwelcome yet skilful surgery. I say "her," because among the Gnats, as among most of these puncturing insects, it is the females only who attain skill in the phlebotomic art, the males being innocent of any share in it, and being indeed unprovided with the needful implements.

Here is a large specimen, resting with elevated hind legs on the ceiling, and now in alarm off with shrill humming flight to the window. I decapitate her without compunction, as it is but a fair penalty for her murderous deeds; and, as of old the axeman held up "the head of a traitor" to public gaze, so I lay *this* head on the glass of the compressorium for your contemplation.

And before I apply pressure to the glass-plate, devote a moment's attention to the *tout ensemble.* First, the head itself is a hemisphere, almost wholly occupied with the two compound eyes, which present the beautiful appearance of a globe of black velvet, studded with gold buttons arranged in lines crossing each other at right angles. The summit of the head, where the two compound eyes unite, bears a sort of rounded pedestal, the area of which forms the sole part of the head not covered by the organs of vision. On this are placed, side by side, the two antennæ, springing from rounded bulbous bases; they consist of twelve (exclusive of the basal bulb) cylindrical joints, which are beset on all sides with short arched hairs, but have besides a whorl of radiating long hairs surrounding the bottom of each joint. The effect of this is exceedingly light and elegant.

Between these projects a long cylinder, which represents the lower lip (*labium*); it slightly swells towards the tip, where it forms a round, nut-like knob, covered with exceedingly minute papillæ, and no doubt constituting a highly sensitive organ of touch. For the greatest part of its length it is covered with lined scales, and with short arched hairs, like the antennæ, while each side of its base is guarded by a labial palp of three joints.

On applying a graduated pressure, slowly increased to actual contact of the plates (or as near an approximation to it as we can effect), we see first that the nut-like tip of the *labium* expands into two concave leaves, like the bracts of a bud, and displays two pairs of more delicate leaves within them. Then from a groove along the upper side of the *labium*, spring out several filaments of great elasticity and of the most delicate tenuity. One pair of these represent the mandibles; they consist each of a very narrow blade with a stronger back like that of a scythe. Their tip is brought to a most acute point, and the edge in immediate proximity to this is cut into about nine teeth pointing backward: the rest of the edge is smooth, but the whole blade is crossed by a multitude of oblique lines of great delicacy, which may be intended to keep the edge constantly keen.

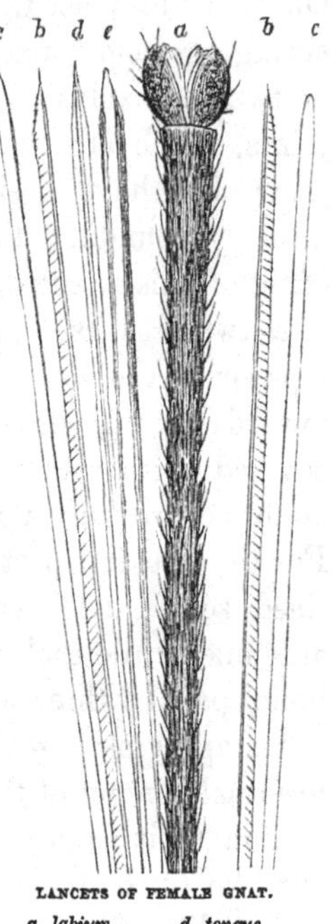

LANCETS OF FEMALE GNAT.

a. *labium.* *b, b.* *mandibles.* *c, c.* *maxillæ.* *d.* *tongue.* *e.* *labrum.*

Next come the *maxillæ*, or lower jaws, horny filaments as long as the former, but still more delicate, constituting simple cutting lancets, with a back and a keen blade, a little widening at the tip.

Besides these there is the *tongue*, consisting of a central rod which is distinctly tubular, and of a thin blade *on each side*, fine-edged and drawn to an acute point. And also the *labrum* or upper lip, an organ having the same general form, but constituting an imperfect tube; a tube, that is to say, from which about a third of the periphery is cut away, so as to serve as a sheath for the tongue, which ordinarily lies within its concavity.

I scarcely know whether this apparatus is not more wonderfully delicate than any we have examined;—even than that of the Flea. And how effective it is you doubtless well know; for when the array of lancets is introduced into the flesh, you are aware that a tumour is left, which by its smart, itching, and inflammation, causes much distress, and lasts many hours. This effect is probably produced partly by the deep penetration of the instruments,—for they are fully one-sixth of an inch in length, and they are inserted to their very base,—and partly by the injection of a poisonous fluid, intended, as has been conjecturally suggested, to dilute the blood and make it more readily flow up the capillary tubes. The channel through which this fluid is injected is probably the tongue, which you see to be permeated by a tube containing a fluid; and the same channel may afford ingress to the diluted blood.

The *labium* does not enter the wound. If you have ever had the philosophic patience to watch a gnat while puncturing your hand, you have observed that the knob at the end of the proboscis is applied to the skin, and that then the organ bends with an angle more and more acute, until at length it forms a double line, being folded on itself, so that the base is brought into close proximity to the skin. Meanwhile the lancets have all been plunged in, and are now sunk into your flesh to their very bottom, while the *labium*, which formed merely the sheath for the whole, is bent up upon itself, ready again to assume its straight form, as soon as the disengaged lancets require its protection.

The tongue of the common Flies (House-fly, Blow-fly, &c.) is an exquisite microscopical object, from its extreme complexity and beauty. You are familiar with the way in which a fly, having alighted close to a drop of tea on the table, applies to it a proboscis with large dilated extremity, and presently licks it all up. You shall now see the curious implement by which this is effected.

The broad portion of the object before us, forming its bottom part, bristling with coarse black hair, is the front of the head of a Blow-fly. From the midst

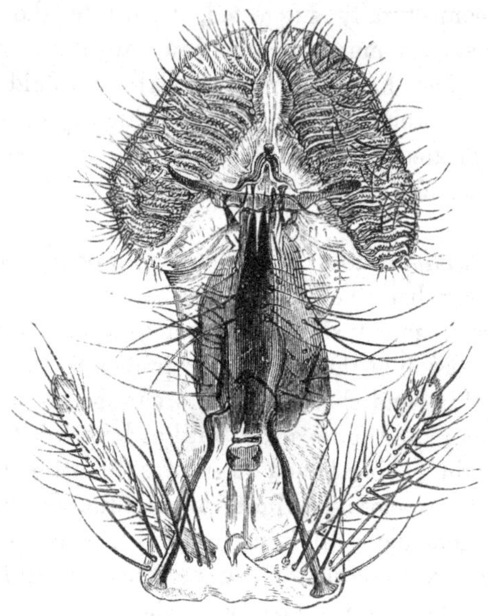

TONGUE OF BLOW-FLY.

of this projects a dark brown mass terminating in two points, and inclosing a narrower and darker object with two long slender roots, dilated at their bases;—this is the pair of *maxillæ* altered and modified into a kind of sheath for the mandibles. On each side projects an elegant club, bristled with coarse black hair, and covered besides with a coat of very minute hairs; these clubs are the maxillary palpi.

But now we come to the terminal part, consisting

of a pair of lobes, together forming a rounded triangle in their outline. This is the dilated and thickened termination of the *labium*, and is the instrument by which the liquids are so rapidly sucked up. It is impossible to describe this beautiful structure intelligibly: and indeed it is not well understood even by those who have devoted their lives to this branch of natural science. The principal feature apparent is a wide clear membrane, through which run with admirable symmetry a series of tubes. These tubes consist of four primary ones, all originating near the centre of the expansion, and radiating thence, two backward towards the two lateral angles of the triangle, and the other two nearly side by side towards its point. From each of these, along its outer side only, branch off the minor tubes, very numerous and close together, going off in a slightly sinuous line direct to the margin, diminishing regularly in their course, and at their extremities curving over, so as to bring their open tips to the surface of the skin.

The construction of these tubes is highly interesting: they are formed, like the air-pipes (*tracheæ*), of a multitude of horny rings; but with this peculiarity, that the rings do not form a continuous spiral, but are separate and distinct; and are moreover imperfect; for each wire (so to speak) does not perform a complete circle, but only about two-thirds of a circle, leaving a blank space; and the tips of the wires end alternately in a fine acute point, and in a rounded fork, like the prongs of a pitch-fork. It has been said that these tubes are modified *tracheæ;* but this fact is by no means obvious to me; for so far from their being connected with the general tracheal system, each of the four main tubes originates in an open centre, and each with an open extremity. I think it likely that they are so many suctorial pipes, through which the fluid to be drunk is pumped up, entering at their minute open tips, and discharging itself into the central cavity, by the open basal extremities of the main tubes.

The most extraordinary modification of jaws, however, is the long spiral tube which is ordinarily coiled up under the face of a Butterfly or Moth, with which it pumps up the sweet nectar of flowers. Many flowers have a deep corolla, and most have the bases of their petals, where the nectar lies, so far from the level of the surface, that probing is necessary to reach it. Bees can enter tubular flowers, and lick their bottoms; and even blossoms that are closed, as the Snapdragon, they know how to force and enter. But Butterflies, with their wide wings, incapable of being folded, cannot enter flowers bodily, and therefore a peculiar apparatus is given them for robbing their contents, as it were, at the doors.

Nothing is easier than to examine this beautiful organ with the naked eye; and much may be learned of its structure by means of a pocket lens. You may thus see in a moment, that it forms a flat spiral of several coils, like the mainspring of a watch; that it runs off to a point, and that this point is double, for it is frequently seen separate a considerable way up. Hence you would probably infer that the organ consists of two equal and consimilar halves, united longitudinally. And so, indeed, it does; and these halves are the representatives of the *maxillæ* or lower jaws of the Beetle, being thus greatly developed at the expense of almost all the other parts. The upper lip and the *mandibles* are discernible only in the form of three most minute plates; the *labial palpi* are large and prominent,—those well-haired points that project in front of the head, one on each side of the spire. This spiral form of the *maxillæ* is called *antlia*.

It is not, however, very easy to fix it in an extended condition on a slip of glass, so as that it shall lie flat throughout its whole length, without injuring the parts or so agglutinating them together, that their structure is concealed or distorted, and in either case unfitted for microscopical examination. The specimen which I have prepared, from the mouth of the Small Garden White Butterfly, is stretched, and fixed in balsam, and will I think show you the structure under a high power very well.

Before we examine it, however, I will cite you the description of one of the most eminent of microscopical anatomists, Mr. Newport. He considers each *maxilla* to be composed of an immense number of short transverse muscular rings, which are convex externally and concave internally, the two connected organs forming a tube. Within each there are one or more large *tracheæ* connected with the *tracheæ* in the head. The inner or concave surface which forms the tube is lined with a very smooth membrane, and extends along the anterior margin throughout the whole length of the organ. At its commencement at the apex, it occupies nearly the whole breadth of the organ, and is smaller than at its termination near the mouth, where the concavity or groove does not occupy more than about one-third of the breadth. In some species, the extremity of each *maxilla* is furnished along its anterior and lateral margin with a great number of minute *papillæ*. These, in *Vanessa Atalanta* (the Red Admiral Butterfly) for instance, form little barrel-shaped bodies, furnished at the free end with three or more marginal teeth, and a larger pointed body in the centre. There are seventy-four of these in each *maxilla*, or half the proboscis. Mr. Newport regards them as probably organs of taste. There are also some curious appendages arranged along the inner anterior margin of each *maxilla*, in the form of minute hooks, which, when the proboscis is extended, serve to unite the two halves together, by the points of the hooks in one half being inserted into little depressions between the teeth of the opposite side; sometimes these are furnished with a tooth below their tips.

With all deference for so respectable an authority, I cannot help seeing that such is not the structure of the *antlia* before us. It is evident to me that each half tube is composed of a membrane stretched upon stiff horny semi-rings, doubtless composed of chitine, and certainly not *muscular*. By bringing the outline of the rounded exterior into focus, we see that these rings form sharp ridges; and by tracing them onwards to the attenuated extremity of the organ, we see them gradually give way to transverse lines of interrupted ridgy warts upon the outside of the membrane. The true muscles appear to be indicated by those oblique lines and bands that are seen in the interior, beneath the horny rings.

This specimen shows very distinctly that the two sides are but semi-tubular, and as one pair of the opposing edges are open at each extremity, and the other pair separate throughout, we are able to discern very clearly the array of hooks, by which the edges are united at the will of the animal. No trace of the curious little pointed barrel-shaped *papillæ* is found here, but I have seen it in other examples.

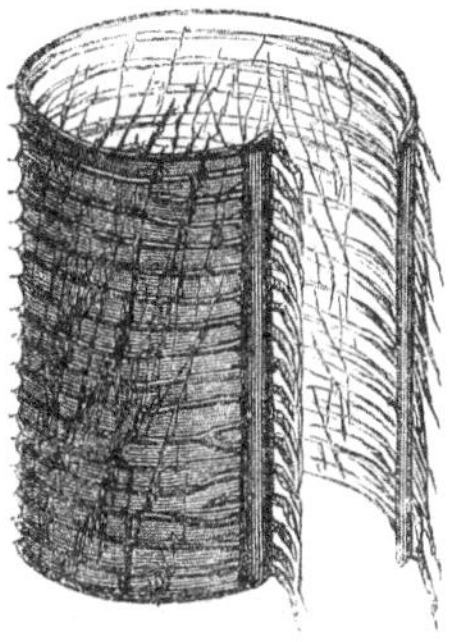

THE SUCKER OF A BUTTERFLY.
A small portion of one half-cylinder.

It seems highly probable, from the observations of the excellent anatomist just named, that the exhaustion of the nectar of a flower, which is effected with great rapidity and completeness, is a process dependent on respiration, and connected with the air-pipe that permeates each division of the sucker.

It will not be a very violent transition, if from the sucking pump of the Butterfly I carry you to the silk-spinner of the caterpillar. Here I have a Silk-worm in the act of commencing its cocoon; by inclosing which in this glass tube, we shall conveniently have the insect at command, and shall be able to view the process under a low magnifying power and reflected light. Now the grey face of the worm is presented to us; and we can see, below the edge of the head-shield, a short broad upper lip, forming two blunt points.

Below this is the pair of strong brown *mandibles*, convex outwardly and concave inwardly, each cut at its broad biting edge into several teeth. Below these are two little points which represent the *maxillæ*, and between them a blunt rounded knob, which is the lower lip (*labium*).

You may also see, on each cheek, close to the base of the *mandible*, a little pit, out of which rises a short columnar organ tipped with two bristles; these columns are the incipient *antennæ*. Outside them you may discern on each cheek, a series of six globes of glass (so they appear) set in the substance of the skin, —five forming a semicircle, and one in the centre; these are "the windows at which the [silkworm's] soul looks through,"—provided he has any soul—in prosaic parlance, his eyes.

Now, having thus introduced the several members of our useful friend's physiognomy to you, let me call your attention to a fleshy wart just beneath the lower lip, and midway between the bases of the two fore legs. This wart terminates in a horny point not unlike a bird's beak, which is perforated, and from the tip of which the glistening yellow filament of silk is ever drawn out, as the caterpillar throws his head from side to side. This pointed wart is the spinning organ; and the thread of silk is, as it issues from the orifice, a fluid gum, which hardens immediately on its exposure to the air. The silk-gum is secreted by the caterpillar in two long blind tubes, which lie twisted and coiled in the interior of the body, occupying nearly the whole space, except that which is taken up by the great digestive canal. These become very slender as they approach the head, and at length terminate in a dilated reservoir, which opens by the little pointed wart which you have just seen.

Many caterpillars are able to suspend themselves at pleasure by means of the thread which they are spinning, lengthening it and "stopping it off," at will. This latter operation they perform (though they cannot recal the thread when once it has issued) by means of an angular point formed by the two slender tubes at their junction in the reservoir; thus compressing the thread of gum, and so preventing any more from issuing. The gum is perfectly colourless in the reservoir, but as it issues forth becomes coated with a varnish, which is secreted in the same organ, and which is poured out at the same time. In the case of the common Silkworm, this varnish imparts to the silk that brilliant yellow hue which it generally possesses, and which, as the varnish is soluble, can be easily discharged from it in the manufacture.

CHAPTER X.

INSECTS: THEIR EARS AND EYES.

A VERY wide field of observation, and one easily cultivated, is presented by the organs of sense in the insect races, and in particular by those curious jointed threads which proceed from the front or sides of the head, and which are technically called *antennæ*. These may sometimes be confounded with the *palpi*, examples of which organs we have been lately looking at; for in a carnivorous Beetle, for instance, both *palpi* and *antennæ* are formed of a number of oblong, polished hard joints, set end to end, like beads on a necklace. And it is probable there may be as much community in the function as in the form of these two sets of appendages; that both are the seats of some very delicate perceptive faculty allied to touch, but of which we cannot, from ignorance, speak very definitely. It is likely, indeed, that sensations of a very variable character are perceived by them, according to their form, the degree of their development, and the habits of the species.

It is not impossible, judging from the very great diversity which we find in the form and structure of these and similar organs in this immense class of beings, compared with the uniformity that prevails in the organs of sense bestowed on ourselves and other vertebrate animals,—that a far wider sphere of perception is open to them than to us. Perhaps conditions that are appreciable to us only by the aid of the most delicate instruments of modern science may be appreciable to their acute faculties, and may govern their instincts and actions. Among such we may mention, conjecturally, the comparative moisture or dryness of the atmosphere, delicate changes in its temperature, in its density, the presence of gaseous exhalations, the proximity of solid bodies indicated by subtle vibrations of the air, the height above the earth at which flight is performed, measured barometrically, the various electrical conditions of the atmosphere; and perhaps many other physical diversities which cannot be classed under sight, sound, smell, taste, or touch, and which may be altogether unappreciable, and therefore altogether inconceivable, by us. It is probable, however, that the *antennæ* are the organs in which the sense of *hearing* is specially seated; a conclusion which has long been conjecturally held, and which is confirmed by some observations recently made on the analogous organs in the *Crustacea*, which I will allude to more particularly presently.

The forms which are assumed by the *antennæ* of Insects are very diverse; and I can bring before you only a very small selection out of the mass. One of the most simple forms is that found in many Beetles, as in this *Carabus* for example. Here, each *antenna* is composed of eleven joints, almost exactly alike and symmetrical, each joint a horny body of apparently a long-oval shape, polished on the surface, but not smooth, because covered with minute depressed lines, and clothed with shaggy hair. There is, however, a slight illusion in the appearance: it seems as if the dividing point of the joints were, as I have just said, at the termination of the oval, but when we look closely we see that the summit of each oval is, as it were, cut off by a line, and by comparing the basal joints with the others, we see that this line is the real division, that the summit of the oval really forms the bottom of the succeeding joint, and that the constricted part is no articulation at all. The first, or basal joint (called the *scapus*), and the second (called the *pedicella*), differ in form from the rest, here but slightly, but often considerably. The whole of the remaining joints are together termed the *clavola*.

You may see a considerable diversity of figure and of aspect generally in this tiny Weevil, which may be accepted as a representative of a great family of Beetles, the *Curculionidæ*. The manner of their insertion strikes us at first sight as peculiar, as is in fact the aspect of the whole head. Instead of a thick substantial solid front, with powerful widely-gaping jaws, such as we saw in the *Carabs*, here projects from between the eyes a long rod-like proboscis, as long as the whole animal besides, curving downwards, and carrying at its very extremity a minute mouth, with all the proper apparatus of lips, jaws, and palpi. Moreover, the *antennæ* are planted on the two sides of this beak, about its mid-length; and they are curiously elbowed, each projecting horizontally at a right angle to the beak for a considerable distance, and then with a sharp angle becoming parallel to it for the remainder of their length. So that, supposing the terminal half of the beak to be broken off just behind the insertion of the *antennæ*, the whole would compose the letter T. Now, the first bend of this angle is composed of a single joint, the *scapus*, which is, in this family, greatly lengthened; and then the two or three final joints are much thicker than all the others, and are as it were fused together into a large oval knob, called the *club*.

Now, a word or two in explanation of this very singular form of head and head-organs. The larva or grub stage of these insects is destined to be passed in the interior of fruits and seeds; the individual which we have been examining (*Balaninus nucum*) was born one morning in August, in the interior of a hazel-nut. Its parent had chosen a suitable nut, just then when it is set for fruit, and as yet green and soft; and had with her proboscis, or rather with her jaws at its tip, as with a gimlet, bored a tiny hole through the yielding shell into the very interior; then turning round, and inserting the extremity of her abdomen with its ovipositor, she had shot an egg into this dark cavity. The juices poured forth at the wound soon healed the orifice; the nut grew; and presently the egg became a little white grub. He then rioted in plenty; prolonged his darkling feast

> "From night to morn, from morn to dewy eve;"

—'twas all "dewy eve" to him, by the way, for no ray of light saw he, till that prosperous condition of existence was done. No wonder he grew fat; and fat those rogues of nut-weevils always are, as you well know. Well, when the nut fell, in October, the kernel was all gone, completely devoured, and our little highway-robber was ready for his winter sleep: he gnawed a fresh hole through the now hard shell, made his way out, and immediately burrowed into the earth, where he lay till June; then became a pupa, and emerged just what you see him, a long-snouted beetle like his mother, in the beginning of August.

Such is his "short eventful history;" and you now see that the long beak is formed entirely with reference to this economy; it is an auger fitted to bore holes into shell-fruits, through their envelopes, for the reception of eggs.

There is a very extensive family of Beetles known as *Lamellicornes*, because the antennal joints are singularly flattened and applied one over the other like the leaves of a book (*lamella*, a leaf). Here is a very common little Chafer found on the droppings in pastures (*Aphodius fimetarius*), in which the last three joints constituting the *club* of the *antenna*, are of an ovate form, and flattened, so as to lie one on another quite close, like three oval cakes; and being connected only at one end of the long axis, they open and shut at the pleasure of the animal, like a long pocket memorandum-book of three leaves.

But this structure is seen to still greater advantage in the much larger Cockchafer, so abundant in May in some seasons. For here the joints composing the club are much more numerous (seven in the male, six in the female), and they are proportionally longer and thinner, and therefore more leaf-like. The insect widely expands them, evidently to receive impressions from the atmosphere; when alarmed, they are closed and withdrawn beneath the shield of the head, but on the first essay towards escape, or any kind of forward movement, the leaves are widely opened, and then after an instant's pause to test the perceptions on the sensorium, away it travels.

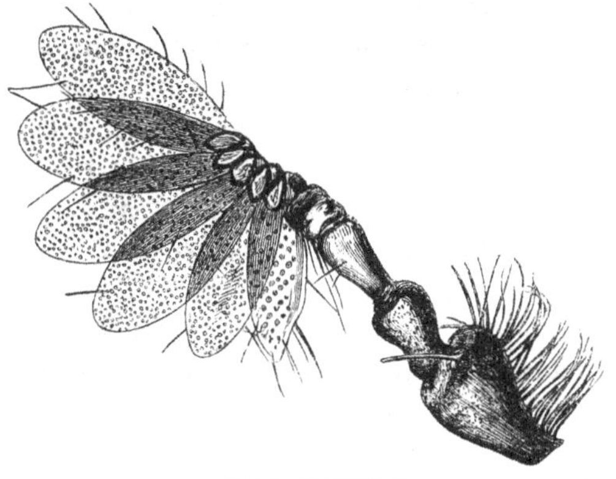

ANTENNA OF COCKCHAFER.

In some Beetles each joint of the series has one of its outer angles more developed than the other, and so produced as to make, with the rest of the joints, a saw-like edge; you may see an example in this Click-beetle or Skipjack (*Elater*), but many members of the same family show the same structure in a far higher degree, the angle being drawn out in a long slender rod, which (with its fellows) imparts to the antenna the appearance of a comb.

But much more curious and beautiful are the antennæ of many Moths, which often resemble feathers, particularly in the group *Bombycina*, of which the

Silkworm is an example; and in the male sex, which displays this structure more than the female. But I will show you a native example.

This is the antenna of a large and handsome, and not at all uncommon moth,—the Oak Egger (*Lasiocampa quercûs*). It consists of about seventy joints, so nearly alike in size and outline, that the whole forms an almost straight rod, slightly tapering to the tip. Each joint, however, sends forth two long straight branches, so disposed that the pair make a very acute angle, and the whole double series of seventy on each side, form a deep narrow groove. These two series of branches, being perfectly regular and symmetrical, impart to the antennæ the aspect of exquisite feathers.

It is, however, when we examine the elements of this structure in detail, using moderately high powers of enlargement, that we are struck with the elaborateness of the workmanship bestowed upon them. Each of the lateral branches is a straight rod, thick at its origin, whence it tapers to a little beyond its middle, and then thickens again to its tip. Here two horny spines project from it obliquely, one much stouter than the other, at such an angle as nearly to touch the tip of the succeeding branch.

Besides this, each branch is surrounded throughout its length with a series of short stiff bristles, very close-set, projecting horizontally (to the plane of the axis of the branch), and bent upwards at the end candelabrum-fashion. The mode in which they are arranged is in a short spiral, which makes about forty-five whorls or turns about the axis; at least in the branches which are situated about the middle of the *antenna;* for these diminish in length towards the extremity, bringing the feather to a rather abrupt point.

The entire surface of the branch gleams under reflected light with metallic hues, chiefly yellows and bronzy greens; which appear to depend on very minute and closely applied scales that overlap each other. The main stem of the feather,—that is, the primary rod or axis,—is somewhat sparsely clothed with scales of another kind, thin, oblong, flat plates, notched at the end, and very slightly attached by means of a minute stem at the base,—the common clothing-scales of the *Lepidoptera,*—specimens of which we have before examined.

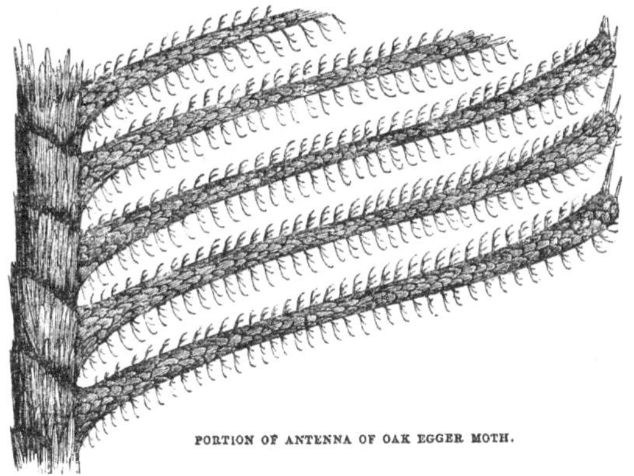

PORTION OF ANTENNA OF OAK EGGER MOTH.

We may acquire some glimpse of a notion why this remarkable development of antennæ is bestowed upon the male sex of this moth, by an acquaintance with its habits. It has long been a practice with entomologists, when they have reared a female moth from the chrysalis, to avail themselves of the instincts of the species to capture the male. This sex has an extraordinary power of discovering the female at immense distances, and though perfectly concealed; and will crowd towards her from all quarters, entering into houses, beating at windows, and even descending chimneys, to come at the dear object of their solicitude. Collectors call this mode of procuring the male "sembling," that is, "*assembling*," because the insects of this sex assemble at one point. It cannot be practised with all insects, nor even with all moths; those of this family, *Bombycidæ*, are in general available; and of these, none is more celebrated for the habit than the Oak Egger. The very individual whose antenna has furnished us with this observation was taken in this way; for having bred a female of this species the evening before last, I put her into a basket, in my parlour. One male, the same evening, came dashing into the kitchen; but yesterday, soon after noon, in the hot sunshine of August, no fewer than four more males came rapidly in succession to the parlour window, which was a little open, and, after beating about the panes a few minutes, found their way in, and made straightway for the basket, totally regardless of their own liberty.

It must be manifest to you that some extraordinary sense is bestowed upon these moths, or else some ordinary and well-known sense in extraordinary development. It may be smell; it may be hearing; but neither odour nor sound, perceptible by our dull faculties, is given forth by the females; the emanation is far too subtile to produce any vibrations on our sensorium, and yet sufficiently potent, and widely diffused, to call these males from their distant retreats in the hedges and woods. I think it highly probable that the great increase of surface given to the antennæ by the plumose ramification we have been observing, is connected with the faculty; perhaps every bristle of the spiral whorls is a perceptive organ, constructed to vibrate with the tender undulations that circle far and wide from the new-born female. Surely the ways of God in creation, as well as in moral government, are "past finding out!"

The male Gnat presents in its antennæ a pair of plumes of equal beauty, but of a totally different character. The pattern here is one of exceeding lightness and grace, as you may see in this specimen Each antenna is essentially a very slender cylindrical stem of many joints (about fourteen); at each joint springs out a whorl of fine hairs of great length and delicacy, which radiate in various directions (not, however, forming a complete circle), curving upward like the outline of a saucer, supposing the stem to be inserted into its centre. The length of these hairs is so great, that the diameter of their sweep equals, if it does not exceed, the whole length of the antenna.

In the tribe of two-winged insects, which we term, *par excellence*, Flies (*Muscadæ*), the antennæ are of peculiar structure. The common House-fly shall give us a good example. Here, in front of the head, is a shell-like concavity, divided into two by a central ridge. Just at the summit of this project are the two antennæ, originating close together, and diverging as they proceed. Each antenna consists of three joints, of which the first is very minute, the second is a reversed cone, and the third, which is large, thick, and ovate, is bent abruptly downwards immediately in front of the concavity. From the upper part of this third joint projects obliquely a stiff bristle or style, which tapers to a fine point. It is densely hairy throughout; and is more beset with longer hairs, on two opposite sides, which decrease regularly in length from the base, making a wide and pointed plume.

Such are a few examples of what are presumed to be the *ears* of Insects; let us now turn our attention to their *eyes.* And we can scarcely select a more brilliant, or a larger example, than is presented by this fine Dragon-fly (*Æshna*), which I just now caught as it was hawking to and fro in my garden. How gorgeously beautiful are these two great hemispheres that almost compose the head, each shining with a soft satiny lustre of azure hue, surrounded by olive-green, and marked with undefined black spots, which change their place as you move the insect round!

Each of these hemispheres is a compound eye. I put the insect in the stage-forceps, and bring a low power to bear upon it with reflected light. You see an infinite number of hexagons, of the most accurate symmetry and regularity of arrangement. Into those which are in the centre of the field of view, the eye can penetrate far down, and you perceive that they are tubes; of those which recede from the centre, you discern more and more of the sides; while, by delicate adjustment of the focus, you can see that each tube is not open, but is covered with a convex arch of some glassy medium, polished and transparent as crystal. There are, according to the computations of accurate naturalists, not fewer than 24,000 of these convex lenses in the two eyes of such a large species of Dragon-fly as this.

Every one of these 24,000 bodies represents a perfect eye; every one is furnished with all the apparatus and combinations requisite for distinct vision; and there is no doubt that the Dragon-fly looks through them all. In order to explain this, I must enter into a little technical explanation of the anatomy of the organs, as they have been demonstrated by careful dissection.

The glassy convex plate or facet in front of each hexagon is a *cornea*, or *corneule*, as it has been called. Behind each cornea, instead of a *crystalline lens*, there descends a slender transparent pyramid, whose base is the *cornea*, and whose apex points towards the interior, where it is received and embraced by a translucent cup, answering to the *vitreous humour.* This, in its turn, is surrounded by another cup, formed by the expansion of a nervous filament arising from the ganglion on the extremity of the optic nerve, a short distance from the brain. Each lens-like pyramid, with its vitreous cup and nervous filament, is completely surrounded and isolated by a coat (the *choroid*) of dark pigment, except that there is a minute orifice or *pupil* behind the *cornea,* where the rays of light enter the pyramid, and one at the apex of the latter, where they reach the fibres of the optic nerve.

Each *cornea* is a lens with a perfect magnifying power, as has been proved by separating the entire compound eye by maceration, and then drying it, flattened out by pressure, on a slip of glass. When this preparation was placed under the microscope, on any small object, as the points of a forceps, being interposed between the mirror and the stage, its image was distinctly seen, on a proper adjustment of the focus of the microscope, in every one of the lenses whose line of axis admitted of it. The focus of each cornea has been ascertained by similar experiments to be exactly equal to the length of the pyramid behind it, so that the image produced by the rays of light proceeding from any external object, and refracted by the convex cornea, will fall accurately upon the sensitive termination of the optic nerve-filament there placed to receive it.

The rays which pass through the several pyramids are prevented from mingling with each other by the isolating sheath of dark pigment; and no rays, except those which pass along the axis of each pyramid, can reach the optic nerve; all the rest being absorbed in the pigment of the sides. Hence it is evident, that as no two corneæ on the rounded surface of the compound eye can have the same axis, no two can transmit a ray of light from the very same point of any object looked at; while, as each of the composite eyes is immoveable, except as the whole head moves, the combined action of the whole 24,000 lenses can present to the sensorium but the idea of a single, undistorted, unconfused object, probably on somewhat of the same principle by which the convergence of the rays of light entering our two eyes gives us but a single stereoscopic picture.

The soft blue colour of this Dragon-fly's eyes,—as also the rich golden reflections seen on the eyes of other insects, as the Whameflies, and many other *Diptera*,—is not produced by the pigment which I have alluded to, but is a prismatic reflection from the *corneæ*.

You would suppose that, having 24,000 eyes, the Dragon-fly was pretty well furnished with organs of vision, and surely would need no more; but you would be mistaken. It has three other eyes of quite another character.

If you look at the commissure or line of junction of the two compound eyes on the summit of the head, you will see, just in front of the point where they separate and their front outlines diverge, a minute crescent-shaped cushion of a pale-green colour, at each angle of which is a minute *antenna*. Close to the base of each *antenna* there is set in the black skin of the head that divides the green crescent from the compound eyes, a globose, polished knob of crystal-like substance, much like the "bull's-eyes" or hemispheres of solid glass that are set in a ship's deck to enlighten the side-cabins. On the front side of the crescentic cushion there is a third similar glassy sphere, but much larger than the two lateral ones. What are these three spherules?

They are eyes, in no important respect differing from the individuals which compose the compound masses, except that they are isolated. The shining glassy hemisphere is a cornea of hard transparent substance, behind which is situated a spherical lens, lodged in a kind of cup formed by an expansion of the optic nerve, and which is surrounded by a coloured pigment-layer.

You may study these simple eyes, or *stemmata* as they are called, in many other insects, though they are not so universally present as the compound eyes. On the forehead of the Honey-bee they are well seen, as three black shining globules, placed, as in the Dragon-fly, in a triangle.

CHAPTER XI.

CRABS AND SHRIMPS.

It is always interesting to trace the varied forms and conditions under which any particular function is performed; and particularly to mark, in creatures very remote from us in the scale of being, the organs devoted to the senses which are so requisite to our own comfort. We have already seen some of these diversities, in examples taken from the classes *Mollusca* and *Insects;* and will now examine some more, as they appear in the *Crustacea.*

If you look at the head of a Crab, a Lobster, or a Prawn, you will see that it is furnished with jointed antennæ, like that of Insects; but whereas in insects there is never more than a single pair, in the creatures of which I am speaking there are two pairs. In the Prawn you may suppose, at first sight, that there are four pairs; but that is because the internal antennæ terminate each in three many-jointed bristles, in structure and appearance exactly like the bristles of the outer pair, two of the three being nearly as long as the outer, while the third is short. In the Lobster, the internal are two-bristled, both bristles rather short, while the external are very long. In the Flat-crabs each pair is simple, the inner minute, the outer long. In the great Eatable Crab each pair is very small, and they are dissimilar.

Now, taking the last-named animal as the representative of his class, let us examine one of his inner antennæ first. It consists of a jointed stem and a terminating bristle; the latter furnished with small hairs common to the general surface of the body, and with long, delicate, membranous filaments (*setæ*), often improperly called *cilia*, which are larger, and much more delicate in structure than the ordinary hairs.

The basal joint is greatly enlarged: if it be carefully removed from its connexion with the head, and broken open, it will be found to inclose in its cavity a still smaller chamber, with calcareous walls of a much more delicate character than the outer walls. This internal cell is considered by Mr. Spence Bate to be a *cochlea*, from its analogy, both in structure and supposed use, to the organ so named in the internal ear of man and other vertebrate animals. It is situated, as has been said, in the cavity of the basal joint of the internal antenna, and is attached to the interior surface of its wall farthest from the median line of the Crab. It has a tendency to a spiral form, but does not pass beyond the limits of a single convolution.

If this interior cell does indeed represent the *cochlea* of more highly-constructed ears,—to which it bears some resemblance, both in form and structure,—then it seems to identify, beyond dispute, these inner or upper antennæ as the organs of hearing.

Now with this conclusion agrees well the manner in which the living animal makes use of the organs in question. The Crab always carries them erect and elevated; and is incessantly striking the water with them, with a very peculiar jerking action, now and then vibrating, and, as it has been called, "twiddling" them. These antennæ, therefore, appear to be always on the watch:—let the animal be at rest, let it be feeding, no matter, the superior antennæ are ever elevated and on constant guard.

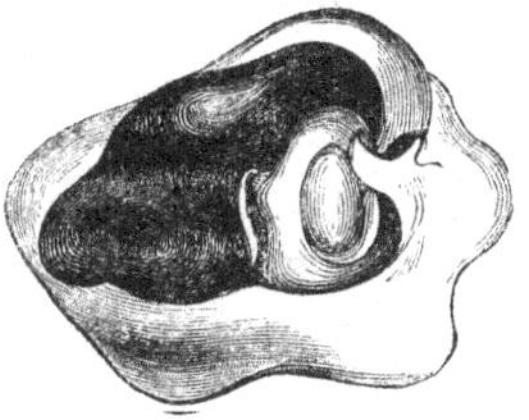

EAR OF CRAB, FROM BEHIND.

The lengthened and delicate setæ with which they are furnished, are, moreover, peculiarly adapted to receive and convey the most minute vibratory sensations from the medium in which they are suspended; and, on the whole, it seems to be satisfactorily settled by Mr. Spence Bate (to whose excellent memoir* I am indebted for these explanatory details) that the inner antennæ are real *ears*.

Having thus taken our Crab by the ears, we will endeavour next to tweak his nose. But stay, we must find it first. We turn our horny gentleman up, and in his flat ancient face we certainly discern little sign of a nasal organ. Our friend Mr. Bate must assist us again. He will tell us to look at the outer or lower antennæ. We will look accordingly, magnifier in hand, while he makes it clear to us that these are a pair of noses.

* Annals of Nat. Hist. for July, 1855.

Each of these organs is formed of a stem consisting in general of five joints, and a filament of many minute joints. In the Prawn and the Lobster all the five joints of the stem are distinct; but in the Crab the whole are, as it were, soldered together into a compact mass, with difficulty distinguishable into their constituent articulations; while in some species their position can be indicated only by the presence of the olfactory operculum.

This important little organ varies in its construction in the different families of *Crustacea.* In the Crab it is a small moveable appendage, situated at the point of junction between the second and third joints; it is attached to a long calcareous lever-like tendon, at the extreme limit of which is placed a set of muscles, by which it is opened and closed; to assist in which operation, at the angle of the operculum most distant from the central line of the animal are fixed two small hinges. When the operculum is raised, the internal surface is found to be perforated by a circular opening protected by a thin membrane.

In the Prawn, Shrimp, and Lobster, there is no operculum, but only the orifice covered by a membrane, which is placed at the extremity of a small protuberance, and is not capable of being withdrawn into the cavity of the antenna, as in the Crab.

In the latter animal, the little door, when it is raised, exposes the orifice in a direction pointing to the mouth; and where there is no door, still the direction of the opening is the same, inwards and forwards, answering to the position of the nostrils in the higher animals. In each case it is so situated that it is impossible for any food to be conveyed into the mouth without passing under this organ; and there most conveniently the animal is enabled to judge of the suitability of any substance for food, by raising the little door, and applying to the matter to be tested the sensitive membrane of the internal orifice.

Thus it is concluded that this lower or outer pair of antennæ are the proper organs of smell, as the upper and inner are of hearing.*

The eyes, though constructed on the same general principles as those of Insects, yet present some particulars worthy of your notice. In the Crabs and Lobsters they consist of numerous facets, behind each of which is a conical or prismatic lens, the round extremity of which is fitted into a transparent conical pit, corresponding to a vitreous body, while the conical extremity of these lenses is received into a kind of cup, formed by the filaments of the optic nerve. Each of these filaments, together with its cup, is surrounded by pigment matter in a sheath-like manner. To see this structure would require anatomical skill; but you may here examine with a low power portions of the cornea, or glassy exterior, of the eye of a Crab and of a Lobster. In the former, you see that the facets into which the cornea is divided are hexagonal, like those of most Insects, but in the latter they are square.

* Op. cit.

But Crustacea have a far greater faculty of circumspection than insects have; for besides the extensive convexity and numerous facets of their eyes, these organs are placed at the extremity of shelly footstalks, which are themselves moveable on hinges, capable of being projected at pleasure, and of being moved in different directions, and of being packed snugly away, when not in active use, in certain grooves hollowed out expressly for them in the front margin of the shell.

If ever you should chance to meet with the exotic Crustacea of the genera *Corycœus* and *Sapphirina*, you would see a form of eye of a quite remarkable and unique character. It is described by Dana in the following terms:—

" A pair of simple eyes consisting of an internal prolate lens, situated at the extremity of a vermiform mass of pigment, and of a large, oblate lens-shaped cornea. The cornea is connected intimately with the exterior shell of the front or the under side of the head, and the two corneæ are like spectacles adapted to the near-sighted lenses within; their size is extraordinary, being often one-third of the greatest breadth of the body in *Corycœus*. The lens and the cornea are often very distant from each other, being separated by a long clear space. The external surface of the cornea is spherical; but the inner is conoideo-spherical, or parabolic. The texture is firm, and when dissected it breaks or cuts like a crystalline lens. The true lens is always prolate, with a regular contour, excepting behind, where it is partly penetrated by the pigment. The pigment is slender, vermiform, of a deep colour, either red or blue, but at its anterior extremity usually lighter, and often orange or yellow." *

We might find much more both instructive and amusing in examining microscopically the structure of the higher Crustacea; but we will now dismiss them in order to discuss some of the lower forms, many of which are so minute that their whole bodies may be watched with ease performing all the functions of life, while confined under our eye, on the stage of the microscope. I refer to the tiny active little creatures known as Water-fleas, which are abundant in both fresh and salt water.

In this jar of fresh water which has been standing in the window for weeks, you may see among the green filaments of *Chara* many little atoms which scuttle hither and thither with a rapid succession of short leaps. These belong to the genus *Cyclops*, and are Crustacea, belonging to the order ENTOMOSTRACA.

By the aid of a glass tube which I stop at one end with my finger, I will endeavour to catch one. It is no easy matter, as you see, for the instant the end of the tube is brought near to one, he takes the alarm and leaps nimbly away before I can make the water rush in by withdrawing my finger from the other end. But I have one at length.

Here it is:—a minim of life not more than a sixteenth of an inch in length, looking something like

* Rep. on Crust. p. 1026.

a pellucid egg, furnished with long antennæ, with five pairs of branching feet, and a long tail terminating in bristles. But its parts and organs must not be dismissed in this summary way; we must look at them in detail.

And first of all, in the very midst of his forehead, like that obscene giant* after whom our tiny atom is named,—he bears a single eye that glares like a ruby. It would need no vast beam of olive-wood sharpened and heated in the fire, and "twirled about" by the united strength of five heroes, to "grind the pupil out;" for though brilliant and mobile, it is far too minute to be touched by the tip of the finest needle. Yet it is elaborately constructed; for it consists of a number (not very large) of simple eyes placed beneath a common glassy cornea. Several muscle-bands are attached to this compound organ of vision, and are arranged so as to form a cone, of which the eye is the base; these give the eye a movement of rotation upon its centre, which may be distinctly seen.

All the limbs, including both pairs of antennæ, two pairs of foot-jaws, five pairs of feet, and a pair of tail-lobes, are furnished at each of their many joints with tufts of long hairs; these appear to act the part of paddles, as the active little animal strikes the water vigorously with all its limbs, for the purpose of progression, and also for the creation of currents in the fluid, which currents subserve a double object,—the bringing constant supplies of water to be respired, and floating atoms of food to be devoured.

* Odyss. IX.

In this individual, which is a female, the antennæ are nearly equal in size throughout their length; but in the male, the middle joints of the upper pair are remarkably enlarged, forming a large swelling, followed by a sudden contraction, the first part of which is hinged. All of the true feet, and the second pair of foot-jaws, are divided to the base into two equal branches, so that the animal seems to possess no fewer than twenty-six limbs: each of which being many-jointed, and each joint, as I have observed, being set with delicately plumose hairs, the whole effect is most elegantly light and feathery.

On each side of the slender tail (more correctly, the abdomen) you see an oval bag connected with the body by an excessively slender thread of communication, and filled tensely with pellucid globose bodies. Like John Gilpin, of equestrian fame, when

> "He hung a bottle on each side
> To keep his balance true,"

our little natatory harlequin "carries weight." But these bags are filled with eggs, a temporary provision for their due and proper exposure to the water, while yet they are protected from enemies. They are developed only at certain seasons, when the eggs having attained a given amount of maturity in the ovary, are transferred through the exceedingly slender tube into these sacs, and are there carried about by the mother until the young are hatched, when the curious receptacles, being no longer needed, are thrown off, and speedily decay.

Here is a second form. It is named *Lynceus*, and

is nearly as common as the *Cyclops* in our stagnant pools. Essentially its structure is the same, but it has this peculiarity, that its body is enclosed within a transparent shell, which is thin and flattened sidewise, and through whose walls all the movements and functions of its parts are distinctly visible. The shell is broadly ovate in outline, comes to a sharp edge above, but is open all along the lower half of its circumference—as if two watch-glasses had been soldered together, edge to edge, and then a portion of the semicircumference had been ground away, so as to leave a thin but long entrance. Through this narrow orifice the limbs are protruded for locomotion, and through it the surrounding water finds its way in currents, bringing oxygen to be respired and food to be devoured.

The translucent shell descends in front into a sharp long beak, below which are seen the organs of the mouth, two pairs of foot-jaws, beset with fine bristles. At the origin of the beak is the eye, consisting, as we saw in the Cyclops, of several lenses, enveloped in a common cornea, the whole forming a moveable organ of a blue-black hue. Just behind this, at the very highest part of the shell, is a little colourless bladder-like vesicle, which constantly maintains a rapidly alternate contraction and dilatation. This is the heart, and this motion circulates the blood.

Below this, there is seen a great translucent irregular mass of flesh, evidently comprising many viscera, which winds along from one end of the shell to the other, nearly occupying its entire area, but not in connexion with it at the hinder part, as we see by its free movements there, where it curves round, and bending beneath terminates in a blunt tail, armed with two strong hooks, which can at pleasure be thrust down through the narrow orifice of the shell, and become partially straightened by being forcibly thrown backward. This great central mass is mainly occupied by the alimentary canal, in which food in various stages of assimilation may at all times be seen, and in which the interesting function of digestion can be witnessed throughout, from the first seizure of the atom and its mastication by the jaws, to the discharge of the effete remains.

The individual before us does not carry at this time eggs in the process of development; but the deficiency is supplied by a *Daphnia* which is playing

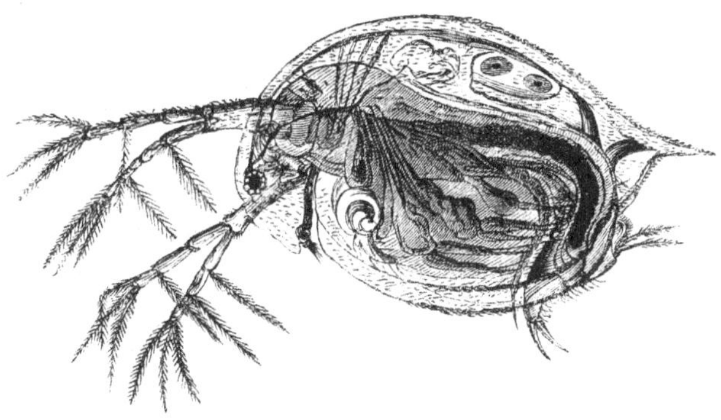

DAPHNIA.

about in the same drop of water. Here you perceive, between the arched outline of the shell and the sinuous outline of the free soft body, an open space of some size, which constitutes a receptacle, in which the eggs are deposited as they are laid, and in which they remain not only until the little animals are hatched, but until they have acquired a sufficient maturity to swim about and get their independent living.

This receptacle—in which you may see five or six eggs—is freely open to the surrounding water, which enters the slit edge of the shell, behind the tail. Perhaps you wonder why the eggs are not washed out by the respiratory currents; they are in fact maintained in their position only by a slender tongue-like projection from the back of the parent, which appears to have that special object. When, however, the young are ready for freedom, the mother has but to depress her body a little more than ordinary, when the door is opened, and the young easily slip from the receptacle into the open water.

These tiny odd-looking sprawling things that you see moving about by quick jerks in the same drop of water, are the young recently hatched. They are quite unlike their parent, having as yet no bivalve shell, no abdomen, and only three pairs of limbs. The body is a transparent plate, resembling the bowl of a spoon in form, but ending in two points which carry pencils of bristles. The large dark eye is conspicuous in front, and the six jointed and bristled limbs radiate from the centre, projecting stiffly on all sides. The second and third pair are seen to be double, each giving off a branch, which is pencilled with bristles like the principal stem.

We have not yet done with these tiny Water-fleas. The sediment at the bottom of this jar of water is quite alive with a host of nimble atoms, some of which you may see crawling up the sides of the glass. They are quite distinct from either of the kinds we have been examining, not only in details of structure, which is more identical indeed than it seems at first sight, but in habit; for whereas they shoot to and fro through the water with great force and rapidity, these can scarcely swim at all; or, if they do, it is with comparative slowness and much apparent effort; though over the smooth side of their glass dwelling, or upon the stems of water-plants, they glide along with much ease and elegance, by the quick vibrations of their pencilled feet.

The form we are now contemplating is distinguished by the name of *Cypris*, a genus which contains a good many British species. It is more completely inclosed in a shell than even the *Lynceus*, and its envelope more truly resembles the shell of a bivalve Mollusk, for the valves are open for more than three-fourths of their circumference; while the portion of the back that is united is sufficiently elastic to allow of some degree of expansion, thus answering the purpose of a hinge.

Now look at the elegant little creature. Its most prominent feature is its two pairs of antennæ, one projecting forwards and curved upwards, the other downwards. Both consist of several transparent joints, and are tipped with long clear bristles; but the pencils which tip the upper pair are specially graceful, being as long as the whole shell, exceedingly slender, beautifully curved, and so transparent that they seem formed of spun glass.

Another peculiarity is that there seems to be but one pair of legs, which terminate each in a hooked spine. You now and then see these awkwardly thrust out from beneath the hinder part of the shell, but locomotion is principally effected by the pencilled antennæ.

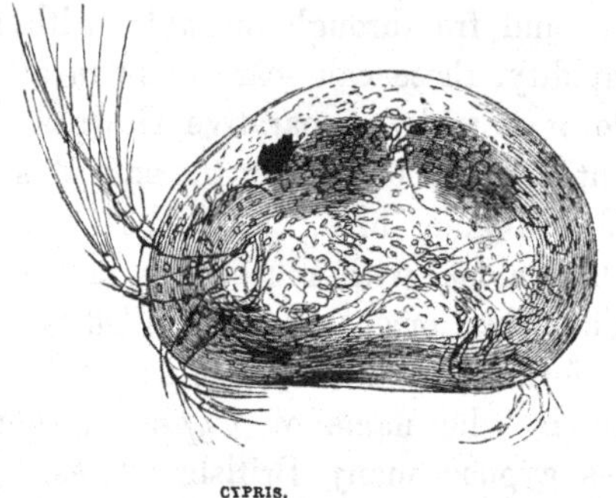

CYPRIS.

There is, however, a second pair of legs, but these do not usually make their appearance outside the shell, being curved backwards to sustain the ovaries.

About thirty years ago an Irish naturalist, Dr. J. Vaughan Thompson, announced a discovery, which, oversetting conclusions previously received by all, caused no little dissent and opposition, and gave rise to a lengthened and wide spread controversy. A very minute crustaceous animal was known, as inhabiting the open sea, to which the name of *Zoea* had been given. It had sessile eyes, and was remarkable for having a long spine projecting from the face, and a similar one standing up from the centre of the back. Another form was known, which constituted the genus *Megalopa:* in which the body was broad, the eyes stalked, and the abdomen projecting behind. This was also small, but somewhat larger than the preceding.

Nobody suspected that these were other than independent forms of animal life, distinct from each other, and equally distinct from every known genus of *Crustacea* besides. It was supposed that no animal of this class underwent metamorphosis,—or that change of form in different periods of life which distinguishes Insects; but that these creatures retained through life the general shape, slightly modified by development of parts and organs, which they each displayed when hatched from the egg.

But these conclusions were quite set aside by the brilliant discovery of Thompson, that *Zoea* and *Megalopa* were the same animal in different stages of existence; and that, moreover, both were but the early states of well-known and familiar forms of larger *Crustacea*, which therefore undergo a metamorphosis as complete as that by which the caterpillar changes to a chrysalis, and the chrysalis to a butterfly, and in every essential point parallel to it.

In the Cove of Cork this naturalist met with a considerable number of *Zoeas*, which he kept in captivity. Some of these passed into the *Megalopa* form, which in turn changed to the most abundant of all our larger *Crustacea*, the common Shore-crab (*Carcinus mœnas*). "Thus, in its progress from the egg to its final development, the Crab was proved to pass through two temporary conditions, which had previously been regarded as types, not of genera only, but of different families; and both strikingly dissimilar from the group to which, in its perfect state, it belongs."

I have not myself examined the transformations of this species; but, as they have been well worked out, and as the animal is so abundant everywhere on the coast that you may easily verify what has been observed, I will cite you the elaborate account of Mr. R. Q. Couch of Penzance, who has investigated the subject with great skill, zeal, and success.

Having procured some specimens of the Shore-crab laden with eggs, just ready for shedding, he goes on to say,—"these were transferred to captivity, placed in separate basins, and supplied with sea-water; and in about sixteen hours I had the gratification of finding large numbers of the creatures alluded to above swimming about with all the activity of young life. There could be but little doubt that these creatures were the young of the captive Crabs. In order, however, to secure accuracy of result, one of the Crabs was removed to another vessel, and supplied with filtered water, that all insects might be removed; but in about an hour the same creatures were observed swimming about as before. To render the matter, if possible, still more certain, some of the ova were opened, and the embryos extracted; but shortly afterwards I had the pleasure of witnessing, beneath the microscope, the natural bursting and escape of one precisely similar in form to those found so abundantly in the water. Thus, then, there is no doubt that these grotesque-looking creatures are the young of the *Carcinus mœnas;* but how different they are from the adult need hardly be pointed out any further than by referring to the figure. When they first escape they rarely exceed half a line in length. The body is ovoid, the dorsal shield large and inflated; on its upper edge and about the middle is a long spine, curved posteriorly, and rather longer than the diameter of the body, though it varies in length in different specimens; it is hollow, and the blood may be seen circulating through it. The upper portion of the body is sap-green, and the lower semi-transparent. The eyes are large, sessile, and situated in front, and

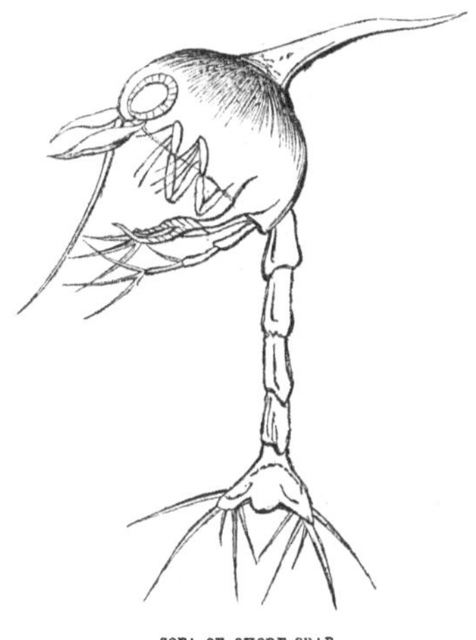

ZOEA OF SHORE-CRAB.

the circumference of the pupil is marked with radiating lines. The lower margin of the shield is waved, and at its posterior and lateral margin is a pair of natatory feet. The tail is extended, longer than the diameter of the shield; and is composed of five equal annulations, besides the terminal one; its extremity is forked, and the external angles are long, slender, pointed, and attached to the last annulation by joints. Between the external angles, and on each side of the median line, are three lesser spines, also attached to the last ring by joints. Between the eyes, and from near the edge of the shield, hangs a long, stout, and somewhat compressed appendage, which, as the animal moves, is reflexed posteriorly between the claws. Under each eye is another appendage, shorter, and slightly more compressed. The claws are in three pairs; each is composed of three joints, and terminates in four long, slender, hair-like appendages. These claws are generally bent on the body, but stand in relief from it. If the animal be viewed in front, the lower margin of the dorsal shield will be found to be waved into three semicircular festoons, the two external of which are occupied by the eyes, and between which the middle one intervenes; the general direction of the claws will be seen to be at right angles to the body. As the young lies inclosed within the membranes of the egg, the claws are folded on each other, and the tail is flexed on them so far as the margin of the shield, and, if long enough, is reflected over the front of the shield between the eyes. The dorsal spine is bent backwards, and lies in contact with the dorsal shield; for the young, when it escapes from the egg, is quite soft, but it rapidly hardens and solidifies by the deposition of calcareous matter in what may be called its skin. The progress of this solidification may be very beautifully observed by watching the circulation in the dorsal spine. When the creature has just effected its liberation from the egg, the blood-globules may be seen ascending to the apex; but as the consolidation advances, the circulation becomes more and more limited in its extent, and is finally confined to the base. These minute creatures, in this early state of their existence, are natatory and wonderfully active. They are continually swimming from one part of the vessel to the other, and when observed free in their native pools, are, if possible, even more

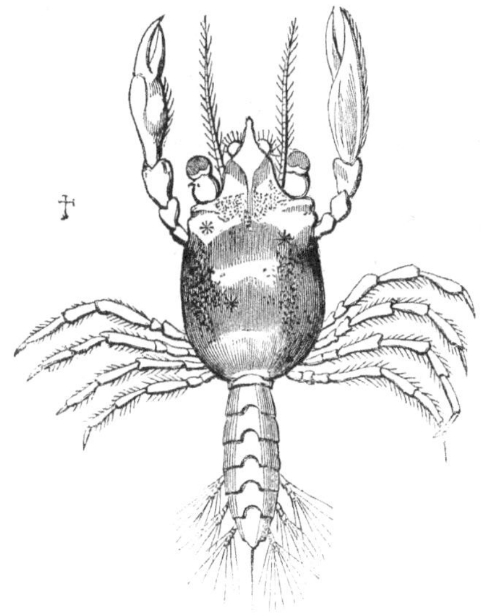

SECOND STAGE OF SHORE-CRAB.

active than when in confinement. Their swimming is produced by continued flexions and extensions of the tail, and by repeated beating motions of their claws; this, together with their grotesque-looking forms, gives them a most extraordinary appearance when under examination. As the shell becomes more solid they get less active, and retire to the sand at the bottom of the vessel, to cast their shells, and acquire a new form. They are exceedingly delicate, and require great care and attention to convey them through the first stage; for unless the water be supplied very frequently and in great abundance, they soon die.

"The second form of transmutation is equally as remarkable as the first, and quite as distinct from the adult animal. In the species now under consideration this second transformation is marked by the disappearance of the dorsal spine; the shield becomes flatter and more depressed, the anterior portion more horizontal and pointed, the three festoons having disappeared. The eyes, from being sessile, are now elevated on foot-stalks, the infra-orbital appendages become apparently converted into antennæ. The claws undergo an entire revolution; the first pair become stouter than the others, and are armed with a pair of nippers," the others being simple; "but the posterior pair are branched near the base, and one of the branches ends in a bushy tuft. The tail is greatly diminished in its relative size and proportions, and is sometimes partially bent under the body, but is more commonly extended. This form is as natatory as the first. They are frequently found congregating around floating sea-weed, the buoys and strings of the crab-pot marks, and other floating substances, both near the shore and in deep water. Their general form somewhat resembles a *Galathea*."*

* Rep. Cornw. Polyt. Soc. 1843.

Thus under Mr. Couch's eye the *Zoea* had changed to a *Megalopa;* and this latter became after a short time a Crab, in which were all the characters that belong to the *order* to which the parent belongs· but not those of the genus, nor even of the family.

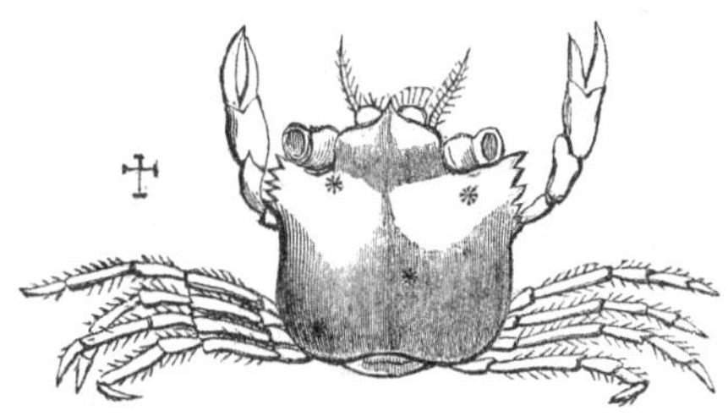

THIRD STAGE OF SHORE-CRAB.

Its form bore a close resemblance to that of the Sargasso Crabs (*Grapsidæ*); for the shield, instead of being large and arched in front, and narrowed behind, was nearly square, while the front was (taking in the eyes) almost straight, the lateral angles much advanced.

This Crab, however, was still very minute; and many sloughings were before it. In the course of these it was destined gradually to attain not only the dimensions of its parents, but also their form. This, however, would be matter of development, rather than metamorphosis: the lateral outlines of the shield would more and more approach each other behind, while the series of points that now belonged to these lateral outlines would become thrown into the front margin, which would by degrees assume an arched form, as you may see in this figure of the adult Crab.

Though I cannot at this moment show you specimens of the *Carcinus* in its earlier stages, yet I have here

ADULT SHORE-CRAB.

a good number of the *Zoeas* of one of those intermediate forms which are the connecting links between the Crabs and Lobsters:—I mean *Galathea*. The adult animal is of a broad squat form, something like what you might suppose a Lobster to be, if it had been flattened between two stones, without being actually destroyed. We have two or three species, one of which is adorned with brilliant scarlet and azure paintings; but I cannot tell to which of them all this infant form belongs.

You perceive that there is a general similarity between these transparent little creatures, and the *Zoea* described by Mr. Couch; but there are great differences in detail. The glassy shield or carapace shoots out in front in a stiff, inflexible, very fragile spine. This is perfectly straight, and nearly thrice the length of the whole shield. It is beset, on various lines on its surface, with short slender spinules jointed to shoulder-like angles, and not serratures. Its interior is perforated by a canal, which dilates and narrows irregularly. The carapace posteriorly is semi-oval, projecting a transparent convex vault far over the part where the abdomen is attached to it, as is seen when the latter bends down. Its extremity gradually tapers into two straight, sub-parallel, stiff spines, about as long as the carapace itself, each terminating in a hooked point.

The abdomen ends in a spinous plate, which is very elegantly lozenge-shaped, and beset with spines. Each of the two latero-posterior edges of the lozenge is cut into six rectangular teeth, and each tooth bears on its hinder face a long spine articulated to it, and most delicately plumose all along its sides. The hindermost pair of spines are short, and are set close together, side by side. Besides these jointed spines, each lateral angle of the caudal lozenge-shaped tail-plate projects into a spine-like tooth.

Though the individuals before us are all in the same state as to the stage of their development, there is some difference in size, indicating, doubtless, a corresponding diversity in age. We will isolate a few of the largest, and put them into a glass trough for microscopical examination.

The largest, during the few minutes which I have

occupied in the process of dipping them out, has undergone a metamorphosis. You observed that, after skipping about the trough for a few moments, it sank quietly to the bottom, where it lay on its back; the next thing that you see is a much more crab-like animal, more opaque, redder, much larger, but lying on its back in the very spot where a moment before you had seen a *Zoea;* while close by it lies the transparent filmy skin which has been cast off.

The new animal is evidently now in its final state, needing only development of its parts, which it would obtain, if in freedom, by successive moults, to acquire the adult form.

If we now submit the exuviæ in detail to a power of 220 diameters, we shall obtain some interesting views of the structure. The slough of the eyes in particular presents one of the most exquisite objects that you can behold. They are somewhat pear-shaped, with the facetted portion well defined. It is the appearance of these facets, varying according as the perfectly hexagonal outline of each or the smooth and glossy convexity comes into focus, that is so peculiarly charming.

Returning now to the examination of one of the living *Zoeas*, you perceive that the three pairs of pencilled limbs do not represent any of the true legs; for the transparency of the integuments allowing the interior to be clearly seen, and the organs of the imago being matured and just ready for sloughing, you discern, with the most beautiful distinctness, the fingered claws (short and stumpy, it is true, as compared with their perfect form in the newly freed imago) folded down upon the breast within the skin, the second pair as large as these, and traces of others beneath them,—all these forming two great projecting lobes, slightly moveable, beneath the thorax of the *Zoea*, and occupying a bulk nearly equal to that of the whole shield.

The circulation of the blood is beautifully clear. The pellucid colourless globules chase each other by starts to and fro, as the eye rests on the outgoing or returning current. It is distinct in some parts where you would scarcely have looked for it; as all over the lozenge plate of the tail, in the interior of the eyes, throughout the posterior spines of the shield, and the frontal spine. But besides, and apparently independent of the circulation, there is a singular fusiform vessel in the latter segments of the abdomen, penetrating the tail-plate, on the ventral side. This vessel, now and then, at irregular intervals, dilates quickly and closes; the wave proceeding upward toward the head, but only for a short distance, and unattended with any impulse to the blood-globules. The nature of this vessel, and its use in the economy of the infant Crab, I can in no wise explain.

CHAPTER XII.

BARNACLES.

You cannot have wandered among the rocks on our southern or western coasts, when the tide is out, without having observed that their whole surface, up to a certain level (often very precisely defined), is roughened with an innumerable multitude of little brownish cones. If you have ever thought it worth while to examine them with more care, you have seen that, crowded as they are, so thickly that frequently they crush each other out of their proper form and proportions, they are all constructed on the same model. Each cone is seen to be a little castle, built up of stony plates that lean towards each other, but which leave an orifice at the top. Within this opening, provided the castle be tenanted by a living inhabitant, you see two or three other pieces joined together in a peculiar manner, which are capable of separating, but which, when brought together, effectually close up all ingress.

Perhaps you have never pushed your investigations farther than this, having a courteous respect for the feelings of the inmate, which has prevented your intruding on a privacy so recluse. But I have been less considerate; many a time have I applied the steel chisel and hammer to the solid rock, and having cut off some projecting piece or angle, have transferred it, all covered with its stony cones, to the interior of a glass tank of sea-water, for more intimate acquaintance with the little builders at leisure.

These are Barnacles (*Balanidæ*). Such a colony I have now in my possession, which I will submit to you, for they present a beautiful and highly interesting spectacle, when engaged in their ordinary employment of fishing for a subsistence. And not only so, but I have living specimens of a much larger and finer species than the common one,—the *Balanus porcatus,* whose castle stands an inch or more in height. The structure, however, and habits are pretty much the same in both.

Without disturbing the busy fishers, then, just take your seat in front of this tank, and with a lens before your eye, watch the colony which is seated on that piece of stone, close to the glass side. From one and another, every instant, a delicate hand is thrust forth, and presently withdrawn. Fix your attention on some one conveniently placed for observation. It is now closed; but in a moment, a slit opens in the valves within the general orifice, displaying a black lining with pale blue edges; it widens to an oval; the pointed valves are projected, and an apparatus of delicate curled filaments is thrust quickly out, expanding and uncurling as it comes, to the form of a fan; then in an instant more the tips of all the threads again curlup, the threads collapse, and the whole apparatus is quickly withdrawn and disappears beneath the closing valves. The next moment, however, they re-open, and the little hand of delicate fingers makes another grasp, and so the process is continually repeated while this season of activity endures.

Now, by putting this specimen into a glass trough, and placing it under a low power of the microscope, we shall see what an exquisite piece of mechanism it is. The little hand consists of twenty-four long fingers, of the most delicate tenuity, each composed of a great number of joints, and much resembling in this respect the antennæ of a Beetle. These fingers surround the mouth, which is placed at the bottom of the sort of imperfect funnel formed by their divergence. They resolve themselves into six pairs of arms, for each one is branched from the basal joint, dividing into two equal and similar portions. Those nearest the mouth are the shortest, and each pair increases regularly in length to the most distant, which are the central pair when the hand is extended. Each division of each of this longest and most extensile pair comprises, in the specimen before us, thirty-two joints, while the shortest consists of about ten, the intermediate ones being in proportion; so that the whole apparatus includes nearly five hundred distinct articulations, a wonderful provision for flexibility, seeing that every joint is worked by its own proper system of muscles.

Moreover, every separate joint is furnished with its own system of spinous hairs, which are doubtless delicate organs of touch, since it has been established that the hairs with which the shelly coats of *Crustacea* are studded, pass through the substance of the latter, and communicate with a pulpy mass, richly supplied with nerves, which lines the shell.* These hairs project at a more or less wide angle from the axis of the finger-like filament, and are graduated in length; and what is very striking, as illustrating the exquisite workmanship of the Divine hand, the hairs themselves are compound structures; for under a high power they seem to be composed of numerous joints, —an illusory appearance probably, what look like joints being rather successive shoulders, or projections and constrictions of the outline, — while each shoulder carries a whorl of finer spines, lying nearly close to the main hair, and scarcely deviating from its general direction. This barbed structure of the hairs is chiefly seen towards their attenuated extremities.

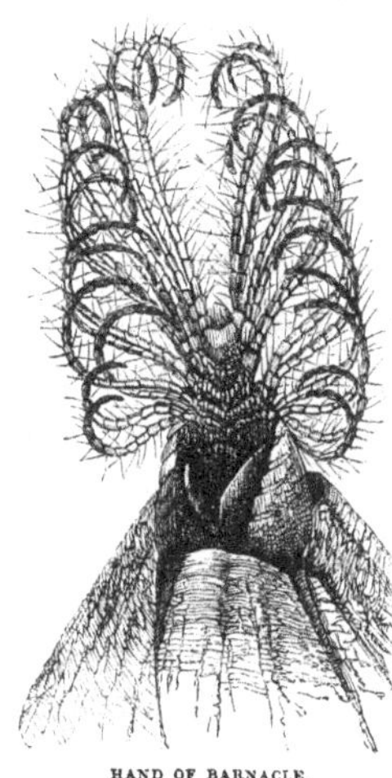

HAND OF BARNACLE.

And now do you ask,—What is the object of this elaborate contrivance, or rather series of contrivances? I answer,—It is the net with which the fisher takes his food—it is his means of living. You have seen that the animal has no power of pursuing prey: he is immoveably fixed to the walls of his castle, which is immoveably fixed to the solid rock. He is compelled therefore to subsist on what passes his castle, and on what he can catch as he sits in his doorway and casts his net at random.

You saw, also, with what a regular perseverance

* Proc. Royal Society, ix. 215.

the casts were made; and now that you have examined in detail the construction of the net, you are prepared to appreciate its fitness for the work assigned to it. Its extreme flexibility, produced by the number of its joints, enables the fingers of the hand, or the threads of the net (which you will), to stretch out or to curl up alternately, while the number of the divergent fingers enables the animal to grasp a comparatively large bulk of water in those curling organs. These, then, form a sieve; the water passes through the interstices of the fingers, while the tiny atoms of solid matter, or the equally minute animalcules that constitute the food of the Barnacle, are sifted out, and detained by the fingers, which curling inward carry whatever is captured to the mouth.

But see how greatly the perfection of the instrument is enhanced by the projecting hairs with which every one of the numerous joints is beset. These, standing out at right angles (or nearly so) to the direction of the finger, meet their fellows from the joints of the next finger, and crossing their points, fill the interstices with an innumerable series of finer meshes, —meshes of such delicacy that it is next to impossible that any organized body inclosed in the given area should escape.

But there is more in them than merely this minute and wide-spread ramification. They are, as we have seen, organs of touch; so that the net has not only the mechanical power of capture, common to an inanimate cast-net which a human fisher uses, but is endowed with the most exquisite sensibility in every part. The slightest contact of an animalcule in the inclosed water with one of those thousands of sensitive hairs communicates instantly an impression to the sensorium, and a consciousness of the fact to the Barnacle; who is thus, without doubt, able with the quickness of thought to close the fingers together at that part, and thus secure the victim.

To make use of the prey thus secured, the Barnacle is furnished with a mouth, which can be protruded into a sort of wart, and is provided with a distinct lip bearing minute palpi, and three pairs of jaws, of which the two outer are horny and toothed, while the innermost is soft and fleshy.

Fixed and immoveable as the Barnacles are in their adult and final stage, they have passed by metamorphosis through conditions of life in which they were active roving little creatures, endowed with the power of swimming freely in the wide sea. In this condition they present the closest resemblance to familiar forms of *Crustacea*, as you will perceive when you examine some specimens of the larvæ that I am able to show you.

I have in one of my tanks an individual of the fine and large Barnacle, *Balanus porcatus*, which, for several days past, has been at intervals throwing out from the orifice of its shell dense clouds of atoms, which form compact columns reaching from the animal to the surface of the water. One of these cloudy columns, when examined with a lens, is seen to be composed of thousands of dancing creatures, resembling the Water-fleas that we lately examined. They maintain a vivacious motion, and yet at the same time keep their association and the general form of the column.

Taking out a few of the dancing atoms, and isolating them in this glass stage-cell, we see that they have exactly the figure, appearance, and character of the young of the common *Cyclops*, so that you would, without hesitation, if you knew nothing of their parentage, assign them to that well-known genus. Their movements are almost incessant, a series of jerking progressions performed by quick but apparently laborious flappings of the limbs, right and left together. They occasionally rest from their exertions for a few moments, but seem to have no power of alighting on any object.

But in order to obtain a more precise idea of the structure of this tiny creature, we must manage to restrain its liberty a little, by applying gentle pressure with the compressorium, just sufficient to confine it without hurting it. The body is inclosed in a broad carapace, shaped much like a heraldic shield, but very convex on the back, and terminating behind in a slender point or spine, which is cut into minute teeth along the edges. Below this shield is seen the body, with three pairs of legs, a great proboscis in the middle pointing downwards and backwards, and the anal fork, which consists of a bulbous base and two diverging points, which project behind under the spine of the shield.

The legs are exclusively swimming organs; they have no provision for grasping, no claws or hooks, nor do they appear to be capable of being used for crawling on the ground or for climbing among the sea-weeds. They are fringed along one edge with long and stout, but somewhat flexible spines, of which those that are nearest the trunk seem more rigid, and are directed more at right angles to the limb than the rest. The legs are formed of many imperfect joints, and the second and third pairs are double from the basal joint outwards, while the first pair are simple. In the fore part of the body a large eye

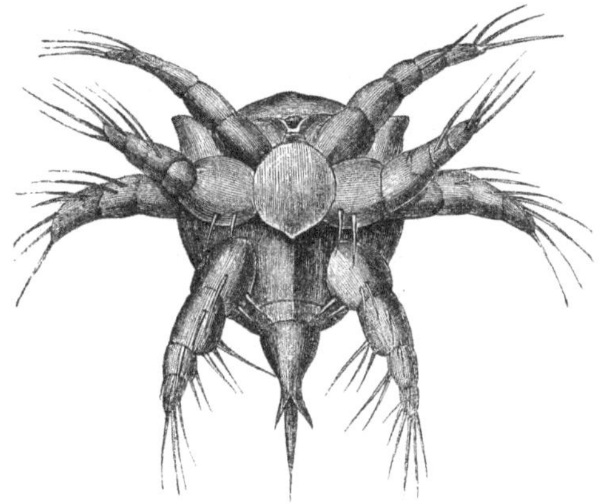

YOUNG OF BARNACLE.

is placed, deep-seated, which is of a roundish form, and is intensely black both by reflected and transmitted light. On the summit of the forehead are placed a pair of thick flexible horn-like organs, which are abruptly bent in the middle, and which I believe represent the first pair of antennæ. This then is the first stage of the Barnacle,—the form under which it appears when it is hatched from the egg.

Among the multitudes which have been evolved during these last few days, and which are now swimming at large in the tank, we may be able to detect some that have passed through their first stage, and having moulted their skin have attained a more advanced form. Here is one, which by its superior size seems to have made some progress towards maturity.

Yes, here are more. These are evidently in their second stage. There is an increase in length; for whereas the former was only $\frac{1}{100}$th of an inch in length, these have attained to a length of $\frac{1}{70}$th of an inch. Yet this increase is observable in no other dimension than that of total length, and this is due to the development of the terminal spine of the shield, which is now much produced, and cut into minute teeth. The anal fork is also attenuated, lengthened, and bent abruptly downward at the base, where it is very mobile; another bend in the middle throwing the extremity into the horizontal again. The delicately membranous pouch-like proboscis is more clearly seen beneath the breast, the extremity of which is directed backwards. In front of this organ there are two decurved very mobile bristles, set on pedicles, the whole closely resembling the internal antennæ in the higher *Crustacea.* The lateral horns or external antennæ appear to terminate in a very delicate brush of hairs, which does not seem to be protrusile.

The little animals in this state swim, generally, back downward; though they frequently assume a perpendicular position, both direct and reversed. I see them now occasionally resting on sea-weeds and *Diatomaceæ*, though the limbs seem even worse fitted than before for crawling, since the spines or bristles with which they are fringed are much increased in length, especially on the third pair.

A specimen nearly twice as large as these last affords us an opportunity of tracing the Barnacle to another point of its transformations. The modifications are chiefly in the proboscis and the anal fork. The former now points directly downwards, is furnished with a pair of minute spines on its anterior side, and with a terminal hook; while its posterior side is set with strong vibrating cilia. The anal fork is greatly increased in dimensions, has its edges armed with spines articulated to its surface, and is marked with longitudinal lines which resemble corrugations. The under-surface of the body is also much corrugated transversely.

In the first moult the spine of the shield was greatly increased, the size of the body itself remaining stationary; in the second moult the ratio is reversed, the body has largely increased, but the spine is nearly *in statu quo.*

We cannot follow the metamorphosis any farther by personal observation, but from the researches of others, and especially of Mr. Darwin, we know that other stages have to be passed before the final fixed condition is attained. As yet no appreciable advance has been made, by either of the two moultings which we have traced, from the free, jerking, dancing Water-flea that was first hatched, towards the sessile Barnacle inclosed in its shelly cone of several valves, and firmly fixed to the solid rock; and we are yet at a loss to imagine how such a change can be effected.

Nor is the matter apparently helped by the next moult; for though there now ensues a great change of form, it does not seem to resemble the adult Barnacle much (if at all) more than before. If described without reference to its parentage, it would still be considered an Entomostracous Crustacean, or Water-flea, but removed to another Tribe. It represents, in fact, a *Cypris*;* the body with its fringed limbs being now included within two convex valves, like those of a mussel or other bivalve shell, either united by a hinge along the back, or rather soldered together there, so as only to allow a slight opening and closing by the elasticity of their substance. The fore part of the head is now greatly enlarged, as are also the antennæ, which project from the shell. The single eye is separated into two, which are large, and attached to the outer arms of two bent processes which are placed within the body, in the form of the letters U U. The legs are increased by the addition of two pairs, and these are doubly bent in a zig-zag form, and can be protruded from between the valves.

It is a highly curious fact that the infant Barnacle has thus passed through two distinct types of animal life, the *Cyclops* and the *Cypris*. These are not one type in different stages, as might be reasonably presumed. The young of *Daphnia* and of *Cyclops* are so much alike, that it would be natural to presume the young of *Cypris* to be of the same form; in which case, we should have in the young Barnacle merely the first and second stages of *Cypris*. But it is not so. *Cypris* does not pass through the *Cyclops* form at all; for, according to Jurine, the young

* See figure on page 220.

when hatched have the appearance of the perfect animal, though varying a little in the shape of their shells.

It is in this second form, which may be considered the pupa of the Barnacle, that the animal quits its free roving life, and becomes a fixture for the remainder of its days. And this is a most wonderful process; so wonderful, that it would be utterly incredible, but that the researches of Mr. Darwin have proved it incontestably to be the means by which the wisdom of God has ordained that the little Water-flea should be transformed into a stony Acorn Barnacle.

Having selected a suitable place for fixing its residence,—such as those massive rocks which sustain the impetuous billows on our sea-worn coasts, —the great projecting antennæ manifest a new and unprecedented function. Glands situated at their base secrete a tenacious glue, which, being poured out in great profusion, cements the whole front of the head to the rock, including and concealing the antennæ themselves. The cement rapidly *sets* under water, and the animal is henceforth immoveable.

It now moults its skin once more. Another great change takes place; the bivalve shell is thrown off, as are also the eyes with their bent supports, and it is seen to be a true Barnacle, though as yet of minute dimensions and with its valves in a very rudimentary condition. It is now the representative of a third type among the Crustacean forms, for it is in effect a Stomapod; such as the Opossum Shrimp (*Mysis*), for example, with the shield composed of several pieces, stony in texture, on account of the great development of their calcareous element, and so modified in form as to make a low cone, the legs (become the *cirri*, or what I have above called the "fingers") made to perform their movements backwards instead of forwards, and the whole abdomen reduced to an almost invisible point.

Marvellous indeed are these facts. If such changes as these, or anything approaching to them, took place in the history of some familiar domestic animal,—if the horse, for instance, was invariably born under the form of a fish, passed through several modifications of this form, imitating the shape of the perch, then the pike, then the eel, by successive castings of its skin; then by another shift appeared as a bird, and then, glueing itself by its forehead to some stone, with its feet in the air, threw off its covering once more, and became a foal, which then gradually grew into a horse;—or if some veracious traveller, some Livingstone or Barth, were to tell us that such processes were the invariable conditions under which some beast of burden largely used in the centre of Africa passed,—should we not think them very wonderful? Yet they would not be a whit more wonderful in this supposed case than in the case of the Barnacle, in whose history they are constantly exhibited in millions of individuals, and have been for ages,—even in creatures so common that we cannot take a walk beneath our sea-cliffs, without treading on them by hundreds!

CHAPTER XIII.

SPIDERS AND MITES.

SPIDERS, I am sure, are not favourites with you. With the exception of the poor prisoner in the Bastille, who had succeeded in taming a Spider,—the only creature besides himself that inhabited his dungeon,—I do not think I have ever heard of any one who loved or admired Spiders *morally*. Yet, physically, we may find much to admire in them, as not a few naturalists have done before us; there are men who have devoted their lives to the study of this unamiable race, and who have discovered in them the same wondrous skill, and the same perfect adaptation of organ to function, of structure to habit, that mark all God's works, whether we think them pretty or ugly, amiable or repulsive.

I am going to show you some of these pieces of mechanism. Remember that the whole tribe is sent into the world to perform one business,—they are commissioned to keep down what would otherwise be a "plague of flies." They are fly-butchers by profession; and just as our beef- and mutton-butchers have their slaughter-house, their steel, their knives, their pole-axe, their hooks, so are these little slaughterers furnished with nets and traps, with caves, with fangs, and hooks, and poison-bags, ready for their constant work. They have, in fact, nothing else to do: their whole lives are spent in slaughtering,—with the exception of rearing fresh generations of slaughterers,—and I suppose they think, and are intended to think, of nothing else.

I was one day in an omnibus, in the corner of which sat a butcher. Presently a man got in, whose blue gingham coat indicated the same trade. He seated himself opposite the other, and the two were soon in conversation. "Do you know Jackson?" says A. "No," says B; "where does *he* slaughter?" The reply gave me a new idea; he evidently considered that "slaughtering" was the only occupation worthy of a man, and therefore the only one worthy of man's thought. Spiders are just the same. If an *Epeira* met a *Clubiona*, probably the first interchange of civilities would be something like—"Where do you slaughter?"

"No one," says Professor Rymer Jones, "who looks at the armature of a Spider's jaws can mistake the intention with which this terrible apparatus was planned. 'Murder' is engraved legibly on every piece that enters into its composition." But surely the Professor is rather severe. I do not think this paragraph was written on an autumn morning, when the flies had driven him out of bed prematurely early, by incessantly alighting on his nose; nor on coming home from a summer evening's walk through the marsh, when clouds of singing and stinging gnats had been the only objects of cognisance to sight, hearing, and feeling. If so, he would have been ready to pronounce "killing no murder," and have blessed the slaughtering Spiders as pursuing a most praiseworthy and useful occupation. Circumstances change opinions.

We will not then touch the moral question; but just look at this apparatus from the head of one of our common Spiders (*Clubiona atrox*), a long-legged and swift species, that builds a compact cloth-like web in our out-houses, with a gallery open at each end for retreat in danger. The specimen is a part of the slough or cast skin, which you may always find in the neighbourhood of such a web; and it is particularly suitable for examination, because it is sloughed in the most perfect condition; every part, the fangs, the palps, the legs, with all their joints, the corneæ of the eyes, the entire skin with every hair,—all are here, and all *in situ*, with a cleanness and translucency which it would require much skill in dissection to obtain, if we captured a living Spider for our purpose.

There are in front of the head two stout brown organs, which are the representatives of the antennæ in insects; though very much modified both in form and function. They are here the effective weapons of attack. Each consists of two joints: the basal one, which forms the most conspicuous portion of the organ, and the terminal one, which is the fang. The former is a thick hollow case, somewhat cylindrical, but flattened sidewise, formed of stiff chitine, covered with minute transverse ridges on its whole surface, like the marks left on the sand by the rippling wavelets, and studded with stout coarse black hair. Its extremity is cut off obliquely, and forms a furrow, the edges of which are beset with polished conical points, resembling teeth.

To the upper end of this furrowed case is fixed by a hinge-joint the fang, which is a curved claw-like organ, formed of hard chitine, and consisting of two parts, a swollen oval base, which is highly polished, and a more slender tip, the surface of which has a silky lustre, from being covered with very fine and close-set longitudinal grooves. This whole organ falls into the furrow of the basal joint, when not in use, exactly as the blade of a clasp-knife shuts into the haft; but when the animal is excited, either to defend itself or to attack its prey, the fang becomes stiffly erected.

By turning the object on its axis, and examining the extreme tip of the fang, we see that it is not brought to a fine point, but that it has the appearance of having been cut off slant-wise just at the tip; and that it is tubular. Now this is a provision for the speedy infliction of death upon the victim; for both the fang and the thick basal joint are permeated by a slender membranous tube, which is the poison duct, and which terminates at the open extremity of the former, while at the other end it communicates with a lengthened oval sac where the venom is secreted. This of course we do not see here, for it is not sloughed with the exuviæ, but retained in the interior of the body; but in life it is a sac, extending into the *cephalo-thorax*,—as that part of the body which carries the legs is called,—and covered with spiral folds produced by the arrangement of the fibres of its contractile tissue.

When the Spider attacks a fly, it plunges into its victim the two fangs, the action of which is downwards, and not from right to left, like that of the jaws of Insects. At the same instant a drop of poison is secreted in each gland, which, oozing through the duct, escapes from the perforated end of the fang into the wound, and rapidly produces death. The fangs are then clasped down, carrying the prey, which they powerfully press against the toothed edges of the stout basal piece, by which means the nutritive fluids of the prey are pressed out, and taken into the mouth, when the dried and empty skin is rejected. The poison is of an acid nature, as experiments performed with irritated spiders prove; litmus-paper pierced by them becoming red as far around the perforations as the emitted fluid spreads.

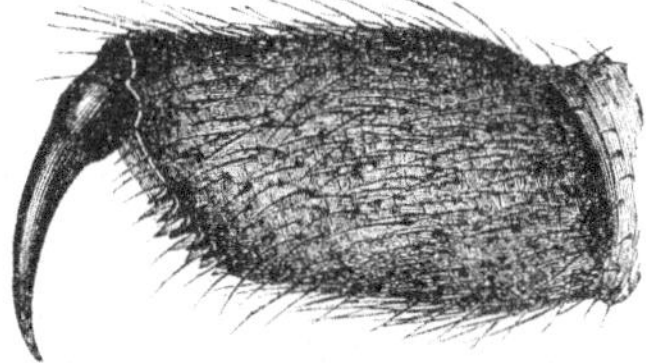

FANG OF SPIDER.

In the slough, the upper surface of the cephalothorax is always detached as a thin plate, convex outwardly, concave inwardly. As it is upon the front portion of this division of the body that the eyes are situate, the slough displays these with great clearness and beauty beneath the microscope. Here you may see them. The whole slough from its thinness is semi-pellucid, but the eyes transmit the light with brilliance, not however as if they were simple round holes, because you can discern very manifestly a hemispherical glassy coat, by which it is refracted.

It is, however, when we examine the forehead of a living or recently killed spider, that we see the eyes to advantage. In this example of the same species (*Clubiona atrox*), you see them, like polished globes of diamond, sunk into the solid skin of the

EYES OF SPIDER.

head. Their form is unimpeachably perfect, and the reflection of light from their surface most brilliant.

The arrangement of these lustrous eyes is worthy of attention. They are generally eight in number in Spiders, but their relative position varies so much, as to afford good characters by which naturalists have grouped them in genera. In the *Clubiona* which we have been examining, they are placed in two nearly straight transverse rows on the forehead; but as this surface is convex, it follows that the axis of every eye points in a different direction from that of its fellows. In *Epeira*, on the other hand,—represented by our great Garden Spider so commonly seen in the centre of its perpendicular web, on shrubs and in corners of our gardens,—the four middle eyes form a square, and the two lateral ones on each side are placed in contact with each other.

It is interesting to remark that their arrangement is not arbitrary, but is ancillary to the varying instincts and wants of the different kinds. On this subject I will quote to you what Professor Owen says:—"The position of the four median ones is the most constant; they generally indicate a square or trapezium, and may be compared with the median *ocelli* in hexapod insects. The two, or the two pairs of lateral *ocelli* may be compared with the compound eyes of insects; the anterior of these has usually a downward aspect, whilst the posterior looks backwards; the variety in the arrangement of the *ocelli* of Spiders always bears a constant relation to the general conformation and habits of the species. Dujés has observed that those Spiders which hide in tubes, or lurk in obscure retreats, either underground, or in the holes or fissures of walls or rocks, from which they only emerge to seize a passing prey, have their eyes aggregated in a close group in the middle of the forehead, as in the Bird-spider, the *Clotho*, &c. The Spiders which inhabit short tubes, terminated by a large web exposed to the open air, have the eyes separated, and more spread upon the front of the cephalo-thorax. Those Spiders which rest in the centre of a free web, and along which they frequently traverse, have the eyes supported on slight prominences which permit a greater divergence of their axes; this structure is well marked in the genus

Thomisa, the species of which lie in ambuscade in flowers. Lastly, the spiders called *Errantes*, or wanderers, have their eyes still more scattered, the lateral ones being placed at the margins of the cephalothorax."*

The shining hemisphere (or nearly a sphere) is in each case covered with a thick cornea, a continuation of the skin, perfectly transparent, and throwing off its outer coats successively in the process of moulting, like that of the rest of the body. The centre of its inner surface is deeply excavated for the reception of a crystalline lens, which is globular in form, and which rests behind on the front surface of a hemispherical vitreous body, without sinking into it. The space between this body and the sides of the lens forms a ring-like channel which is filled with an aqueous humour, and into this projects a circular process of the thick pigment-coat, which corresponds to the choroid, thus defining the pupil of the eye, and at the same time confining the lens to its proper situation. The margin of this pigment-ring may be considered as an iris, and is of various colours, as red, green, or brown in those species which are active by daylight, while it is black at the back of the eye. The nocturnal species have no dark pigments, but are furnished with a curtain (*tapetum*), which reflects a brilliant metallic lustre, and makes the eyes of these Spiders glare in the twilight, like those of cats.

It will be interesting to compare with this range of eyes, the same organs in a kindred animal, the common Harvestman (*Phalangium cornutum*). Here in the centre of the cephalo-thorax rises a short pillar,

* Comp. Anat. (Ed. 2) p. 451.

which is crowned with two rows of conical points, with polished black tips. On each side of the pillar is a large black eye, hemispherical in form, and brilliantly glossy, exactly resembling, indeed, those which we have just examined. There are, however, only this single pair which thus look out laterally, exactly like the eyes of Birds. There is, indeed, a speck on each side of the thorax, considerably removed from the eye-pillar, just above the origin of the first pair of legs, which has been mistaken for an eye; but it is truly a spiracle, or breathing hole.

There are many other points of interest about this Harvestman, such as the conical spines which stud the head, body, and limbs; the multitude of small bead-like joints into which the foot (*tarsus*) is divided; and in particular the hammer-like form of the modified antennæ, which bend abruptly downwards, and have pincer-tips. These are highly curious, and you may examine them at your leisure; but for the present we will return to our Spiders.

Ever since those mythic times when Arachne contended with Minerva for supremacy in needlework, and was changed, for her pains, into a spider, our little spinners have been famous (*Spider* = Spinne) for their matchless achievements in thread. And still their industrious art is plied everywhere around us—in our chambers, in our windows, in our cellars, in our walls, in our gardens, in waste and desert places, and even under water. But you shall hear what Professor Owen says on the degree and mode in which Spiders exercise their singular secreting faculty, which "varies considerably in the different species. Some, as the

Clubionæ, line with silk a conical or cylindrical retreat, formed, perhaps, of a coiled-up leaf, and having an outlet at both extremities, from one of which may issue threads to entrap their prey. Others, as the *Segestriæ*, fabricate a silken burrow of five or six inches in length, in the cleft of an old wall. The *Mygale cementaria* lines a subterraneous burrow with the same substance, and manufactures a close-fitting trap-door of cemented earth, lined with silk, and so attached to the entry of the burrow as to fall down and cover it by its own weight, and which the inmate can keep close shut by means of strong attached threads.

"The arrangement of Spiders by M. Walckenaër into families, characterised by their habits, places the principal varieties of their webs in a very concise point of view.

"The *Cursores*, *Saltatores*, and *Laterigradæ*, make no webs: the first catch their prey by swift pursuit; the second spring upon their prey by insidious and agile leaps; the third run, crab-like, sideways or backwards, and occasionally throw out adhesive threads to entrap their prey. The *Latebricolæ* hide in burrows and fissures, which they line with a web. The *Tubicolæ* inclose themselves in a silken tube, strengthened externally by leaves or other foreign substances. The *Niditelæ* weave a nest whence issue threads to entrap their prey. The *Filitelæ* are remarkable for the long threads of silk which they spread about in the places where they prowl in quest of prey. The *Lapitelæ* spin great webs of a close texture, like hammocks, and wait for the insects that may be entangled therein. The *Orbitelæ* spread abroad webs of a regular and open texture, either circular or spiral, and remain in the middle, or on one side, in readiness to spring upon an entangled insect. The *Retitelæ* spin webs of an open mesh-work, and of an irregular form, and remain in the middle or on one side, to seize their prey. Lastly, the *Aquitelæ* spread their silken filaments under water, to entrap aquatic insects.

"The silken secretion of Spiders is not applied only to the formation of a warm and comfortable dwelling for themselves, or of a trap for their prey; it is often employed to master the struggles of a resisting insect, which is bound round by an extemporary filament, spun for the occasion, as by a strong cord. It forms the aëronautic filament of the young migratory brood. It serves to attach the moulting *Hydrachna* to an aquatic plant by the anterior part of the body, when it struggles to withdraw itself from its exuvium. Lastly, a softer and more silken kind of web is prepared for the purpose of receiving the eggs, and to serve as a nest for the young." *

The silk with which these various fabrics are constructed is a thick, viscous, transparent liquid, much like a solution of gum arabic, which hardens quickly on exposure to the air, but can meanwhile be drawn out into thread. So far, it agrees with the silk of the silkworm and other caterpillars; but the apparatus by which it is secreted, and that by which it is spun, are both far more complex and elaborate than those of the latter. Generally speaking, there are three pairs of spinnerets, or external organs, through which the

* Comp. Anat. (ed. 2), p. 458.

threads are produced, but in some few cases there are only two pairs, and in others, as the Garden Spiders (*Epeira*), the hindmost pair seem to be united into a single spinneret. These are always situated at the hinder extremity of the body, and I will show them to you presently. First, however, I will describe the internal apparatus—the source of the threads.

The glands which secrete the gummy fluid are placed in the midst of the abdominal viscera, and in some instances—as in the female of *Epeira fasciata*, a species which makes a remarkably large web—they occupy about a quarter of the whole bulk of the abdomen. About five different kinds of these glands may be distinguished, though they are not all present in every species. The *Epeiræ*, however, present them all.

In this genus there are:—1. Small, pear-shaped bags, associated in groups of hundreds, and leading off by short tubes, which are interlaced in a screw-like manner, and open in all of the spinnerets. 2. Six long twisted tubes, which gradually enlarge into as many pouches, and then are each protracted into a very long duct, which forms a double loop. 3. Three pairs of glandular tubes, similar to the preceding, but which open externally through short ducts. 4. Two groups of much branched sacs, whose long ducts run to the upper pair of spinnerets. 5. Two slightly branched blind-tubes, which terminate by two short ducts in the middle pair of spinnerets.

It is not very easy to examine the spinnerets with a microscope, so as to make out their structure. If we confine the Spider in a glass cell, it is so restless that the least shock or change of position will cause it to move to and fro; and, besides, when it does become quiescent, the spinnerets are closed in towards each other, so that we cannot see their extremities. By selecting a specimen, however, recently killed, such as this *Clubiona*, we may discern sufficient to enable us to comprehend their construction.

Looking, then, at the abdomen from beneath, we see the three pairs of spinnerets clustered together close to the extremity. The pair most forward are shaped somewhat like barrels, whose free ends bend over toward each other. They are covered with stiff black hairs, and just within the margin of what may be called the *head* of the barrel (for it is cut off horizontally, with a sharp rim), there is a circle of very close-set, stiff, whitish bristles, which arch inwards. The whole flat surface of the "head," within this circle of bristles, is beset with very minute horny tubes, standing erect, which are the outlets of the silk-ducts, that belong to this pair.

Behind this first pair are seen the middle pair, almost concealed, however, from their shortness and smallness, and from the approximation of the first and third pairs. We can discern that they are more teat-like than the preceding, terminating in a minute wart, which is prolonged into a horny tube. The whole teat is set with similar tubes, which are larger and longer than those of the first pair. Finally, the third pair resemble palpi, for each consists of two lengthened joints, and are bluntly pointed. The spinning tubes in these are limited, as it appears to me, to one or two at the extreme end of each spinneret, the whole surface besides being covered with the ordinary long bristles. Strictly speaking, however, they are three-jointed, for all the spinnerets spring from wart-like sockets, which may be considered as basal joints; and as the circlet of bristles in the first pair doubtless indicates a short joint, sunken as it were within the preceding, this pair is likewise three-jointed; the middle pair appears to be but two-jointed.

The minute horny tubes are themselves composed of two joints, the basal one thick, the terminal one very slender, and perforated with an orifice of excessive tenuity, through which the gum oozes at the will of the animal, as an equally attenuated thread. On our *Clubiona*, the number of tubes in all the spinnerets is about three hundred; but in the Garden Spider (*Epeira*) they exceed a thousand.

This remarkable multiplicity of the strands with which the apparently simple and certainly slender thread of the Spider is composed, has attracted the attention of those philosophers who seek to discover the reasons of the phenomena they see in nature. The explanation was first suggested, I believe, by Mr. Rennie,* but it has been amplified with much force by Professor Jones, in the following words:—

"A very obvious reflection will here naturally suggest itself, in connexion with this beautiful machinery; why, in the case of the Spider, it has been found necessary to provide a rope of such complex structure, when in so many Insects a simple, undivided thread, drawn from the orifice of a single tube, like the thread of the silkworm, for instance, was sufficient for all

* Insect Architecture, 337.

required purposes. And here, as in every other case, it will be found, on consideration, that a complicated apparatus has been substituted for a simple one only to meet the requirements of strict necessity. The slow-moving Caterpillar, as it leisurely produces its silken cord, gives time enough for the fluid of which it is formed to harden by degrees into a tenacious filament, as it is allowed to issue by instalments from the end of the labial pipe; but the habits of the Spider require a very different mode of proceeding, as its line must be instantly converted from a fluid into a strong rope, or it would be of no use for the purposes it is intended to fulfil. Let a fly, for example, become entangled in the meshes of a Spider's web; no time is to be lost; the struggling victim, by every effort to escape, is tearing the meshes that entangle it, and would soon succeed in breaking loose did not its lurking destroyer at once rush out to complete the capture and save its net, spun with so much labour, from ruin. With the rapidity of thought, it darts upon its prey, and before the eye of the spectator can comprehend the manœuvre, the poor fly is swathed in silken bands, until it is as incapable of moving as an Egyptian mummy. To allow the Spider to perform such a feat as this, its thread must evidently be instantaneously placed at its disposal, which would have been impossible had it been a single cord, but being subdivided into numerous filaments, so attenuated as we have seen them to be, there is no time lost in the drying; from being fluid they are at once converted into a solid rope, ready for immediate service."*

* Nat. Hist. of Anim. ii. 339.

No doubt you have often admired the exquisite regularity of those Spiders' webs which are called geometric; that of our abundant Garden Spider, for instance. You have observed the cables which stretch from wall to wall, or from bush to bush, in various directions, to form the scaffolding, on which the net is afterwards to be woven; then you have marked the straight lines, like the spokes of a wheel that radiate from the centre to various points of these outwork cables, and finally the spiral thread that circles again and again round the radii, till an exquisite net of many meshes is formed.

But possibly you are not aware that these lines are formed of two quite distinct sorts of silk. It has been shown that the cables and radii are perfectly unadhesive, while the concentric or spiral circles are extremely viscid. Now the microscope, or a powerful lens, will reveal the cause of this difference; the threads of the cables and radii are perfectly simple, while the spiral threads are closely studded with minute globules of fluid, like drops of dew, which, from the elasticity of the thread, are easily separated from each other. These are globules of viscid gum, as is easily proved by touching one or two with the finger, to which they will instantly adhere; or by throwing a little fine dust over the nets, when the spirals will be found clogged with dirt, while the radii and cables remain unsoiled. It is these viscid threads alone that have the power of detaining the vagrant flies which accidentally touch the net.

The diversity in the secreting organs already alluded to, as well as in the spinnerets, is no doubt connected with this difference in the character of the silk; and it is worthy of remark, that this diversity is greatest in such Spiders, as the *Epeiræ*, which spin geometric nets.

Immense is the number of globules of viscidity that stud the spiral circles of one of these nets. Mr. Blackwall, the able and learned historian of the tribe, has estimated that as many as 87,360 such pearly drops occurred in a net of average dimensions, and 120,000 in a large net of fourteen or sixteen inches diameter; and yet a Spider will construct such a net, if uninterrupted, in less than three quarters of an hour.

Scarcely less admirable is the ease and precision with which the little architect traverses her perpendicular or diagonal web of rope; a skill which leaves that of the mariner who leaps from shroud to backstay in a ship's rigging immeasurably behind. To understand it, however, in some measure, look at this last joint of one of the feet of our well-used *Clubiona.* It is a cylindrical rod, ending in a rounded point; every part of its surface is studded with stiff, rather long, horny bristles, which, springing from the side, arch inward towards the point. Now this array of spines effectually prevents a false step, for if any part of the leg, which is sufficiently long, only strike the thread, the latter is certain to slip in between the bristles, and thus to catch the leg. But more precision than this is requisite; especially when we observe with what delicacy of touch the hinder feet are often used to guide the thread as it issues from the spinnerets, and particularly with what lightning-like rapidity the larger net-weavers will, with the assistance of these feet, roll a dense web of silk around the body of a helpless fly, swathing it up, like an Egyptian mummy, in many folds of cloth, in an instant.

Look, then, at the extreme tip of the ultimate joint. Two stout hooked claws of dark horny texture are seen proceeding from it side by side, and a third of smaller size, and more delicate in appearance, is placed between them, and on a lower level. The former have their under or concave surface set with teeth (eighteen on each in this example), very regularly cut, like those of a comb, which are minute at the commencement of the series near the base of the claw, and gradually increase in length to the tip. These are doubtless sensible organs of touch, feeling

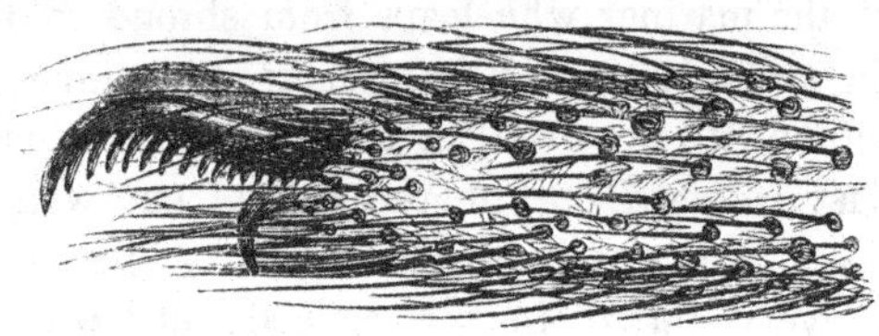

CLAWS OF SPIDER.

and catching the thread; and they, moreover, act as combs, cleansing their limbs, and probably their webs, from the particles of dust and other extraneous matter which are continually cleaving to them.

There are Spiders in the sea also. I can show you one which is sufficiently common on the southern shores, sprawling and crawling sluggishly among the filamentous seaweeds and branching flexible zoophytes. Here it is, *Nymphon* by name.

Its most prominent characteristic is the excessive slenderness of all its parts, but especially its eight legs, which are exceedingly lengthened, comprising each eight joints, and no thicker than the finest thread. On the other hand, the body is reduced to a minimum; the abdomen, which in the Spiders and Harvestmen of the land is so bulky as to constitute the chief volume of the animal, is here so minute that you will have some difficulty in finding it at all; it is, in fact, that tiny atom of a point that projects between the hindmost pair of limbs. The thorax, indeed, is a little more developed; but even this has scarcely any appreciable breadth or thickness, being scarcely more than the extended line formed by the successive points of origin of the limbs.

The head, however, is distinct and well furnished. It is crowned with a short column, much as in the Harvestman, on the summit of which are placed four black eyes, set in square; these, under the magnifying power which we are applying to them, gleam like diamonds, the light being highly refracted through them. It is the high refractive power of these eyes, as of those which we have lately been examining, which makes them appear black; for, as I have explained, they are really transparent lenses, covered with polished corneæ, and furnished with the other essentials requisite for the transmission of the rays of light to the optic nerve, or, as in this case, direct to the brain.

In front, you see the head projects into a stout oval or cylindrical proboscis, terminating in a small mouth and stout jaws, and furnished at the sides with a pair of spine-like palpi, and a pair of pincer-claws (modified antennæ) somewhat resembling the nippers of a Crab or Lobster.

Such is the outward form of this tiny speck, the whole body of which scarcely equals in dimensions a quarter of an inch of sewing cotton. And now I will beg your attention to the singular manner in which digestion is carried on in this atom, which you will discern plainly enough through its brown but translucent skin. If you look carefully at either of the long, many-jointed legs, you will see that it is permeated by a central vessel, the walls of which contract periodically with a pulsation closely resembling that of a heart, by which granules or pellucid corpuscles, floating in a clear fluid, are forced forward. There is no uniformity in the direction of the pulsatory waves; sometimes, as in the limb we are watching, they proceed from the body towards the extremity; but, in some of the others, we shall probably find, even at the same moment, that the waves have a retrograde course; and this contrariety may occur in two contiguous limbs on the same side of the animal. By continuing our observation for some minutes, we shall find also that its force is varying and uncertain; strong and regular at one time, weak and vacillating at another, and sometimes even quite intermitted, or, at least, quite imperceptible.

By selecting a limb in which the movements are strong, you may trace the vessel to its termination in a blind sac in the last joint but one of the limb; and then follow it up to its junction with a great vessel which runs longitudinally through the trunk, of which all the vessels that permeate the limbs are branches, and whence the circulating globules all proceed. This great vessel is the stomach, and this circulation is the provision for dispersing the nutritive properties of the food to all parts of the system. There is in these humble and simply organised animals no proper blood, or, at least, none included in a system of arteries and veins; but the products of digestion are carried to the most distant parts of the body, through this extraordinary development of the stomach or intestine (both in one), and through this series of blind canals, by means of their own irregular contractions, aided by the muscular movements of the body and limbs.

You would scarcely forgive me if I took no opportunity of showing you the Cheese-mite, that first object of wonder to every child that looks through a magnifying glass. And no place so suitable for its introduction, as in connexion with its cousins, the Spiders and Harvestmen. Well, fortunately, we need not search far for specimens; for here, in the cavity of this almost defunct skeleton of a cheese, we can find as many millions as you can reasonably desire to select from. Here is a fat one; we'll take him.

You can see with a pocket lens that it has a plump, polished, oval body, of a pellucid white hue, and eight short red legs; but for more than this we must go to the tube. Look at him now, as he lies on his back, helplessly sprawling and throwing his feeble legs about, in the live-box.

His oval body is divided by a transverse furrow

into thorax and abdomen, like a Beetle's; and there is another division between the head and thorax, wherein it differs from the Spiders. The first two pairs of legs are separated by an interval from the last two pairs; they are all of a translucent pale red hue, as is also the head: each consists of seven short joints, the last of which has a sort of heart-shaped pad, something like a horse's hoof, and a single hooked claw, which works against its sole.

The structure of the head cannot be seen satisfactorily otherwise than by crushing the Mite in the compressorium; a process which, when we remember how many thousands we crush down in our oral compressorium every time we eat ripe cheese, needs not excite much compunction. We must put a drop of water between the plates, in order to wash away the opaque granules which will escape from the bodies of the animals, when the skin, and all the solid parts, will be left beautifully clear and distinct. Moreover, by putting half-a-dozen specimens in at once, we shall secure them pressed in various aspects, and be pretty sure of some perfectly flat and symmetrical.

I have one under such conditions; the parts of the mouth nicely expanded, and the whole well displayed. Now for a high power; for, to discern this properly, we cannot do with less than 600 diameters.

Viewed from beneath, we see a broad *labium*, nearly square, divided at the tip into two blunt points, with a sharp notch between them. The two lateral edges are, as it were, buttressed by the pair of palpi, which are thick, and consist of four joints each; these are distinguished by the bristles at each joint, though the whole are united; soldered, as it were, to the sides of the lip.

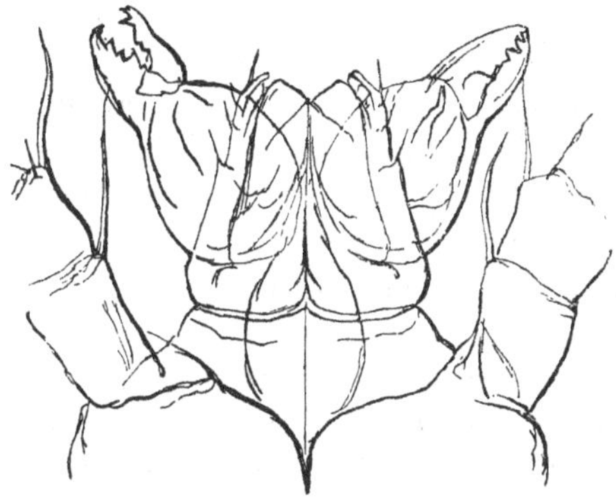

HEAD OF CHEESE-MITE.

The upper portion of the mouth is formed by two stout *mandibles*, which are jointed to the front of the head, and can be either widely expanded, or brought together so as to form a covering to the *labium*. They are pincer-form, like the claws of a crab, the two fingers being strongly toothed on their opposing surfaces. They thus form effective prehensile instruments. These mandibles can be advanced separately or together, and the whole head can be elevated or depressed.

In the water of ponds we may frequently see, playing among the sub-aquatic vegetation, bright-coloured Mites; sometimes rich velvety green, sometimes purple, but more commonly brilliant scarlet; often curiously marked with sinuous patterns or spots of black. They swim rapidly and evenly by means of rapid rowings with their legs, which are thickly fringed with long hairs. I have one here, which seems to be the *Hydrachna histrionica.* It is a little, flat, circular, cushion- or cake-like creature, scarlet, with four clouds of black on its back, about one-sixth of an inch in diameter. You may notice the effective oars which the legs form by means of their thick fringe of hair, and in particular the power which the hind pair possess by reason of the enormously dilated hip-joint, affording space for broad and powerful muscles.

The structure of the mouth differs greatly from the same parts in the Cheese-mite. The palpi here are long and perfectly free throughout; the fourth joint is long and slender, and is curiously hollowed at the end to receive the terminal joint, which forms a short claw, and which falls down upon the former. The mandibles, too, are not pincers, but consist each of a thick joint, cut off obliquely behind, like the nib of a pen, while the other extremity is blunt and broad, and bears a strong curved claw; the lip is oval, and cleft in the middle, and is wedged in between the bases of the first pair of legs.

CHAPTER XIV.

WHEEL-BEARERS.

I MUST now introduce you to a class of animals peculiarly microscopic, since without our marvel-showing instrument they are wholly beyond the sphere of human cognizance. Yet they have been ever since its invention favourite objects with the microscopist; and I am free to confess that, among all the classes of animated beings, this of the *Rotifera* has been my own special delight. Their numerous and varied forms, often of remarkable symmetry and elegance, their swiftly-revolving wheels, their vigorous and sprightly motions, their curious habits and instincts, their complex organization, and the ease and correctness with which this is discerned through their tissues, which have the transparent brilliance of the purest crystal—all combine to impart a charm to the Wheel-bearers, which makes the observer hail their appearance in his drops of water with pleasure, and linger over them with unwearied delight.

The peculiarity which specially characterises them is the presence of certain organs called *cilia*, and their arrangement in such a manner that their motion gives to the observer the impression that two toothed wheels are placed on the front of the animal, which are in rapid revolution on their axes. This was believed to be the real fact by the earlier microscopists, though they were utterly unable to conceive how such a movement could consist with parts maintaining an organic connexion between themselves. It is, however, an optical illusion, depending on the nature of ciliary movement, which therefore I must first endeavour to explain to you.

Cilia are organs which play a very important part as instruments of locomotion, as well as of other functions, in all the lower forms of animals, and in the early stages of some of the higher forms. They are found also characterising the lowest forms of vegetable life, giving to them the means of spontaneous locomotion, which renders them liable to be mistaken for animals. They consist of prolongations of the fleshy tissue into long and very delicate hairs, which are endowed with a special faculty of motion. This consists of a bending down in a given direction to a certain extent of flexure, followed by a rapid resuming of the perpendicular; which is however immediately succeeded by like bendings and straightenings in alternate gradation. The simplest condition of this movement is that in which a single cilium only exists, by whose successive lash-like beats upon the surrounding water, the animal is rowed along as a boat through the sea. But far more commonly cilia are arranged in rows, or in many series of rows, in which case the bending and straightening of the individual cilia do not occur otherwise than in strict and orderly relation to each other. For instance, one cilium in a given row begins to bend, the one next to it then begins, then the third, then the fourth, and so on, all precisely in the same direction, all in precisely the same time, all with precisely the same force, and all to precisely the same extent. It follows, that before the first has completed its beat and resumed the erect position, three or four others are in various degrees of flexion, regularly graduated; and that if the eye could look laterally at such a row of cilia suddenly arrested and fixed as they were, we should see their tips tracing a wavy line instead of a straight one. Moreover, since the bending of any cilium brings its tip nearer to its successor than it was before, and this approximation increases the farther the flexure proceeds, it follows that at the bottom of each wave the tips of the cilia overlap their successors, while the spaces perpendicularly above their bases are left more open by the removal of their points.

Hence, in microscopical observation of ciliated animals, though the individual cilia are too minute to be discerned while still, we can readily discern the increased density (and therefore opacity) of the bottom of a wave, contrasted with the increased openness (and therefore clearness) of the summit. So that the optical effect is that of an alternate succession of dark and light spots blending into each other.

But as no cilium in the series is for two successive moments in the same degree of flexure, and as both it and all its predecessors and successors are ever urging on their perfectly timed and regulated course, the waves are never fixed, but always gliding on with a swift but beautifully even rapidity. And as it is with the waves, so it is with their optical effect upon the eye; the black and white spots, or rather the black spots with blank intervals, appear to be constantly chasing each other in ceaseless race.

You are then prepared to take a peep at this beautiful *Brachionus pala*. A cup of elegant form, swelling at the sides and narrowing a little at the mouth, has one side of its rim furnished with four spines, the middle pair of which are very slender, sharp, and needle-like; the other side of the rim is undulated, but not toothed. The bottom of the cup terminates in two broad blunted points,—when seen directly in front; but a lateral view considerably modifies the whole form. Then you see that the back of the cup is much more swollen, the belly-edge being nearly straight, and that this latter descends much lower than the dorsal line, the bottom being as it were cut away obliquely and slightly hollowed. Between the two bottom points, there is a round opening, for an object which we shall see presently. Such is the shell, or *lorica*, as it is technically called, which is of a rather stiff, elastic consistence, of a horny (chitinous) texture, and of the most glassy transparency, permitting us to trace every vessel, every organ, and every function of the animal within the shell with perfect distinctness. The little creature is of unwonted dimensions in its class, for it is one thirty-sixth of an inch in length. Hence it is just visible as a white speck moving in the water, to the unassisted eye, while a pocket lens reveals its beautiful form.

Within this translucent shell you see a confused mass of moving viscera, a multitude of irregular sacs and bands, lying over each other, whose crowding, changing, and vanishing lines distract the attention, and prevent you from making out anything definitely. But a waved outline of hyaline flesh is protruding from the rim of the shell, and now, having reached beyond the level of the spine-points, it rapidly unfolds into three broad flattish lobes, and in an instant each of the two lateral ones is crowned by a wheel of dark points in rapid horizontal revolution. Is not this a charming sight? Round and round go the wheels, forming two perfect crowns, which rotate with uninterrupted and unceasing course, smooth and regular, which we can compare with nothing else than the crown-wheel of a watch, if allowed to run down.

Now these are examples of ciliary action. Though at first it is almost impossible to persuade oneself that there is not an actual rotation of parts, yet this is only an illusion, as I have already explained. The waves alone move, the cilia themselves retaining their position unchanged except that they alternately bend and erect themselves. It may assist your idea of this motion to advert to a field of corn over which a smart breeze is blowing. You see that waves chase each other across the field; but your reason, and indeed your observation, tells you, that this appearance is produced entirely by the alternate bending and rising of the ears of corn, which are of course stationary.

The beauty and wonderfulness of these ciliary wheels are so striking, especially when one sees them for the first time, that for a while we see nothing else: we cannot take our eye off from them. But when you have a little satisfied your sense of seeing, you may examine other points of interest in this charming little animal.

The cilia are remarkably stout and long in this genus, but on the middle lobe of the front there are other processes of the same character; but still stouter. These too are not properly vibratile, at least they do not make circular wheels: ordinarily, they project like stiff erect bristles, or converge towards each other.

Between the two middle spines the shell is cut into a deep notch, out of which protrudes, when the wheels are expanded, a curious little organ, consisting of a couple of fleshy tubes, the one sheathed in telescopic fashion within the other, and bearing at its tip a pencil of bristles, which can in turn be sheathed. This organ doubtless represents the united antennæ of Insects.

But, you ask, what is that much more conspicuous organ that is alternately thrust out and drawn back at the bottom of the shell, and that is so nimbly whisked about in all directions, looking, with its numberless transverse wrinkles, and its little fingers at the tip, so like an elephant's trunk in miniature? This is the creature's foot; the only one he has; and as I said the little tubular telescope represents the two antennæ fused into one, so we must consider that this flexible member represents all the six pairs of an Insect's legs united, or perhaps, more philosophically, one of the pairs, the rest being obsolete. It must not be considered as a tail, not only from its function, which is decidedly that of locomotion, but also from its position on the ventral side of the intestinal orifice. It is a curious organ, capable of great elongation, or, at the will of the animal, of entire retractation within the abdomen; and this in an instant: while, as you observed, it is flung about, and dashed from side to side, and bent hither and thither with a sort of insane energy. The means by which these movements are performed, you may easily discern in several pairs of muscular bands which run throughout its whole length, their upper insertions being placed high up on the interior of the shell, where, during contraction, you may see them swollen into thick bulbs.

The foot terminates in two short conical fingers or toes, which can be drawn in or extended, widely separated or brought into contact, at pleasure. By means of these the animal has the power of mooring itself, even to the smooth surface of glass; and that so firmly that from them it can stretch itself in all directions by turns, now and then shaking itself to and fro with sudden violence, as if irritated, yet without letting go its foot-hold.

While thus anchored, the action of the ciliary wheels produces considerable whirlpools in the surrounding water, as you will see very distinctly when we have recourse to a curious but simple expedient, first invented by Gleichen, and since much used by Ehrenberg, of mixing some colouring matter with the water in which the animal is. I take a little carmine with a wetted hair-pencil, as if I were going to colour a drawing, and allow a small portion of the pigment to diffuse itself in the water which is in the live-box, then, putting on the cover, I quickly replace the whole on the stage, and re-find my little Brachion: and now I again submit it to your observation.

The whole field is now filled with scattered granules of irregular form and size, of a dark-red hue. These are the particles of carmine floating in the water; particles of alumina, that is to say, stained with cochineal. They are in motion, and their movement is more energetic the nearer they are to the little animal, which is rotating vigorously in the midst of them. They describe two great circles, concentrical with the two wheels of the Brachionus, and it is easy to see that their rotations are the cause of the movement. The motion of the cilia communicates itself to the surrounding water, and produces circular currents, into which every floating atom within a certain distance is drawn, and in which it then continues to whirl round with a rapidity which increases as it approaches the centre of rotation.

But the *Brachionus* suddenly lets go its foot-hold, and a surprising change takes place. No more currents are made in the water, but the animal itself glides swiftly away head foremost with an even course, revolving on its axis as it goes. What is the immediate cause of its movement? The ciliary action which before produced vertical currents.

In order to explain this, let me suggest to you a homely comparison. Suppose you see a boat on a still lake, and in it a man pulling a pair of oars. He pulls vigorously, but the boat does not move an inch, and you perceive that she is fast moored; a rope holds her to a post on the bank. But does the man's rowing produce no effect? O yes; the successive strokes of his oars upon the water have communicated motion to the fluid, and a strong current is made on each side of the boat, in a direction opposite to that in which he strives to row her forward, the force of which is felt to a distance proportionate to the vigour and continuance of his pulling. The reason of this is that the boat is fixed, and all the force of the impact is spent on the water.

But now another man approaches the post, and unties the rope. Instantly the boat glides ahead, and continues to do so, urged by the repeated strokes of the oars, whose effect on the water in making currents is now slight and imperceptible. The reason of this is that the water is now the fixed body (or nearly such), and the force of the impact is mainly spent on the moveable boat.

The *Brachionus* is the boat, its cilia are the oars, and its foot is the rope. As long as this last maintains its hold, the whole force of the ciliary stroke is spent on the water, and currents are the result; but as soon as this hold is broken, the force acts on the animal (=boat), which is thus rowed rapidly forwards.

The use of the cilia in this latter case is obvious. They enable the little animal to rove about at its wayward will; and doubtless motion is as pleasant and necessary to it as to the fish in the sea, or the bird in the air. But what is the object of their vigorous rotation, when the animal chooses to maintain a firm hold with its foot? What is the use of rowing a boat, if you do not choose to let go the painter?

To solve this enigma, let us search up our little Brachion once more; he will not roam long before he settles soberly again. Yes, here I have him moored. Now, mark carefully the vortices, which are so vigorously circling around the animal's front, and you will perceive that the movement is not a strictly circular one, but that each whirlpool has an outlet close to the cilia; for the accumulated and condensed particles of pigment, after many rotations, pass off in an united stream between the two crowns, and go away horizontally in a line from the ventral side of the front. That is to say, each vortex pours off its accumulation at a point on the inner side of the ciliary circle, and the two streams, uniting, pass off from the lip of the shell, to be drawn in again, however, by and by, when the centrifugal force is exhausted.

Now this stream passes immediately over the mouth, which is an opening in the flesh of the front, forming a deep cleft on the ventral side, the lips of which, as also the whole interior of the tube, of which it is the orifice, are richly covered with cilia. A certain portion of the atoms are thus arrested by these cilia, and are hurled by their vibrations down this gulf. Yet not all, nor nearly all; for the lips appear to possess the sense of taste, or of some modification of touch, which enables them to refuse or receive the atoms presented to them, so that only such particles pass down the throat as are selected for food. Some of the atoms of pigment are admitted, and one of the most pleasing sights connected with these animals, is to watch the swallowing of coloured food, its reception into the singular sunken mouth, where the great powerful jaws act upon it: thence its dismissal through the gullet, where certain glands pour upon it their secretions, into the stomach, where other glands, answering to a liver, change it, and thence into the intestine and rectum, until its indigestible portion is discharged through the cloacal orifice.

The object of the mingling of colour with the water in which these and similar animals are held for observation, was the tracing of the phenomena of digestion. And, indeed, it renders the whole process beautifully distinct; for, from the transparency of the tissues, the presence of the coloured pellet is everywhere recognisable, since it retains its form and hue under all its changes, clearly revealing to us the shape, dimensions, and directions of the various canals through which it passes; here and there diffusing throughout the viscus in which it is held a beautiful roseate hue, more or less deep, without, however, losing its own definite outline.

Let me now direct your attention to the organs devoted to the seizing and mastication of the food. And the more because the form of these organs in the *Rotifera* is quite peculiar,—quite unlike what is found in any other class of animals, though the parts are essentially the same as those which we have already seen as entering into the mouth in Insects.

Removing the carmine-stained water, I put into the live-box a drop from a vase very rich in organisms of many kinds. Among these you see very numerous the mulberry-like clusters of that beautiful green creature, *Syncrypta volvox*, which is now pretty generally considered a plant, though from its spontaneous motion, swimming evenly along, revolving on its axis as it goes, you would be inclined to agree with earlier observers in thinking it an animal. These appear to be favourite morsels with the Brachion: one has already been devoured, and is quite visible in the alimentary canal, its brilliant green hue shining out through the translucent viscera and tissues. Others are approaching, and two or three are just now drawn into the vortex of the ciliary current. It is amusing to see the manœuvres which the *Brachionus* makes to take his prey. I say *manœuvres*, for there really seem to be perception and intelligence. The mode in which it directs its ciliated flaps towards the spot where a *Syncrypta* is whirling, or suddenly stretches forward to the extent of the long foot, as if it would seize the prey by force, seems to indicate a cognizance of its proximity; as do also, still more, the manner in which it depresses the lip-like lobe of the rotatory organ on one side, when the prey is in the vortex on that side, and the eager haste with which it shrinks down into its shell the instant the little mulberry drops at length into the throat.

BRACHIONUS.

But now comes the tug of war; the black, millstone-like jaws open wide and stretch forward to grasp the little victim (which is still distinctly visible through the transparent tissues): they touch the globular envelope, but cannot quite grasp it. The Brachion redoubles its efforts; the jaws gape vigorously, but can only scrape the sides of the little globe, which at every touch slips away, the expanse of the jaws being not quite sufficient to embrace it.

At last the little animal becomes indignant; the jaws no more endeavour to grasp, but with a very distinct and sudden upward jerk throw out the prey, which until now has been retained and pressed downward by the contraction of the sides of the sensitive throat. Strange to see, the little *Syncrypta*, after all its imprisonment and rough handling, is no sooner free than it whirls merrily away, revolving as it pursues its even ciliary course, just as if no interruption of its freedom had occurred.

Meanwhile, however, better success attends the Brachion's hunting; for a smaller globe has sunk into the throat, and passes with a gulp into the mouth, between the gaping jaws, which instantly close upon it, and, working vigorously, bruise it down with a hammer-like action upon a sort of central table. After this process has gone on for a few minutes, the green mass, less perfectly defined than before, slips through a narrow postern-gate, along a short narrow alley, into the digesting stomach.

But what sort of a mouth is this? It is inclosed within the tissues of the body, not very far from its centre, so that no part of it comes into contact with the external water, or even approaches any part of the superficies of the body. It has been usual to call the great hemispheric bulk in which the symmetrical hammers work so vigorously, a gizzard, but it is a true mouth, and the hammers are true jaws.

This form of mouth is termed a *mastax;* it consists of a dense but transparent muscular mass, forming three lobes at its lower part, deeply cleft at the front of its ventral side, where the passage which I have called the throat, but which is more correctly designated the *buccal funnel*, enters. Within this muscular bulb are placed two bent organs like hammers, called *mallei*, and a third central table, called the *incus*. The mallei approach each other dorsally, while the incus is placed towards the ventral side, its stem pointing obliquely away from the centre.

Each malleus consists of two portions, united by a free but powerful hinge-joint. The lower joint (*manubrium*) is shaped somewhat like a shoulder-blade; and the upper joint (*uncus*) is set on at nearly a right angle to it, but is capable of considerable change of direction by means of its hinge. It consists of five or six finger-like teeth, connected by a thin web of membrane.

The *incus* also consists of several distinct pieces. The principal are two stout *rami*, resting on what appears, when you look at the back or belly of the animal, to be a slender foot-stalk (*fulcrum*). But when you get a lateral view the foot-stalk is seen to be only the edge of a thin plate, to the upper edge of which are jointed the rami, in such a manner that they can open and close, like the blades of a pair of shears. Each ramus is a thick, three-sided piece, with the upper side hollow, and the inner flat, and in contact with that of its fellow, in a state of repose. The uncus of each malleus falls into the concavity of its corresponding ramus, and is fastened to it by a stout triangular muscle, which allows some freedom of motion.

Many muscles are inserted into various parts of these organs, and into the walls of the mastax, which impart various and complex motions to all the parts. Thus, as we have seen, they are adapted to the various functions of mouth-organs, those of grasping, holding, bruising, and chewing food.

The mallei correspond with the mandibles of Insects; and the rami of the incus with the maxillæ; while the walls of the mastax with the two edges of its orifice correspond with the mouth, with its labrum and labium.

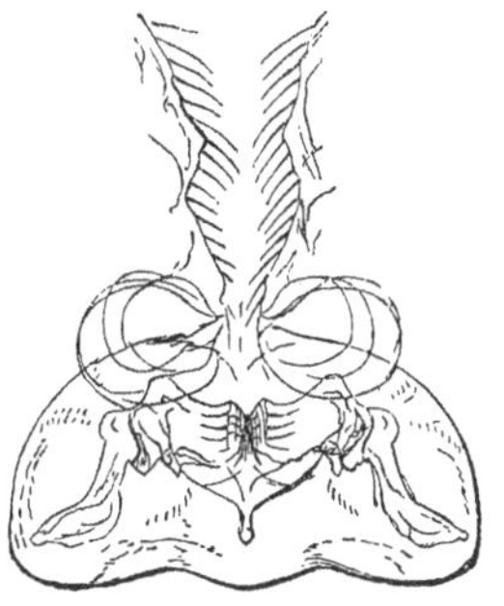

MOUTH OF BRACHIONUS.

It is true we are somewhat startled to find a mouth placed far down within the cavity of the breast; but there are other forms in this class, some of which I may be able to show you, where the mastax has essentially the same structure, in which it is placed at the front margin of the body, from which the jaws can be freely protruded. The difficulty will seem less if you weigh the following considerations:—

The integument in the *Rotifera* is very flexible, and, especially in the frontal regions, is extremely invertible. In those genera in which the mouth apparatus can be brought into contact with the external water, it is ordinarily, to a greater or less degree, retracted within the body, by the inversion of the surrounding parts of the exterior, while, in those genera in which it is permanently enclosed, analogy requires us to consider the condition as induced by a similar inversion, but of permanent duration. If we imagine the head of a soft-bodied Insect-larva retracted to a great degree (as is done partially by many Dipterous larvæ), the skin of the thoracic segments would meet together in front, around a purse-like opening, which would be the orifice of such a buccal funnel as exists in most *Rotifera.* In the latter, it is the normal condition; in the former, it is merely accidental and temporary.

We need not devote any more minute consideration to the digestive apparatus in our little Brachion, but there are some other points in its structure which are worth noticing. In the central line of the body, just above the mouth, as you see the animal in a dorsal view, there is a square speck of a rich crimson hue, the edges of which, when we view it under reflected light, glitter and sparkle like a precious stone. But when we obtain a perfectly lateral view, we perceive that the situation of this gem-like speck is considerably nearer the dorsal side of the shell than the mouth, and that it forms a wart-shaped prominence on a large turbid mass which occupies the whole front portion of the animal. By comparison of this organ with the corresponding parts in other genera, there is every reason to infer that this turbid mass is an enormous brain, the nervous matter being in a very diffuse condition; and that the ruby seated on it is an eye, consisting of a crystalline lens, and a layer of crimson pigment beneath it.

The oval bodies that you see attached to the hinder part of the shell are eggs. Most of the females that we meet with carry one or more, sometimes to the number of six or seven. The specimen we are examining had two at first, one on each side the foot-orifice; but just now a third was excluded,—an operation which occupied but an instant,—and this took its place beside the former two, so that we now see three. These eggs are generally carried by the parent until the young are hatched. The oldest of these three is nearly ready for hatching, and if you watch awhile you will see the birth of the young. At first exclusion, the egg which was seen some time before in the ovary, as a semi-opaque mass, of well-defined but irregular shape, immediately assumes a form perfectly elliptical, and its coat hardens into a brittle shell. This is so transparent that the whole process of maturation can be watched within the shell. The yelk is at first a turbid mass, in which are many minute oil-globules. Soon it divides into two masses, then into four, then into eight, sixteen, and so on, by the successive cleavage of each division, as fast as it is made, till these divisions are very numerous. Then we begin to see spontaneous movements; the outline of the young separates in parts from the wall of its prison; folds are seen here and there, and fitful contractions and turnings take place. Soon an undefined spot of red appears, which gradually acquires depth of tint and a definite form, and we recognise the eye. Slight waves are seen crossing one end of the egg; these become more and more vigorous and rapid, and at length we see that here is the situation of the frontal cilia. The mastax appears, and the jaws, and soon the latter begin to work;

though it must be only by way of practice, for it is hard to imagine what they can yet find to masticate.

All these phenomena have successively appeared in the egg we are now watching; and at this moment you see the crystalline little prisoner, writhing and turning impatiently within its prison, striving to burst forth into liberty.

Now a crack, like a line of light, shoots round one end of the egg, and in an instant the anterior third of the shell is forced off, and the wheels of the infant Brachion are seen rotating as perfectly as if the little creature had had a year's practice. Away it glides, the very image of its mother, and swims to some distance before it casts anchor, beginning an independent life. At the moment of the escape of the young, the pushed-off lid of the egg resumes its place, and the egg appears nearly whole again, but empty and perfectly hyaline, with no evidence of its fracture, except a slight interruption of its outline, and a very faint line running across.

This is a female young: the male is totally unlike the female, and is very much smaller. We can always tell whether an egg is going to produce male or female young, by the great difference in its size, the female being more than twice the bulk of the male egg. All of one brood are of the same sex; we never see a *Brachionus* with male and female eggs at the same time. What is very strange, is, that the male has no shell, no spines, no mouth, no jaws, no stomach, no intestines; no ciliary wheels; its cilia, which are very long and powerful, being arranged in one circle round the whole front. Its movements are exceedingly fleet.

Perhaps you are tired of *Brachionus*, and are ready to cry out, "Ohe! jam satis!"* Well, then, I will turn him off, and show you another elegant little creature, the Whiptail (*Mastigocerca carinata*). I have here in a bottle some stalks of the Water-Horsetail (*Chara vulgaris*), which I obtained in a pond a few weeks ago. These I examine in this way. Taking hold of the *Chara* with a pair of pliers, I pull it partially out of water, and, allowing it to rest on the neck of the bottle, I cut off with a pair of scissors, or with a penknife on my nail, about one-fourth of an inch of the tips of three or four leaves, which adhere together by their wetness. These tips I place in the live-box, with a drop of water, and having separated them with a needle, I put on the cover, and examine them with a triple pocket lens; holding up the box perpendicularly, not opposite the light, but obliquely, so that the field is dark; but the light reflected and refracted by the animalcules shows them out beautifully white and distinct, even the minute ones. The forms and some characters of the middling and larger can be quite discerned thus; for example, the slender tail of the one I am now going to show you, I can thus see. The position of any particular individual to be examined being thus marked, it is readily put under the object-glass of the microscope. I have found these leaves very productive of the more stationary animalcules, the *Rotifera* especially.

It was in this way I this morning found the pretty and delicate little Whiptail, which I am going to make the subject of our evening's study. It is inclosed in a glassy shell (*lorica*) of a long oval form, from which

* "O dear! quite enough of this!"

rises on the front half of the back a thin ridge, which in the middle has a height nearly equal to half the diameter of the body, but tapers off at each end. Its base is corrugated with wrinkles. This is not set on symmetrically, but leans over considerably to the right side. Its basal portion is hollow, and is continuous with the general cavity of the shell, for we sometimes see portions of the viscera in its interior.

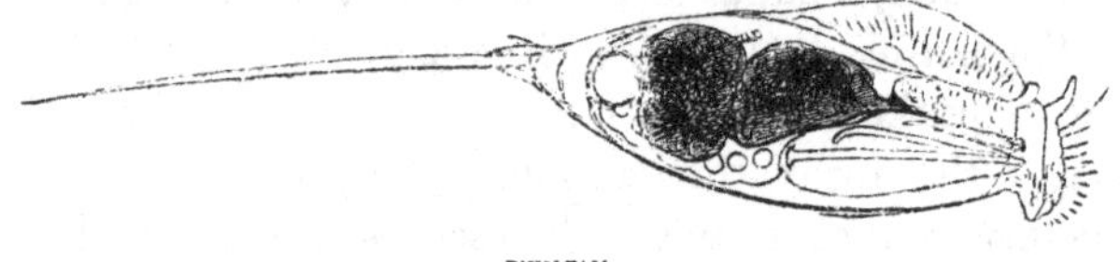

WHIPTAIL.

The head of the animal is rounded, and divided into several blunt eminences or lobes, which are set with cilia; these rotate constantly, but irregularly and feebly, and do not make manifest wheels, as *Brachionus* does. A small antenna projects from the back of the head, capable of being erected or inclined. A long brain descends along the base of the ridge, carrying a bright and rather large crimson eye set like a wart on its interior angle

Instead of the flexible and contractile foot of *Brachionus*, the Whiptail has a single horny spine of great slenderness, and exceeding in length the whole body. This spine probably represents not the foot, but one of the toes at the end of the foot. For it is attached to a very short foot, in the midst of two or three bract-like spines, one of which, longer than the rest, and distinctly moveable, probably represents the other toe undeveloped. The long spine is set on by a proper joint, a globose bulb being inserted into a socket, which allows it free motion, in all directions except backward. The socket itself is contained in a second joint, the basal part of which is inserted at some distance within the aperture of the lorica. This articulation is formed by an infolding of the skin, but is permanent in its position.

The most remarkable circumstance connected with this elegant little animal is the unusual form of the dental apparatus, which differs so immensely from that of *Brachionus*, that we should never recognise them as being the same organs, if we had not numerous intermediate links, which by insensible gradations connect the two remote forms.

The *mastax* is a somewhat slender sac, much produced in length, and with the component lobes greatly and irregularly developed. The *incus* has a *fulcrum* of great length and slenderness, a straight rod with a dilated foot. The *rami* are small, and pincer-shaped, but with the angles greatly produced. The *mallei* have long, slender, incurved *manubria*, and simple *unci*.

But the remarkable circumstance is the non-symmetrical character of the apparatus. The left side is much more developed than the right. The left angle of the *incus* descends to a greater distance than the right; and its extremity is dilated into an expansion, with several irregular points, to which muscular threads are attached. The *ramus* also of the same side is larger than its fellow; so with the *mallei*. The *mánubrium* of the right is comparatively short, very slender, and of uniform thickness; with a long, slender, rod-like *uncus*, doubly bent in the middle. The left is much longer, irregularly swollen, clubbed at the articulation, and bearing a thick, curved, knotted *uncus*, which terminates at a point not precisely opposite the tip of its fellow. These circumstances, combined with the unsymmetrical character of the dorsal ridge, of the foot-spine, and of some other organs, render this genus a highly curious one to the naturalist.

The little Whiptail is as lively in its motions as it is elegant in its form. When swimming, it glides with considerable swiftness through the water, turning frequently on its course, and often partially revolving on its long axis. When inclosed, as is often the case, by two fragments of the filamentous *Chara*, it travels along the sides of its inclosure, nibbling, as it goes, the floccose and sedimentary deposits on the surfaces of the leaves. The long spine-foot is commonly carried inertly after it; when the animal suddenly turns, of course the tail is bent at the basal joint, but it is not habitually whisked about, as is the tail of *Brachionus*, nor is it so much used as a support or turning point. The animal has the power of so using it, however, and of adhering with considerable force to the glass of the box or the side of a phial, by its point.

We have hitherto looked at our *Rotifera* by transmitted light, and their crystalline transparency renders them beautiful objects when thus exhibited. But we will now look at the Whiptail by the direct light of the sun upon it, condensed, but not to a burning point, by the bull's-eye lens.

It now possesses a peculiar beauty of another character. The body generally is colourless as a vase of glass, but reflects the rays brightly from its polished surface. An advancing egg in the ovary is opaque white, as is the front part of the mastax; the stomach and intestine filled with vegetable matter are of a yellow-green; the rotating head appears of a pale blue, and the eye shines out as a speck of opaque vermilion.

With the dipping-tube I will now take up a drop of water from the bottom of the *Chara*-jar, allowing a little of the loose sediment to flow in also. This is a random cast; we know not what we may get, though we are pretty sure to catch something. Now then for the examination. Ha! here is the curious Skeleton Wheel-bearer, *Dinocharis pocillum*;—nay, several of them.

This genus is remarkable for possessing true joints in the foot; not merely telescopic inversions of the skin, but permanent articulations with swollen condyles, resembling those of the antennæ of a beetle. Hence the Skeleton has great freedom and precision of motion; using the tips of the long toes as a fixed point, it throws its body hither and thither to a great distance, with remarkable agility. These joints admit of forward and lateral flexure, but you never see the body brought backward beyond a perpendicular position, the swelling of the terminal portion of each articulation precluding further motion in that direction; just as the joints of our knees and elbows permit bending in one direction but not in the other.

This is another indication that these divisions are true joints; and I direct your attention to the point, because the fact helps to indicate that this class of animals has its proper affinities with the ARTICULATA, which has been denied by most naturalists.

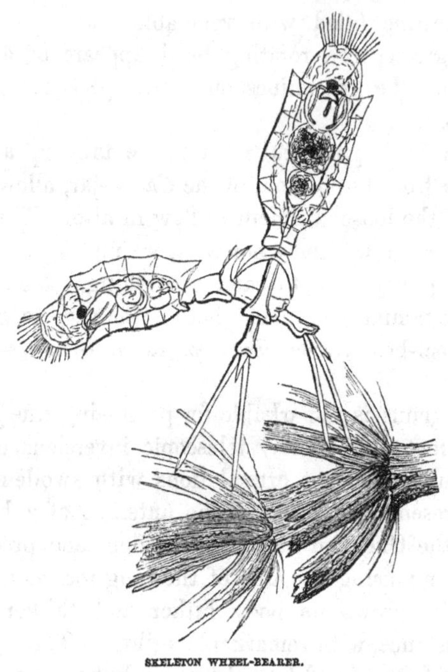

SKELETON WHEEL-BEARER.

The form is curious. Elevated at the summit of a long foot, consisting of three joints, which surmount two unusually lengthened and slender toes, is a vase-shaped lorica, which is three-sided. Its surface is covered all over with minute points, very closely set, so that it resembles shagreen; besides which it forms numerous sharp ridges, which run across transversely. The two sides run off into thin lateral wings, which come to a sharp edge; the back angle also forms a ridge, but less sharp and thin. In front, the shell, or lorica, is as it were cut off abruptly, like the rim of a goblet, but out of this rises a second column, connected with the rim by an elastic membrane, which allows some freedom of motion. This column is widely divided in front and behind, and rises to a point on each side. When the rotatory front is withdrawn, these points approach and meet, closing the orifice; but when the head is protruded they are widely separated.

Internally, we see the usual viscera contained in so narrow a cavity that we are ready to suppose the walls of the lorica unusually thick; this is, however, an optical illusion, dependent on its dilatation into those angular wings already noticed. The cavity penetrates into them; for in one of these specimens I see those curious convoluted threads that are believed to be connected with respiration, within the lateral wings. The stomachs are generally full of green and brown food, but they will not imbibe carmine.

Let us look, however, a moment longer at the singular foot. Between the first and second joints there are two projecting spines; these differ much in different individuals as to their length, slenderness, and direction; sometimes being quite short, at others as long as the toes; generally, they arch downwards, but occasionally they stand out straight, or even curve upwards. In some specimens these spines appear to be processes of the first joint, but in others we can see distinctly that they belong to a little intermediate piece between the first and second *apparent* joints. Between the two toes, on the hinder aspect, projects from the last joint a small spine, which is perhaps the rudiment of a third toe, since we find that number in some genera of this class. The whole foot, including the toes, is rough with the shagreen-like points that cover the lorica.

You have already noticed the rapidity and fitful irregularity which the long and many-jointed foot confers upon the movements of this curious little form. From the toe-tips, as a point of adhesion, it throws its body to and fro, or from side to side, in a peculiar manner. The toes are sometimes sprawled out, like the legs of an expanded pair of compasses, and sometimes the joints of the foot are suddenly bent in zig-zag fashion, and then as abruptly straightened. The animal swims gracefully, but only with moderate swiftness, the rotatory crown of cilia being small, though forming the usual vortices when the animal is moored; while thus swimming, the toes are gracefully stretched behind, nearly in contact with each other. It is lively in its motions, but these seem performed without any ostensible object; we do not often see it attempt to eat, or nibble at any substance.

I think we never find the Skeleton except among the sediment at the bottom of the water in which it is kept; among which also we frequently see the remains of defunct specimens—the skeleton of the Skeleton; this itself makes a pretty object: the lorica, with its points and ridges, the thoracic column, the foot with its joints and spines, and the toes, being perfectly preserved, and rendered even more clear than during life, because of the removal of all the soft internal parts by decay, and by the efforts of those little scavengers, the smaller species of infusorial animalcules. These quickly find their way into the interior of any dead animal with a shelly case, as a Wheel-bearer, a Water-flea, or an Insect, and soon devour every particle of soft flesh, cleaning out the case in the most tidy manner.

Here is a tiny subject which will test your powers of observation, and possibly your patience, in satisfactorily defining its structure, partly on account of its swift motion and irregular leaps, and partly on account of its extreme transparency. It is a crystalline cup, somewhat like the body of a wine-glass, without any foot, but bearing many flat sword-shaped processes, which, proceeding from the breast, commonly lie flat on each side, down the body, the points projecting below. These are evidently stiff and highly elastic, and their use is manifest to any one who sees the creature in active motion. It swims with a rapid gliding progress, head foremost, but at almost every moment it makes a sudden forcible jerk or leap backwards or to one side, and that so quickly that the eye often cannot follow it in the transition. The organs by which these jumps are effected are the long breast-spines, which are suddenly thrown out in various directions, and they may frequently be seen extended the instant after a leap. When we consider that the creature is jerked often four or five times its own length, through so dense a fluid, we shall perceive how strong the muscular action must be which moves the lever-like spines. The creature is thrown irregularly, often with the side foremost, or the back, or made to perform a somersault in the act. It is probably a sensitiveness to danger or annoyance that prompts these violent leaps; at least, it frequently performs them, after a momentary examination of any floating matter with which its course brings it into contact.

The rotatory organs, the source of the common gliding motion, are not very large or conspicuous; the cilia appear to be set all along the brow. The eye is very noticeable: it is placed near the front, and seems to be of a deep bluish-black hue.

I have not, however, as yet introduced the nimble little stranger by name. We may call it familiarly the Sword-bearer, but Professor Ehrenberg has named it *Polyarthra platyptera*.

This eminent authority on all that concerns these minute forms has placed the species among those which are destitute of a horny lorica or shell. But he is certainly in error here, for, as you may see, there is manifestly a stiff lorica, which covers the back and sides, but which gapes widely in the middle of the under side, throughout its length. From the lateral points, however, a membrane may be seen for a short distance, which doubtless protects the viscera from actual exposure.

The sword-like fins appear to be twelve in number, arranged in groups, or bundles, of three each; one bundle being set on each side of the dorsal, and one on each side of the ventral aspect, at about one-fifth of the entire length from the frontal points. These are all that we can ordinarily count; but I have seen more: one day, while examining a specimen that presented a vertical aspect to me—*end-on*, to speak familiarly—the fins being all expanded, I saw with perfect distinctness a seventh pair, proceeding from the middle of the breast. They are flat, thin, narrow blades, of exceeding delicacy; all distinctly serrated on both edges, the teeth pointing from the base outward; each is strengthened by a central rib. They are jointed independently, on rounded shelly knobs, and are doubtless moved by strong muscles. Under pressure, the knobs and the fins are brought out with beautiful distinctness. Here again we have true jointed limbs.

On the front you may discern a pair of tiny antennæ, each bearing a pencil of very fine bristles. And just below the level of their base, in the centre of the dorsal region, you see the large eye, of a deep red hue, so deep that it frequently looks as if it were actually and intensely black. Just below the eye, apparently, but considerably more towards the ventral aspect, there is a huge mastax, occupying almost half the length of the whole body. The jaws are very simple in their construction, and therefore very instructive, for they contain the same elements as in *Brachionus;* but from their excessive tenuity, and for other reasons connected with the form of the animal, they are calculated to tax to the utmost your perseverance and skill in manipulation to resolve them. They were an enigma to me for years.

The great *mastax* is pear-shaped, pointing obliquely towards the middle of the belly. This form is owing to the great length of the *fulcrum*, and the wide curvature of the *mallei*. The *rami* are very broad, somewhat square at their base, flat, but much arched longitudinally. They open and shut vigorously, with a snapping action, but are not protruded from the front; their whole interior edges come into contact. The *mallei* are simple slender bent rods, apparently

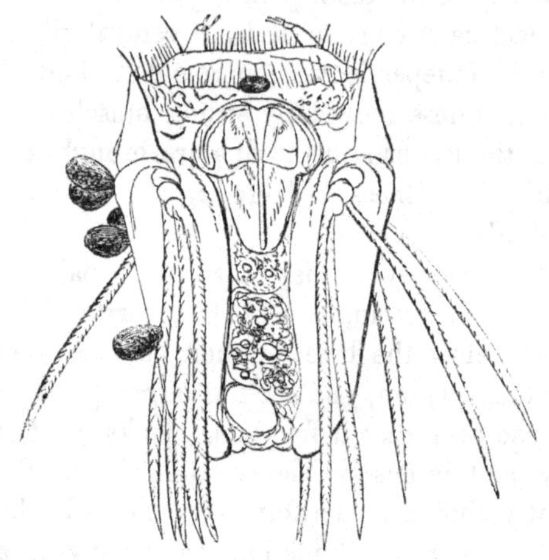

SWORD-BEARER.

without distinct articulation. During life they are thick and irregular in outline, owing to their being invested with dense muscles; as is the whole upper portion of the *mastax*. These muscles conceal or disguise the form and action of the parts during life, but the introduction of a drop of solution of potash into the water instantly dissolves away the fleshy parts, leaving the solid organs, or those composed of chitine, beautifully clear, and fit for observation. Without this aid it would be impossible to resolve the structure of these minute animals.

The little Sword-bearer, like the *Brachionus*, carries its eggs attached to the hinder part of its body, for some time after they are discharged; the minute green oval bodies that you see sticking to the side of this specimen, are not, however, eggs, but parasitic animalcules (*Colacium vesiculosum*), which very frequently infest this species, adhering to various points of the shell, and even to the sword-fins.

What I have now to submit for your examination is one of the rarest species of the class, and certainly not the least singular in its form. It is the Tripod Wheel-bearer (*Actinurus Neptunius*). When fully extended, its length exceeds that of almost every other species, for it reaches about one-twentieth of an inch, but its extreme thread-like slenderness precludes the unassisted eye from taking cognizance of it, as its thickness, even when greatest, is not more than one six-hundredth of an inch.

From this excessive length and tenuity, the appearance of the creature is very remarkable. It may be likened to a cylindrical tube, out of which protrude a great number of draw-tubes from both extremities, principally the posterior one. Those in front terminate in an oval proboscis, which having a sort of finger at its extremity, and two eyes, with an antennal tube projecting obliquely backwards, presents, when viewed laterally, a strong resemblance to the head of a rabbit, the antenna representing the ears. In front, and just below this head-like proboscis, is a double swelling, containing the rotatory organs, which are small and seldom unfolded. The eyes are deep black; probably, as in the last example, a red of great intensity. When the head is withdrawn, the integument is very clearly seen to be introverted. The body consists of one long cylindrical tube, which receives three or four short joints to complete the abdomen; at the dorsal aspect of the extremity of the last of these is the cloaca; at this point the diameter is already very much attenuated; but there are eight or nine more joints which constitute the foot, and these are of extreme slenderness. Towards the extremity, two processes are given off behind, each consisting of a club-shaped piece, with a slender bristle at the tip. The foot terminates in three long, slender, cylindrical, divergent toes, which are flexible, and commonly bent outward; they are equal in thickness, and truncate. These are often retracted in various degrees, even when the foot is otherwise extended.

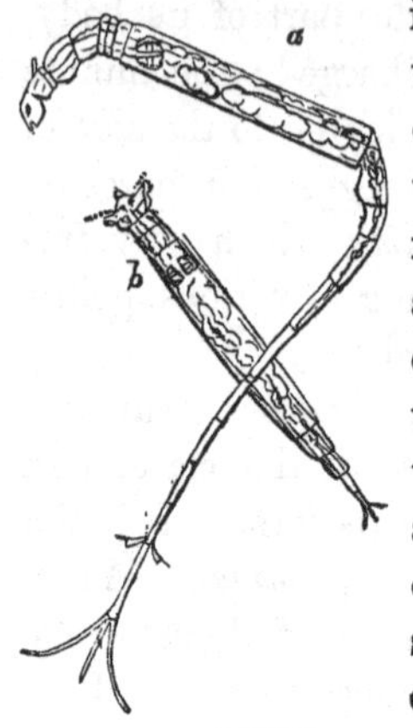

TRIPOD WHEEL-BEARER.

Owing to the slenderness of the body, the viscera are greatly elongated. The mastax, as usual in this family, consists of two hemispheres (each bearing two teeth, set transversely, but converging to the centre); it is situated at a considerable distance from the wheels, and is reached by a long buccal funnel. The digestive canal is a long sac, apparently undivided; it originates directly from the mastax, with, I think, two small basal glands; its posterior extremity becomes gradually tapered to the cloaca. In the specimen we are examining, a small quantity of fæcal matter of a yellowish-brown colour is collected in two small masses, near the extremity. Along the ventral aspect runs the ovary, which in this specimen contains two long oval eggs in advanced development; from their transparent brightness, I suspect the young are produced before birth. I think I can detect a contractile bladder, but am not certain.

The dorsal region of the trunk is marked with strong rugged lines running longitudinally; these look like corrugations of the integument, but I incline to think them the strongly developed muscles for the retractation of the foot. Muscles are seen running through the joints of the foot, until they can no longer be traced, from their tenuity. The viscera can be demonstrated with difficulty, partly owing to the longitudinal muscles, which are so strong and close, and partly from the incessant contraction and elongation of the parts, which drive the internal organs hither and thither. It refuses, you see, to swallow carmine, which might have assisted us.

This singular animal is lively in its motions, especially in the protrusion and retractation of the extremities. These are constantly alternating, and a very curious sight it is to see the immense length of foot suddenly thrust forth from the body, in which it had been completely hidden, the starting out of the horizontal processes, and the diverging of the long toes, as these are successively uncovered. The latter do not seem to be often used as instruments of prehension or adhesion. Indeed the animal does not appear very much given to change of place, but lies in the water, alternately contracting and elongating. Frequently, as the foot is thrust out, the body is made to bend forward so as to form a right angle (see the engraving, in which the animal is thus represented at *a*; *b* represents it when the head is rotating, but the foot is almost wholly withdrawn within the body; in which state the resemblance to a telescope, or to a nest of glass-tubes, is striking).

The last specimen of this class of tiny favourites that I shall show you is one of more than ordinary beauty. It is the Two-lipped Tube-wheel of the Hornwort (*Limnias ceratophylli*). Hitherto we have seen such examples as have the power of freely swimming from place to place at pleasure; but there is a considerable group, of which this is a member, which are permanently stationary, being fixed for life to the leaves or stems of the vegetation that grows under water. The stiff and spinous whorls of the Hornwort (*Ceratophyllum demersum*), that grows commonly in sluggish streams and pasture-pools, is a favourite resort of the species, but it is not confined to any one plant. Here, for instance, it has chosen as the site of its residence the much-cleft leaves of the Water Crowfoot (*Ranunculus aquatilis*); those leaves, I mean, which, growing wholly under the water, are divided into a multitude of slender finger-like filaments, so different from those which float on the surface, and which are merely notched.

You can readily find the Tube-wheels by the aid of a pocket lens, and even with the naked eye when you have seen one or two. By holding up this phial, in which a little plant of the Crowfoot is growing, and searching, with the lens, the window being in front of you, the filaments one by one, you will readily perceive, here and there, little shining objects standing up, or projecting in various directions from the surface of the leaves. The colony is rather numerous in this case, and we shall have no difficulty in selecting our specimens.

On this filament, which I have seized with the joints of a pair of pliers, I can see at least half a dozen of the little parasites. This, then, I will nip off from the plant, and put it with its tiny population into the live-box. Here it is ready for examination.

Several of the animals are in the field of view; but we will look at one at a time. A long narrow tube, slightly widening at the mouth, is affixed by the lower extremity to the slender filaments of water-grass, crowfoot, &c. It is about one-fifty-fifth of an inch in length, pellucid but tinged with brownish yellow. It appears to be of a gelatinous texture, and is covered with extraneous substances, such as decaying animal or vegetable matters, which adhere to its surface. From the mouth of the tube protrudes a transparent colourless animal, the head of which is rounded, with the extremity pursed up. Suddenly it unfolds its flower-like wheel, which consists of two broad nearly circular lobes united, the margin of which is set with strong cilia, much resembling

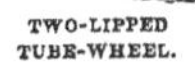
TWO-LIPPED TUBE-WHEEL.

those of the last species. Each cilium appears to be curved, and to be thickened at the middle—the optical expression of the ciliary wave; and the effect of the rotation, as each seems to pursue its fellows around the circular course down the dividing sinuosity, up the opposite sides, and round the margin again, is very striking. The cilia at the front are interrupted between the lobes. In the centre of each lobe is a broad plate, surrounded by a bright ring, and crossed by radiating lines, which also extend towards the ciliated margin; probably these are muscular filaments. The funnel is between the lobes, and leads by a short œsophagus to a bulbous transparent mastax, in which are seen jaws that work on each other. Below this is a long capacious sac, without convolutions or constrictions, but apparently granular in its texture. The alimentary canal is bent upwards through the whole length, terminating in an orifice behind the rotatory organ; for though I have not traced it when empty, I have seen the fæcal matter driven rapidly upward as through a canal, until the mass was discharged just behind the sinuous cleft. On our mixing carmine with the water, the effect is very striking; the particles, whirled round in two circular vortices, are poured in an accumulated torrent through the sinuosity, and over the elevation at the front of the head. We presently perceive a slender line of crimson passing down below the mastax, which indicates a slender stomach-tube there; and, after a while, a little ball of the same pigment accumulates, and is seen resting a little lower down. This then indicates the form and position of the stomach; it must be a very slender canal, terminating in a small rounded bag, at about one-third of the distance from the mastax to the base of the tube. The lengthened sac which you see is the ovary, from which the eggs are discharged into the lower part of the case.

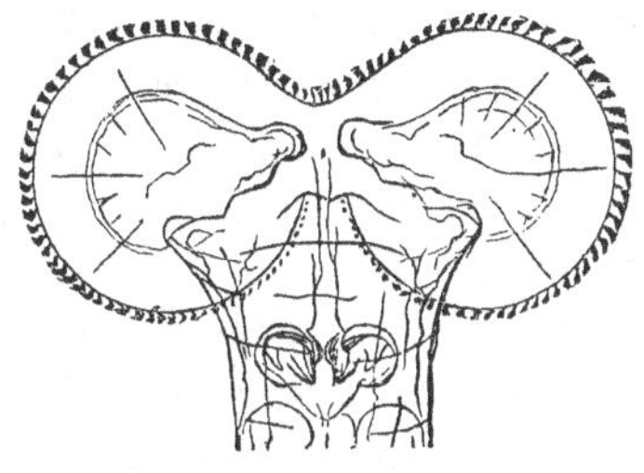

WHEELS OF TUBE-WHEEL.

The mouth needs a little explanation in detail. As you see it, you probably discern little resemblance in its parts to the same organ in *Brachionus*, and yet essentially it is formed of the very same constituents; and as it is very instructive to observe the modifications, in different animals, of a common model of any particular organ, it will be worth while to devote a few minutes' careful observation to this structure before us, especially as it is here seen with more than usual brilliancy and clearness.

The mastax, then, which you see in the centre of the animal, just below the level of the beautiful flower-like wheels, consists as usual of three subglobose lobes; one on each side appropriated to each malleus, and the third descending towards the ventral aspect, which envelops the incus. The mallei are more intimately united to the rami of the incus than in the former type, each uncus forming, with its ramus, a well-defined mass of muscle, enclosing the solid parts, and in form approaching the quadrature of a globe; two flat faces opposing and working on each other. Across the upper surface of the mass the uncus is stretched, as three long parallel fingers arched in their common direction, and imbedded in the muscular substance; their points just reaching the opposing face of the ramus, and meeting the points of the opposite uncus, when closed. The manubrium is much disguised, by being greatly dilated transversely, forming three bow-like loops of little solidity, to the chord of which the fingers are soldered, *not articulated.* The surface of the dense muscular mass displays striæ parallel to the fingers, and, as it were, continuing their number towards their dorsal extremity, becoming fainter till they are imperceptible. These striæ do not disappear when the muscular parts are dissolved by potash; and hence I infer the existence of a delicate investiture of solid substance similar to that of the teeth, &c., enclosing the muscular mass.

The incus, which cannot be separated from the mallei, thus consists of two portions, corresponding to the rami in *Brachionus*, &c., each of which forms the lower part of the quadrantic mass just described. At the ventral extremity they are articulated to a slender fulcrum, which is a little bent downward. The solid framework of each ramus sends off, from its inferior surface, a slender curved process, which is connected with the extremity of the fulcrum.

The action of this apparatus is as follows:—The ciliary vortices produced by the waves of the coronal disk, pass together through the upper sinus, and are hurled in one stream along the centre of the face, nearly to the projecting chin. Here is placed the orifice of the buccal funnel, a perpendicularly descending tube of considerable width, slightly funnel-shaped at the top, the interior surface of which is strongly ciliated. It descends straight upon the mastax, over the part where the unci unite. But just above this point there are two valves projecting from the walls of the tube, also well ciliated. These can be brought into contact, or separated in various degrees, at will, and being very sensitive, they regulate the force of the inflowing current, and doubtless exclude hurtful or useless substances. The current now flows along the two rami of the incus, as I have already described; and, passing between their separated points, descends into the œsophagus, a slender duct opening beneath them, and leading to the digesting stomach.

As this current passes, the manducatory apparatus acts upon the particles of food which it brings in its course. The quadrantic masses approach each other and recede, with a rapid rolling movement, in the direction of the curvature of the mallei; while, at the same time, the rami of the incus open and close their points, rise and sink, and occasionally perform a kind of shoveling action. The points of the fingers of the unci, meeting each other, doubtless pierce and tear the *Infusoria* swallowed, and the striated faces of the quadrantic masses bruise, squeeze, and grind them down.

When the muscular investiture is dissolved away by potash, the essential identity of the whole structure with that of the type already described becomes abundantly evident. Even the mallei, which in some aspects present difficulty, when viewed vertically, are but little changed; the fingers are parallel instead of divergent, and the handle-like character of the manubrium is lost; but three areas, enclosed by loops or carinæ of solid substance, reveal their true nature.

We will now see if we can separate the animal from its tube, so as to examime its lower parts. By a gentle pressure upon the bottom of the tube with the edge of a penknife, which I bring to bear upon it by the aid of this simple microscope, the creature is induced to wriggle out of his case. Replacing now the cover of the live-box, and placing it again on the stage of the compound microscope, we see that the lower part of the body forms a foot analogous to that of a *Brachionus*, covered with ring-like wrinkles, and separated from the body by an abrupt constriction and diminution of the diameter. At the very extremity there is a sort of sucking-disk, by which we may presume the hold of the animal upon the plant is maintained. No organic connexion subsists between the foot and the tube; for the latter is not an essential part of the animal, though absurdly called a lorica by Ehrenberg, but only an accumulation of mucus successively exuded from the body, and thrown off in the form which it possesses by the contractions and other movements of the body.

But see! the poor naked creature is writhing in contortions, which become more and more convulsive and spasmodic: and now it evinces great rigidity in these, till the body has become almost shapeless, portions of the surface being here and there violently forced out into projections, and the foot strongly curled up. The only signs of life that now remain are the occasional fitful workings of the jaws. Are we then to suppose that the shelter of the gelatinous case is needful to its continued existence, or did I inflict a mortal injury upon it when I laid the edge of my penknife upon its lower part to drive it forth? Most probably the latter is the true solution.

Out of the colony that remains, we will now select another specimen, with ripening eggs, in order to watch the development of the young. Here is one with three eggs lying obliquely in the tube, one of which is already showing the impatient movements of the embryo within. Ha! now the egg-shell has burst, and the little creature escapes from its prison, and quickly makes its way to the mouth of the parent-tube. Now it is free, and swims away rapidly, in a giddy headlong manner. It is quite unlike its mother; for its form is trumpet-shaped, resembling that of a *Stentor* with a wreath of cilia around the head, interrupted at two opposite points: the central portion of the head rises into a low cone. There is as yet no trace of the beautiful double-petalled flower.

It has been whirling giddily about the live-box for about a quarter of an hour, but now it begins to manifest tokens of weariness; or rather the time is approaching for it to select a place of permanent sedentary abode. Its motion is sensibly retarded; it now and then adheres to the glass momentarily, by its foot, and moves forward by successive jerks, not proceeding more than its own length at a time, and this apparently with effort. The periods of its remaining stationary become longer, so that you may suppose it finally settled two or three times, before its wanderings are quite over, some shock or alarm sending it off to a little distance again.

At length it wanders no more; its foot holds fast to the glass, and its movements are confined to whirlings round and round on this as a pivot, and to sudden contractions of length. Presently we see a very delicate film surrounding the point of attachment;—the first rudiment of the tube, a ring of mucus thrown off from the skin, and pressed down to the foot by the contractions of the body. Meanwhile, the ciliary crown is dividing itself into two, and now we see already the essential form and appearance of the mature animal, every moment developing its perfection.

CHAPTER XV.

WORMS.

An examination of the diverse modes in which locomotion is performed among animals, and the various organs and modifications of organs that subserve this important purpose, would form no uninteresting chapter in natural history. You have two feet, your dog has four; in the bird, two of these are converted into wings, with which it rises into the air; in the fish all of them are become fins, with which it strikes the water. But it is in the invertebrate classes that we discover the strongest variations. The Poulpe "flops" awkwardly but vigorously along, by the alternate contractions and expansions of the web that unites its arms; the Snail glides evenly over the herbage by means of its muscular disk; the Scallop leaps about by puffs of water driven from its appressed lips; the Lobster shoots several yards in a second by the blow of its tail upon the water; the Gossamer Spider floats among the clouds upon a balloon that it has spun from its own body; the Centipede winds slowly along upon a hundred pairs of feet; the Beetle darts like an arrow upon three; and the Butterfly sails on the atmosphere with those painted fans which are properly "aërial gills." How elegantly does the

Planaria swim by the undulation of its thin body, and the *Medusa* by the pumping forth of the water held within its umbrella! How wondrously does the *Echinus* glide along the side of the tank on its hundreds of sucking-disks! How beautiful, and at the same time how effective, are the ciliary wheels of the *Brachionus.*

I am now going to show you some other examples of travelling machinery in an humble and despised, but far from uninteresting class of animals,—the Worms. Here is an Earth-worm upon the garden-border. With what rapidity it winds along, and now it pokes its sharp nose into the ground, and now it has disappeared! If your eye could follow it, you would see that it makes its way through the compact earth not less easily nor less rapidly than it wound along the surface. If you take it into your hand, you perceive no feet, wings, fins, or limbs of any kind; only this long cylinder of soft flesh, divided into numerous successive rings, and tapering to each extremity. The very snout which you saw enter so easily into the substance of the soil, is no hard bony point, but formed of the same soft yielding flesh as the other parts. And yet with no other implement does the lithe worm penetrate whithersoever it will through the ground. How does it effect this?

The fineness of the point to which the muzzle can be drawn is the first essential. This can be so attenuated that the grains of adherent soil can readily be separated by it, when its action is that of the wedge. The body being drawn into the crevice thus made, the particles are separated still farther. Now another provision comes in; the whole surface of the skin secretes and throws off a quantity of tenacious mucus or slime, as you will immediately perceive if you handle the Worm; this has the double effect of causing the pressed particles of soil to adhere together, and then to form a cylindrical wall, of which they are the bricks, and the slime the mortar; and also of greasing, as it were, the whole interior of the burrow or passage thus made, so that the Worm can travel to and fro in it without impediment; while the fact that the slime is continually poured forth afresh prevents the least atom of earth from adhering to its body. This you have doubtless observed, or may observe in a moment, if you will take the trouble to thrust a spade into the ground and give it two or three shakes. You will presently see on all sides the alarmed Earth-worms coming swiftly to the surface, and will notice how perfectly sleek and clean they are.

But these contrivances are only accessories: we have not yet discovered the secret of the easy movement. The mere elongation of the snout is no explanation of the disappearance of the Worm in the burrow; for you will naturally and reasonably say that this elongation cannot extend beyond a certain limit; and what then? No further progress can be made unless the hinder parts of the body are, by contraction, drawn up towards the elongated front;—but what holds the front in place meanwhile? Why, when the muscles contract, does not the taper, wedge-like muzzle slip back and lose the ground it had gained?

This we will now look at. I take up this Worm and put it in a narrow glass cell, where we may watch its movements. It presently begins to elongate and contract its body vigorously, apparently alarmed at its unwonted position; and the mucus is thrown off in copious abundance. We apply a low microscopic power to it, and catch glimpses, now and again, as it writhes about, of a number of tiny points protruded and retracted with rhythmical symmetry through the skin. Its mobility precludes our discerning much more than that these points are very numerous, that they are arranged in four longitudinal lines, running along the ventral side of the animal,—two lines on each side,—and that in each line there is a point protruded from each of the many rings of which the Worm's body is made up.

In order to see a little more of these organs we must sacrifice a Worm; having killed it, and divided the body in the middle, I cut off, with sharp scissors, a small transverse portion, say two or three rings, and press the fragment between plates of glass. Now, with a higher magnifying power, we discern in the midst of the translucent flesh the points in question. They are not, however, single; but each protrusile organ consists of a pair of transparent, brittle, glassy rods, shaped like an italic *ſ*, of which the recurved points are directed backwards when thrust out from the skin.

The mode in which these assist the progression of the Worm is well described by Professor Rymer Jones. "The attenuated rings in the neighbourhood of the mouth are first insinuated between the particles of the earth, which, from their conical shape, they penetrate like a sharp wedge; in this position they are firmly retained by the numerous recurved spines appended to the different segments; the hinder parts of the body are then drawn forward by a longitudinal contraction of the whole animal—a movement which not only prepares the creature for advancing further into the soil, but by swelling out the anterior segments forcibly dilates the passage into which the head had been already thrust: the spines upon the hinder rings then take a firm hold upon the sides of the hole thus formed; and, preventing any retrograde movement, the head is again forced forward through the yielding mould; so that, by a repetition of the process, the animal is able to advance with the greatest apparent ease through substances which it would at first seem utterly impossible for so helpless a being to penetrate." *

Implements analogous to these are found in most of the animals of the class *Annelida*, to which the Earth-worm belongs. But in this creature you see them in their simplest form: it is to the aquatic Worms that you must look if you wish to see the amazing diversity, complexity, and delicacy of these organs. In these there are one or two pairs of "feet" on each ring, consisting of wart-like prominences, which are perforate and protrusile, and through the middle of which work a number of bristles (*setæ*), arranged in a radiating pencil, something like the hairs of a paint-brush. In this transparent and colourless little Nais from fresh water, you may see their form and arrangement; in complexity they present an

* Gen. Outline, p. 202.

advance upon the Earth-worm, for here there are some seven or eight bristles in each pencil, which radiate in the same plane, and are graduated in length; they are very slender, bent at the tip, and as transparent as if drawn out of spun glass. It is interesting to observe with what lightning-like rapidity they are thrust out and withdrawn in constant succession, as the body is ever lengthening and shortening.

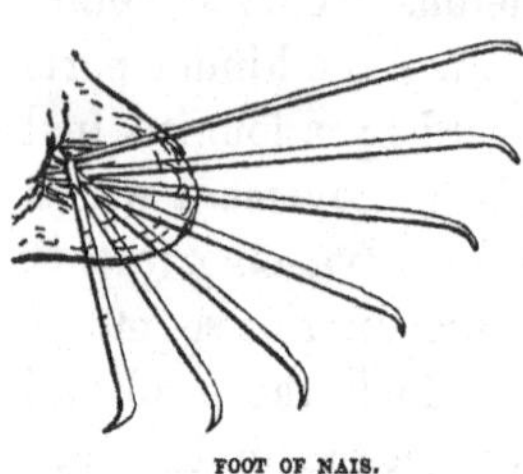
FOOT OF NAIS.

Let us exchange this little freshwater Worm for a marine one. Here is a *Polynöe*, a curious genus, very common under stones at low water on our rocky shores. It is remarkable on several accounts. All down the back we discover a set of oval or kidney-shaped plates, which are called the back-shields (*dorsal elytra*); these are flat, and are planted upon the back by little footstalks, set on near the margin of the under surface: they are arranged in two rows, overlapping each other at the edge. These kidney-shaped shields, which can be detached with slight violence, are studded over with little transparent oval bodies, set on short footstalks, which are perhaps delicate organs of touch. The intermediate antennæ, the tentacles, and the *cirri*, or filaments of the feet, are similarly fringed with these little appendages, which resemble the glands of certain plants, and have a most singular appearance. If we remove the shields, we discover, on each side of the body, a row of wart-like feet, from each of which project two bundles of spines of exquisite structure. The bundles, expanding on all sides, resemble so many sheaves of wheat, or you may more appropriately fancy you behold the armoury of some belligerent sea-fairy, with stacks of arms enough to accoutre a numerous host. But if you look closely at the weapons themselves, they rather resemble those which we are accustomed to wonder at in missionary museums,—the arms of some ingenious but barbarous people from the South Sea Islands,—than such as are used in civilised warfare. Here are long lances, made like scythe-blades, set on a staff, with a hook at the tip, as if to capture the fleeing foe and bring him within reach of the blade. Among them are others of similar shape, but with the edge cut into delicate slanting notches, which run along the sides of the blade like those on the edge of our reaping-hooks. These are chiefly the weapons of the lower bundle; those of the upper are still more imposing. The outmost are short curved clubs, armed with a row of shark's teeth to make them more fatal; these surround a cluster of spears, the long heads of which are furnished with a double row of the same appendages, and lengthened scymitars, the curved edges of which are cut into teeth like a saw. Though a stranger might think I had drawn copiously on my fancy for this description, I am sure, with your eye upon what is on the stage of the microscope at this moment, you will acknowledge that the resemblances are not at all forced or unnatural. To add to the effect, imagine that all these weapons are forged out of the clearest glass instead of steel; that the larger bundles may contain about fifty, and the smaller half as many, each; that there are four bundles on every segment, and that the body is composed of twenty-five such segments; and you will have a tolerable idea of the garniture and armature of this little Worm, that grubs about in the mud at low-water mark.

Should it ever be your fortune to fall in with a species of Sea-mouse (*Aphrodite hystrix*) which inhabits our southern coast a little way from the shore, you may be delighted and surprised with a modification of these organs, which exhibits a more than ordinarily obvious amount of creative forethought and skill. I will describe them in the words of the learned historians of these animals, MM. Audouin and Milne-Edwards:—

"The feet are divided into two very distinct branches, the lower of which is large, conical, of a yellowish-brown hue, and much shagreened on the surface. The upper branch is much less salient than the lower. We observe at the foot of the dorsal shields two bundles of rigid bristles: the one, expanded like a fan and applied upon the shields, is fixed immediately outside the insertion of those organs; the bristles which compose it are awl-shaped, without teeth, slightly curved, and directed inwards and backwards; their colour is a clear brown, with golden reflections. The second bundle is inserted more externally, on a tuberculous footstalk, and points horizontally backwards and outwards. The bristles which enter into its composition are very long, very strong, and terminated by a lance-shaped point, of which the edges are garnished with teeth curved backwards towards the base. These are veritable barbed arrows, having the extremities sometimes exposed, but often concealed in a sheath which is formed of two horny pieces, capable of opening and of closing again upon them.

"The use of these two valves it is not difficult to detect. They protect the points of the arrow, and permit the *Aphrodite* to receive them again into its body unharmed; whereas, without this precaution, the tissues which they traverse would be cut and mangled. But when these weapons are deeply plunged into a foreign body, as into the soft flesh of those animals which annoy the Worm, since the sheath does not penetrate with them but folds back, it follows that their teeth are inserted without any protection, and that on account of their backward direction they can be withdrawn only with great difficulty; thus, in most cases, the dart becomes broken; but the animal is furnished with so great a number, that these losses are scarcely felt, and there remain to it amply sufficient for its defence in all contingences." *

You will have noticed that the learned French zoologists seriously countenance the notion that these exquisitely elaborated organs are weapons of offence. But in this I think they are in error, misled by the resemblance, already alluded to, which they bear to weapons of human construction. The manner in which they act as implements of locomotion has been beautifully demonstrated by Dr. Williams in the Nereidous Worms, of which he observes that in

* Litt. de la France, ii. 71.

nearly all species the feet are constructed with express reference to progression on solid surfaces. In many instances, the bristle is compound, consisting of a staff with a variously armed point or blade jointed to its extremity. "Viewed by the light of mechanical principles, nothing can be so obvious as the reason why the *setæ* in these, as in nearly all other *Annelida*, are jointed. If they consisted of rigid, unbending levers, it is manifest that they would prove most awkward additions to the sides of the animals; if fixed too deeply in the surrounding soil, they would not act at all as levers; if too superficially, the Worm would be compressed in its tube at the moment when the *setæ* of the opposite feet would meet in a straight line. These difficulties are effectually and skilfully obviated by the introduction of a joint or a point of motion on each *seta*. This is one instance among many which the eye of the mechanician would detect in the organization of the *Annelida* in which Nature takes adroit advantage of mechanical principles in the attainment of her ends." *

Look now, in illustration of these principles, at the bristle-feet of this beautiful green *Phyllodoce.* No doubt you have often seen it in the little hollows of our rocky ledges, and especially on beds of young mussels, and probably you have admired the elegant ease with which its lithe and tortuous body writhes and winds, like a bit of green silken cord, in and out among the compactly crowded shells. You have wondered, too, at the difficulty which attends the attempt to take it up, not on account of the rapidity of its

* Rep. on Brit. Annelidæ, 211.

motions, but because of the extreme slenderness and slipperiness of the subject, and of the power which it possesses of insinuating itself into the smallest crevice.

The foot in this genus has but a single branch, and a single pencil of bristles, which is placed between the flat swimming leaf that ornaments each segment and the lower *cirrus*. The bristles are of the compound jointed form, but the joint is fixed in a peculiar manner. The basal portion is drawn out into a very slender long straight shaft, terminating in a knob somewhat resembling the end of a limb-bone. This is slit in one direction to receive the terminal piece, which is shaped somewhat like a lance-head, and is inserted into the slit exactly as a knife-blade is fixed into the haft. The head is in fact a knife-blade, with a thickened back and a very thin edge, which is notched into teeth of the most exquisite delicacy. The blade is slightly curved, and drawn out into a long acute point; and the whole bristle is formed out of an elastic horny substance (probably chitine) that rivals in transparency and brilliancy the purest flint-glass.

I might adduce a vast variety of examples of these organs in the Marine Worms, all of which would charm you by their elegance and by their extreme diversity; but I have other things to show you in this interesting class of animals, which fortunately are so common on all our shores that you will have no difficulty in procuring plenty of specimens for your private observation and study. And if you need intelligent guidance you cannot have a better mentor than Dr. Williams, whose admirable "Report on the British Annelidæ" I have just cited.

Before we dismiss our little *Phyllodoce* to its home in the aquarium, we must try to get a sight of its pretty mouth. The Worms are somewhat wayward in displaying this part of their charms, sometimes exposing it at intervals of a second or two for very many times in succession, at others sullenly keeping it closed; and no efforts that I am aware of on our part will induce the display: we must await their pleasure. It is, in fact, a turning of the throat inside out. In most of the Worms the head is minute, and what seems to be the mouth is but the orifice from which the throat or proboscis is everted. In the *Phyllodoces* this organ is a great muscular sac, in some species equalling in length one-fourth of the whole body.

Ha! there it appears! What a chasm yawns in the under side of the head, as the interior begins rapidly to protrude, turning inside out as it comes forth, like a living stocking, until it assumes the form of an enormous (*comparatively* enormous, of course) pear-shaped bag, the surface of which is beset with a multitude of secreting warts or glands, somewhat like the papillæ which stud the tongue in higher animals! The extremity, which is perforated, is surrounded by a muscle, by means of which it contracts forcibly on whatever it is applied to, and thus holds it firmly, while the re-inversion of the sac drags it into the body to be digested.

But this huge proboscis disappears as rapidly and as wonderfully as it was revealed. Commencing at what is now the outer extremity, which is quickly turned in, the whole swiftly returns to its cavity in the inverse order to that in which it was extended; and now that it is all engulphed, we marvel that so vast a sac can be packed away in so slender a case.

In this instance the armature of the proboscis is feeble; but we have species which are very elaborately armed. There is a minute species of *Lombrinereis*, which commonly appears in our aquaria after they have been some time established, and breeds in vast numbers on the floccose matter that clogs the bottom and sides. In this tiny Worm there is a formidable array of jaws, resembling black hooks, which we may discern through their pellucid tissues snapping and cutting viciously like so many pairs of hooked scissors. Though I have often had this little species in my tanks in copious abundance, I regret to say I cannot find any at this moment for our examination, and shall therefore content myself with translating for you MM. Audouin and Milne-Edwards' description of the jaws as they appear in a closely-allied form, but of far greater dimensions, *Eunice*.

"The proboscis is not very protrusile; when it is withdrawn its external orifice is longitudinal, and the jaws are fixed on each side, all facing the medial line. When it is projected, however, the two margins of the longitudinal cleft become transverse in separating, and the jaws follow the same movement, and diverge in the ratio of their forwardness. A kind of lower lip which is affixed to the under side of the proboscis is composed of two horny blades united towards their front extremity, and prolonged behind into points. The jaws are to the number of seven; three on the right and four on the left; the two upper ones are perfectly alike, and mutually opposed; they are large, narrow, pointed, recurved hook-wise at the tip, and jointed at their hinder ends on a double horny stem shorter than themselves. The second pair of jaws are large, broad and flat, mutually alike, and jointed on the lower side of the first pair; . . . their internal edge is straight and cut into deep teeth. The third pair are small, thin, concave, and notched; they are affixed by their inferior edge outside and in front of the second pair, which they conceal during repose. Finally, the supernumerary jaw, which is found on the left side only, is small, semi-circular, toothed, and placed between the second and third pairs. All these pieces are surpassed by the margin of the proboscis, which is often hard and black."*

From this complex and formidable mouth we will pass to one of quite another form, not less effective, perhaps more formidable, but ordained by the goodness of God to be a most valuable agent in the relief of human suffering. I mean the Medicinal Leech, of which we can readily procure a specimen from our friend the apothecary.

Here it is. There is no protrusile proboscis, but the throat is spacious, and capable of being everted to a slight degree. The front border of the mouth is enlarged so as to form a sort of upper lip, and this combines with the wrinkled muscular margin of the lower and lateral portions to form the sucker. With the dissecting scissors I slit down the ventral margin of the sucker, exposing the whole throat. Then, the edges being folded back, we see implanted in the

* Litt. de la France, ii. 138.

walls on the dorsal region of the cavity three white eminences of a cartilaginous texture, which rise to a sharp crescentic edge; they form a triangular, or rather a triradiate figure.

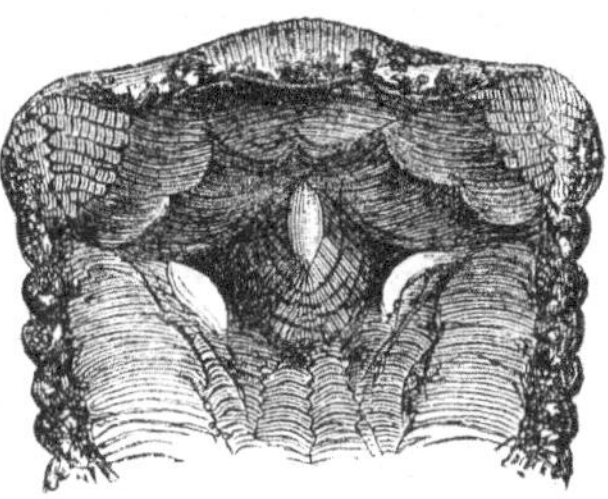
THROAT OF LEECH LAID OPEN.

Now, if you recollect, this is the figure of the cut made in the flesh wherever a Leech has sucked, as it is of the scar which remains after the wound has healed. For these three little eminences are the implements with which the animal, impelled by its blood-sucking instincts, effects its purpose. But to understand the action more perfectly, we must use higher powers.

I dissect out of the flesh, then, one of the white points, say the middle one, and laying it in water in the compressorium, flatten the drop, but use no more pressure than just enough for that. Now I apply a power of 150 diameters, and we will look at it in succession. You have under your eye a sub-pellucid mass, of an irregular oval figure, and of fibrous texture, one side of which is thinned away apparently to a keen edge of a somewhat semi-circular outline. But along this edge, and as it were imbedded into it for about one-third of their length, are set between seventy and eighty crystalline points, of highly refractive substance, resembling glass. These points gradually decrease in size towards one end of the series, and at length cease, leaving a portion of the cutting edge toothless. At the end where they are largest, they are nearly close together, but at length are separated by spaces equal to their own thickness. The manner in which they are inserted closely resembles, in this aspect, the implantation of the teeth in the jaw of a dolphin or crocodile.

But this appearance is illusory. By affixing the little jaw to the revolving needle, we bring the edge to face our eye. It is not an edge at all; but a narrow parallel-sided margin of considerable breadth. And the teeth are not conical points, as they seemed when we viewed them sidewise, but flat triangular plates, with a deep notch in their lower edge. Thus they partly embrace, and are partly inserted in, the margin of the jaw.

Observe now how beautifully this apparatus subserves the purpose for which it is intended. By means of its sucker, the Leech creates a vacuum upon a certain part of the skin, exactly like that produced by a cupping-glass. The skin covered is drawn into the hollow so far as to render it quite tense, by the pressure of the surrounding air. Thus it is brought into contact with the edges of the three jaws, to which, by means of powerful muscles attached to them, a see-saw motion is communicated, which causes the little teeth soon to cut through the skin and superficial vessels, from which the blood begins to flow. The issue of the vital fluid is then promoted by the pressure around, and so

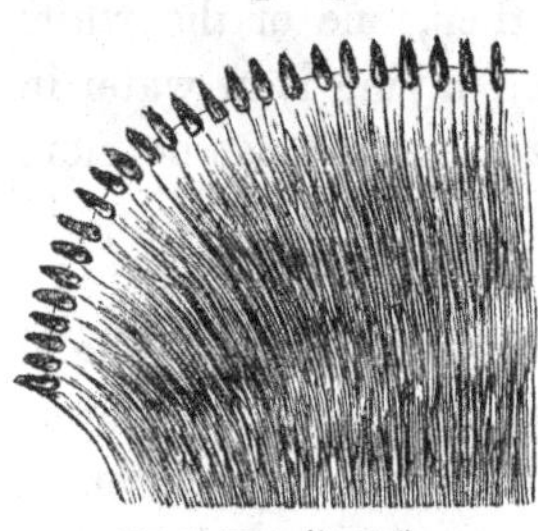
JAW OF LEECH (*in part*).

goes on until the enormous stomach of the Leech is distended to repletion.

It has been suggested that this whole contrivance, with the instinct by which it is accompanied, is intended for the benefit of Man, and not of the Leech. Blood seems to be by no means the natural food of the Leech; it has been ascertained to remain in the stomach for a whole twelvemonth without being digested, yet remaining fluid and sound during the entire period: while, ordinarily, such a substance cannot in one instance out of a thousand be swallowed by the animal in a state of nature. Whether this be so or not,—whether man's relief under suffering were the *sole* object designed, or not, it was certainly *one* object; and we may well be thankful to the mercy of God, who has ordained comfort through so strange an instrumentality.

The progress of marine natural history, as studied in the aquarium, has made our drawing-rooms and halls familiar with a multitude of curious and beautiful creatures which a few years ago were known only, and that very imperfectly, to the learned professors of technical science. Among the forms which embellish our tanks are several species of *Serpula*, and Worms allied to it. The shelly contorted tube which this painted Sea-worm inhabits, and which it has built up itself around its own body, with stone and cement which that body supplied, is well known to you; as is also the curious conical stopper with which it closes up its bottle as with a cork, when safe at home, and the lovely crown of gorgeously coloured fans which it expands when it takes ("the *air*," I was about to say, but rather) the *water*. You are familiar, too, with the lightning-like rapidity with which, while in health and vigour, the *Serpula*, on the slightest alarm, retreats into his fortress, taking care to clap the door to after him. But perhaps you have never had an opportunity of examining the mechanism by which this rapid flight is effected.

As there are two distinct movements performed by the Worm,—the slow and cautious and gradual protrusion, and the sudden and swift retreat,—so there are two distinct sets of organs by which they are performed. Shall I sacrifice one from this fine group to demonstrate the mechanism? Well, then, I carefully break the shelly tube, and extract the Worm uninjured.

Its form is, you perceive, much shorter and more dumpy than you would have supposed from looking at the tube; and it is somewhat flattened, having a back- and a belly-side. On the former there is a sort of shield, the sides of which bear wart-like feet, —about seven pairs in all,—which are perforated for the working of protrusile pencils of bristles, similar in structure and in function to those which we lately examined.

Here is one of the pencils extracted. To the naked eye it is a yellowish body with a satiny lustre; and this effect depends upon the light being reflected from a number of nearly parallel lines,—the staves of the spear-like bristles, which the eye cannot resolve in detail. A drop of the caustic solution of potash cleanses the bundle from the fleshy matter which would otherwise obscure the vision, and now I place it on the stage.

With this power of 400 diameters you see a multitude—some twenty or thirty, or more—of very long, slender, straight rods, of a clear yellowish horny substance, set side by side, like a sheaf of spears in an armoury. Each one merges, at its upper end, into a sort of blade, which is slightly bent, and which tapers

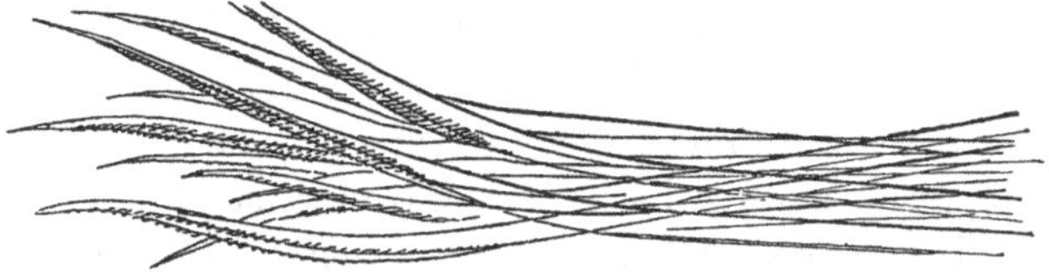

PUSHING-POLES OF SERPULA.

to an exceedingly fine point. But its chief peculiarity is that the blade has a double edge, not like a two-edged sword, the edges set on opposite faces, but on the same face, set side by side, with a groove between them; and each edge is cut with the most delicate and close-set teeth, the lines of which pass back upon the blade, as in our reaping-hooks.

These pencils of spear-like bristles are the organs by which the protrusion of the animal is performed. Their action is manifestly that of pushing against the walls of the interior, which on close examination are seen to be lined with a delicate membrane, exuded from the animal's skin. The opposite feet of one segment protrude the pencils of bristles, one on each side, the acute points and teeth of which penetrate and catch in the lining membrane; the segments behind this are now drawn up close, and extend their bristles; these catch in like manner; then an elongating movement takes place; the pencils of the anterior segments being now retracted, they yield to the movement and are pushed forward, while the others are held firm by the resistance of their holding bristles; thus gradually the foreparts of the animal are exposed.

But this gradual process would ill suit the necessity of a creature so sensitive to alarm, when it wishes to retreat. We have already seen how, with the fleetness of a thought, its beautiful crown of scarlet plumes disappears within its stony fastness: let us now look at the apparatus which effects this movement.

If you look again at this *Serpula* recently extracted you will find, with a lens, a pale yellow line running along the upper surface of each foot, transversely to the length of the body. This is the border of an excessively delicate membrane; and on placing it under a higher power (say 600 diameters) you will be astonished at the elaborate provision here made for prehension. This yellow line, which cannot be appreciated by the unassisted eye, is a muscular ribbon, over which stand up edgewise a multitude of what I will call combs, or rather sub-triangular plates. These have a wide base; and the apex of the triangle is curved over into an abrupt hook, and then this is cut into a number (from four to six) of sharp and long teeth. The plates stand side by side, parallel to each other, along the whole length of the ribbon, and there are muscular fibres seen affixed to the basal side of each plate, which doubtless give it independent motion. I have counted 136 plates on one ribbon; there are two ribbons on each thoracic segment, and there are seven such segments:—hence we may compute the total number of prehensile comb-like plates on this portion of the body to be about one thousand nine hundred, each of which is wielded by muscles at the will of the animal; while, as each plate carries on an average five teeth, there are nearly ten thousand teeth hooked into the lining membrane of the cell, when the animal chooses to descend. Even this, however, is very far short of the total number, because long ribbons of hooks of a similar structure, but of smaller dimensions, run across the abdominal segments, which are much more numerous than the thoracic. No wonder, with so many muscles wielding so many grappling hooks, that the retreat is so rapidly effected!

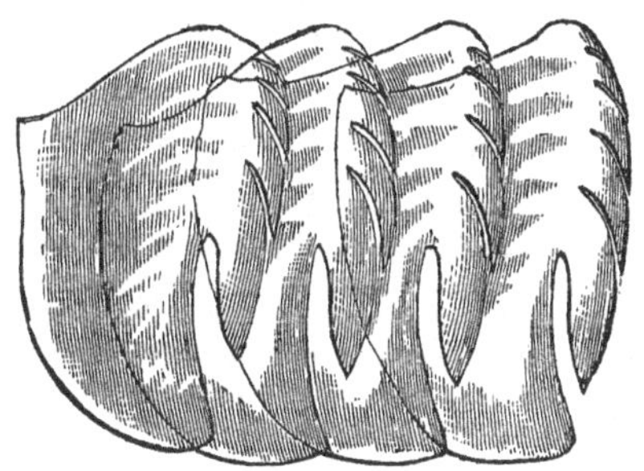

HOOKS OF SERPULA.

CHAPTER XVI.

SEA-URCHINS AND SEA-CUCUMBERS.

PEERING about among the rocks to-day at low-tide, I found, on turning over a large stone, an object which, though familiar enough to those who are conversant with the sea and its treasures, would surprise a curious observer fresh from the fields of Warwickshire. It is a ball, perfectly circular, and nearly globular,—only that its under part is a little flattened,—hard and shelly in its exterior, which is however densely clothed with a forest of shelly spines, each one of which has a limited amount of mobility on its base. On attempting to remove it, I find that it adheres to the stone with some firmness, and that on the exercise of sufficient force, it comes away with a feeling as if something were torn, and I find that a multitude of little fleshy points are left on the stone. Having dropped my prize into a glass collecting-jar of sea-water, I presently see that it is slowly marching up the side, sprawling out on every side a multitude of transparent hands, with which it seems to feel its way, and which are evidently feet also, for on these it crawls along at its own tortoise-pace. And I now see that it is the knobbed ends of some of these feet which were torn away by my forcible act of ejectment, and left clinging to the stone.

It was not the first time that I had seen the Sea-urchin (*Echinus miliaris*); and I might have passed it by with a feeling of satiated curiosity, had I not recollected our evening's amusement. Oh, ho! said I, what a fund of microscopic entertainment is inclosed in this stone box! So I brought it home, and now produce it as the text of our conversazione.

Every part is a wonder; but we must examine each in order. Take the spines first.

As we examine these organs on the animal crawling at ease over the bottom of a saucer of sea-water, using this triple lens, we see that each is a taper pillar, rounded at the summit, and swelling at the base, where it seems to be inserted into a fleshy pedestal, on which it freely moves, bending down in all directions, and describing a circle with its point, of which the base is the centre. Each spine is for the greater portion of its length of a delicate pea-green hue, but the terminal part is of a fine lilac or pale purple. The whole surface appears to be fluted, like an Ionic column, but this is an illusion, as you will see presently.

I now detach one of the spines, cutting it off with fine-pointed scissors as near the base as I can reach. I put it with as little delay as possible into the live-box, and examine it with a high power, say 600 diameters. Look at it. You see the ciliary currents very distinctly; and if you move the stage so as to bring the basal portion into view, you may discern even the cilia themselves, very numerous and short, quivering with a rapid movement. The currents are not longitudinal, but transverse, and somewhat peculiar. The floating atoms which come within their vortex are drawn in at right angles to the axis of the spine, and are presently hurled away in the same plane; forming a circle, whose plane is perpendicular to the direction of the spine. The surface upon which these cilia are set is a transparent gelatinous skin, of extreme tenuity, stretched tightly over the solid portion, which it completely covers, and studded with minute oval orange-coloured grains.

The substance of which the spines are composed is best seen by crushing a few of these organs into fragments. We now see a texture beautifully delicate; they are formed of calcareous substance as transparent as glass, and reflecting the light like that material; hard but very brittle; clear and solid, with a fibrous appearance in some parts, but in others excavated into innumerable smooth rounded cavities which join each other in all possible ways. It is to this structure that the spine owes its strength, its lightness, and its brittleness.

This arrangement of the calcareous deposit in a sort of glass full of minute inter-communicating hollows is very peculiar, but it is invariably found in the solid parts of this class of animals; so that the experienced naturalist, on being presented with the minutest fragment of solid substance, would, by testing it with his microscope, be able at once to affirm with certainty, whether it had belonged to an *Echinoderm* or not. And this uniformity obtains in all the diverse forms which the animals assume, and in all the various organs which are strengthened by calcareous deposits, —Crinoid, Brittle-star, Five-finger, Urchin, Sea-gherkin, or Synapta;—ray, plate, spine, sucker-disk, lantern, pedicellaria, dumb-bell, wheel, or skin-anchor,—whenever we find calcareous matter, we invariably find it honey-combed, and eroded, as it were, in this remarkable fashion.

Dr. Carpenter has described this texture so well, that I shall not apologise for quoting his words to you, especially as you will have an opportunity here of testing their correctness, by personal observation. "It is," he remarks, "in the structure of that calcareous skeleton, which probably exists, under some form or other, in every member of this class, that the microscopist finds most to interest him. This attains its highest development in the *Echinida*, in which it forms a box-like shell, or 'test,' composed of numerous polygonal plates jointed to each other with great exactness, and beset on its external surface with 'spines,' which may have the form of prickles of no great length, or may be stout, club-shaped bodies, or, again, may be very long and slender rods. The intimate structure of the shell is everywhere the same: for it is composed of a network, which consists of carbonate of lime, with a very small quantity of animal matter as a basis, and which extends in every direction (*i.e.* in thickness, as well as in length and breadth), its areolæ or interspaces freely communicating with each other. These 'areolæ,' and the solid structure which surrounds them, may bear an extremely variable proportion one to the other; so that, in two masses of equal size, the one or the other may greatly predominate; and the texture may have either a remarkable lightness and porosity, if the network be a very open one, or may possess a considerable degree of compactness if the solid portion be strengthened. Generally speaking, the different layers of this network, which are connected together by pillars that pass from one to the other in a direction perpendicular to their plane, are so arranged that the perforations in one shall correspond to the intermediate solid structure in the next, and their transparency is such, that when we are examining a section thin enough to contain two or three such layers, it is easy, by properly 'focussing' the microscope, to bring either one of them into distinct view. From this very simple but very beautiful arrangement, it comes to pass that the plates of which the entire 'test' is made up, possess a very considerable degree of strength, notwithstanding that their porousness is such, that if a portion of a fractured edge, or any other part from which the investing membrane has been removed, be laid upon fluid of almost any description, this will be rapidly sucked up into its substance." *

To return, however, to our spine. When we look at it laterally, the appearance is such that we cannot but firmly believe, that it is grooved throughout with straight and deep longitudinal furrows. But if we break off the same spine transversely, and so exhibit it that the broken end shall be presented to the eye, we perceive that there are no grooves; but that the points in the circumference, which seemed to be the summits of the ridges, which are very narrow, are really lower than the intermediate spaces, which we supposed to be the grooves, and that the surface of these spaces is really convex in a slight degree.

* The Microscope, p. 553.

The explanation of these contradictory appearances is easily given. Meanwhile, however, they read an important lesson to the inexperienced microscopist, not to decide too hastily on the character of a surface or a structure, from one aspect merely. So many are the chances of illusion, that the student should always seek to view his subject in different aspects, and under varying conditions of light, position, &c.

It is by making a thin transverse section of a spine, —cutting off a slice of it, to speak in homely phrase, —that we shall demonstrate the structure, which is very beautiful. This is an operation requiring much delicacy and practice, and implements for the special purpose; and hence it is best performed by professional persons, who prepare microscopic objects for sale. You may see such a section, however, on this slide; but I do not know whether the spine belongs to the species we are examining.

The whole central portion is formed of the sponge-like calcareous matter, which, from the variously reflected and refracted rays of light, appears nearly opaque, and of a bluish colour by transmitted light. This structure sends forth radiating points (making longitudinal ridges, of course, in the perfect spine); and it is the opacity of these points (or ridges) which reach the circumference, that gives to the spine the appearance of being fluted. Indeed it would be fluted if this were the entire structure; but the open space left between these projecting radii is filled with the solid glassy matter, having, as we see, a convex surface. This, however, from its perfect transparency, is not seen when we look at the side of the spine, the eye going down to the bottom of the interspace. The spine is, in fact, a fluted column of spongy glass, with the grooves filled with solid glass.

We have not yet seen, however, the beautiful mechanism appropriated to the movement of these spines. You can hardly see this to advantage in the living animal, but here is the entire shelly box of a dead *Echinus*, on which, while for the most part the surface is denuded of spines, a few dozen remain sufficiently attached to show what I wish to demonstrate, viz. the mode of articulation. You observe that the whole globose shell is covered with tiny knobs, differing in size, and not set in very regular, or at least not very obvious order, but showing a tendency to run in lines from pole to pole of the globe. Giving attention to one of the larger of these knobs, under a lens it is seen to be a hemispherical eminence on the shell, the very summit of which is crowned by a tiny nipple of polished whiteness, resembling ivory. Now if we carefully lift one of the still remaining spines from its attachment, which in its present dried state is so fragile that the slightest touch is sufficient for the purpose, we shall note that its base rests on this tiny nipple; and on turning it up, and bringing the magnifying power to bear upon its base, we see that

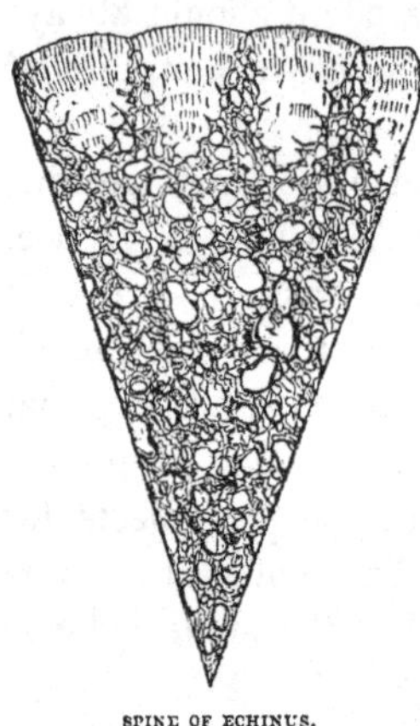

SPINE OF ECHINUS.
Segment of Section.

this is excavated with a hollow, whose dimensions exactly correspond with those of the nipple. It is indeed a true "ball and socket" joint, like that of the human hip or shoulder, and is surrounded by a capsular ligament to keep it in place, the muscles which sway the spine from side to side and cause it to rotate being inserted outside the capsule. Professor Edward Forbes calculates that upon a large *Echinus*, such as this dried specimen of *E. sphæra*, there are more than four thousand spines, every one of which has the structure, the mechanism, and the movements that we have been examining. Well may he say, that "truly the skill of the Great Architect of Nature is not less displayed in the construction of a Sea-urchin than in the building up of a world!"

To return now to our little *E. miliaris*, which has been all this time coursing round and round his saucer, wondering, perchance, at the narrowness and shallowness of the White Sea in which he finds himself. Again we peer, lens to eye, over the bristling surface, and discern, shooting up amidst the spines, and almost as thickly crowded as they, multitudes of the tiny organs which have caused so much doubt and discussion among naturalists. Müller, the great marine zoologist of Denmark, who first discovered them, thought them parasitic animals, living piratically upon the unwilling Urchin, and accordingly gave them generic and specific names. The term *Pedicellaria*, which he assigned to his supposed genus, is that by which modern naturalists have agreed to call them still, though the word is not now used in a generic sense, since it is indubitably established that they are not independent animals, but essential parts of the Urchin itself. Müller described three distinct sorts, and I have added a fourth to the number; they are named *P. triphylla*, *tridens*, *globifera*, and *stereophylla*. They all agree in these particulars:—that each has a long, slender, cylindrical, fleshy stem, through the centre of which runs an axis or rod of calcareous substance; that the base of the stem rests on the skin of the Urchin; that on the summit is placed a head consisting of three pieces, which are capable of being widely opened and of being closed together, at least at their tips; that the edges of these pieces, which come into mutual contact, are furnished with teeth, which lock into each other; that the head-pieces (like the stem) consist of calcareous centres, clothed with flesh; that, besides the opening and shutting of the head, the stem can be swayed from side to side; and that all these movements are spontaneous, and apparently voluntary. It appears that the head-pieces close on any object presented to them, such as the point of a needle, and hold with considerable force and tenacity, so that the *Pedicellaria* may be drawn out of the water without relaxing its grasp.

Looking at one of the first-named kind, the *Pedicellaria triphylla*, of this *Echinus miliaris*, we see that it consists of three broad and thick sub-triangular pieces, jointed into a head, set on a thickish stem of transparent gelatinous fibrous substance, through which a slender core of calcareous matter runs that looks fibrous and blue. The three moveable pieces or blades are convex externally, concave internally;

thin in substance, furnished along their opposing or concave sides with two longitudinal ridges or keels, each of which is cut into the most beautifully fine teeth, so that the edge of each ridge looks like a shark's tooth; the edges of the pieces are also similarly toothed: these shut precisely into each other.

In the larger *E. sphæra*, the head-blades of this kind have one stout central ridge, which is rounded and not toothed. It forms the front of a great interior cavity which opens by two orifices on each side of the column.

The moveable pieces inclose a skeleton of calcareous substance, glassy, colourless, and brittle, in which, according to the plan I have already described, are excavated a multitude of oval cavities which form irregular rows; a central line runs down each piece, that is solid and free from cavities. This calcareous skeleton is encased in a gelatinous flesh, similar to and continuous with that of the stalk.

This is the smallest kind, the head being about $\frac{1}{56}$th of an inch in height.

Considerable modifications are found to exist in the details of each form, in the relative proportions which the parts bear to each other, and so forth; so that two forms, which in their extreme conditions widely differ, mutually approach, and appear to run into each other. This is the case with the present, and with the form which I will now show you.

P. tridens is much larger than any of the other forms, the moveable head being about $\frac{1}{20}$th of an inch in length, and the whole organ about $\frac{1}{8}$th of an inch.

This may be considered as essentially *P. triphylla*, modified by the blades being greatly drawn out in length, and at the same time rendered quite slender, so that they may be called pins; they meet only at the points, where they often cross, the interspaces of the basal parts being open. The inner edges of these are notched with teeth as in *P. triphylla*, of which those near the tips are larger and cut into subordinate teeth of exquisite minuteness.

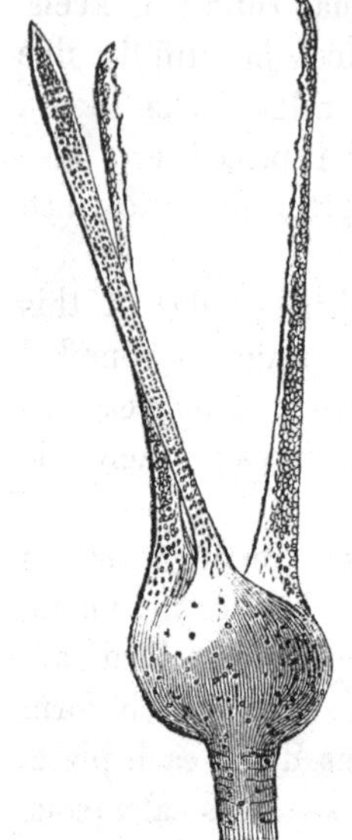
HEAD OF PEDICELLARIA TRIDENS.

We have here an opportunity of seeing that the oval or square markings, which are thickly placed throughout the calcareous substance of the blades, are certainly cavities in it; for in those examples in which the pins, which are very brittle, are broken, the edge of the fracture is not even, but jagged with holes exactly corresponding with the marks in question; so that the structure is the same as that of the spines and of all the other solid parts of the Urchin.

We will now examine some specimens of *P. tridens*, treated with potash, which will enable us to see the calcareous support better. The head-blades expand at the base into three-sided prisms or pyramids, each of the two interior sides of which is indented with a large cavity, leaving a projecting dividing ridge, armed with teeth somewhat remote from each other. The one exterior angle is toothed in a corresponding manner, but the opposite angle appears plain. The angle of one blade-base fits into the cavity of its neighbour; and, so far as I have observed, when the two edges thus overlap, it is the toothed one that is on the outside. Looking from the circumference towards the centre of the head, it is the left angle that is toothed and external, the right being plain and sheathed. This observation, however, applies only to *E. miliaris;* for, in the corresponding organs of *E. sphæra,* both sides of the trigonal base appear untoothed, except close to the bottom, where a deep notch indents each margin.

Viewed from beneath, the head assumes an outline which is rondo-triangular; but yet such that each side of the triangle has a very obtuse projecting angle in the middle, where the blade-bases meet each other. They fit accurately, and each has a deep oblong cavity in its bottom, which does not, as I conceive, communicate with the interior.

By selecting one of these heads, which has been divested of its fleshy parts by immersion in caustic potash, and then well cleansed by soaking in clean water, and placing it under a low power of the microscope—100 diameters, for example—with a dark ground, and the light of the lamp cast strongly upon it by means of the Lieberkuhn, or the side-condenser, we shall have an object of most exquisite beauty. The material has all the transparency and sparkling brilliance of flint-glass, while the elegantly-shaped pins, the perfect symmetry of the prismatic bases, the arch which is lightly thrown across their cavity, the minute teeth of the tips locking accurately into each other, and the oval cavities in the whole structure set in regular rows, and reflecting the light from thousands of points, constitute a spectacle which cannot fail to elicit your admiration.

P. globifera is formed on the same model as *P. triphylla*, but is more globose, and each piece appears to have a deep cleft at the point, which does not extend to the interior side, where a thick ridge runs down from the point to the base. At the summit of this ridge, in each of the three divisions, there is set a strong acute spine, directly horizontally inwards, so that the three cross each other when the blades close, which they do energetically,—a formidable apparatus of prehension! The stem is much more slender than in *P. triphylla*, and the height of the head of one of average size is only $\frac{1}{43}$d of an inch. It is peculiar also in being slender throughout, and in having the knobbed calcareous stalk extending up to the head, which appears to work on it. In each of the other sorts the stalk extends only through a part of the distance, above which the investing fleshy neck becomes wider and empty.

But the internal structure is not quite the same as in the others. The main portion of the head is composed of gelatinous flesh; the calcareous support being reduced to that ridge which runs up the interior side of the blade. This is somewhat bottle-shaped, with a bulbous base, and a long slender neck, with two edges on the inner face, which are armed with horizontal hooked spines, some of which are double, and the whole terminates in a sort of ring, formed by the last pair of spines, which unite into the acute horizontal point that I have already mentioned. The skeleton is filled with oval cavities, like that of the others.

The fourth kind of *Pedicellaria*, which I call *P. stereophylla*, is quite distinct from either of the others. It is very minute, the head being only $\frac{1}{200}$th of an inch in height. The head is a prolate solid spheroid, cut into three segments, exactly as if an orange were divided by three perpendicular incisions meeting at the centre. Thus the blades meet accurately in every part when closed, but expand to a horizontal condition. These are almost entirely calcareous, being invested but thinly with the gelatinous flesh. They are filled with the usual oval cavities, set in sub-parallel arched series.

The head is set on a hollow gelatinous neck nearly as wide as itself, and thrown into numerous annular wrinkles; its walls are comparatively thin, disclosing a wide cavity, apparently quite empty, as the blue calcareous stem extends only half-way from the base to the head. At this point the neck contracts rather abruptly, and continues to the base, but just wide enough to invest the stem.

This sort is confined, so far as I have seen, to the ovarian plates and their vicinity, where they are numerous.

Thus these tiny organs, so totally unlike anything with which we may parallel them in other classes of animals, do not merely afford us amusement, and delight us by their elegance of shape and sparkling beauty, the variety and singularity of their forms, the elaborateness of their structure and the perfection of their mechanism, but excite our marvel as to what can be the object which they subserve in the economy of the creature,—what purpose can be fulfilled by so many hundreds of organs so singular, and scattered over the whole surface of the shelly body.

It is very difficult to answer this question. The only organs with which they can be compared are the singular "birds' heads" in so many of the *Polyzoa*, which we looked at some time ago. But, unfortunately, a like mystery enshrouds the use of those processes, and the only light that we have as yet upon either form is that of dim conjecture. It has been supposed that, in both cases, the function of the prehensile forceps is to seize minute animalcules or floating atoms of food, and pass them to the mouth: but the supposition is involved in great difficulties; as the organs, however fitted for prehension, seem peculiarly unsuited for transmitting objects; besides that the great majority of them are placed very remote from the mouth. I can only repeat the conjecture which I have hazarded in the case of the Polyzoan "birds' heads," viz. that the *Pedicellariæ* are intended to seize minute animals, and to hold them till they die and decompose, as baits to attract clouds of *Infusoria*, which, multiplying in the vicinity of the Urchin, may afford it an abundant supply of food.

There is yet another series of organs which stretch out from every part of the periphery of this living box; scarcely less numerous than either the spines or *Pedicellariæ*, but very different from both. They are what I alluded to just now as the feet. Let us pay a moment's attention to their appearance and action, before we examine their structure.

We see, then, extending from various points of the shelly case of the Urchin, and reaching to twice or thrice the length of the longest spines, slender pellucid tubes, slightly tapering towards their free extremity, which then abruptly dilates into a hemispherical knob, with a flat end. These very delicate organs are extended or contracted at the will of the animal, turned in every direction, waved hither and thither, and evidently have the faculty of adhering very firmly by their dilated tips to any object to which they are applied.

So much we can discern as we watch the creature disporting in this vessel of water; but we will now endeavour to learn a little more about its structure and economy. Selecting for this purpose a sucker which is extended to great length, I snip it across with a pair of sharp scissors, as near the base as I can. Mark the result. The terminal knob which was attached to the bottom of the saucer maintains its hold, but the tube has suddenly shrunk up to a sixth part of its former length, exchanging at the same time

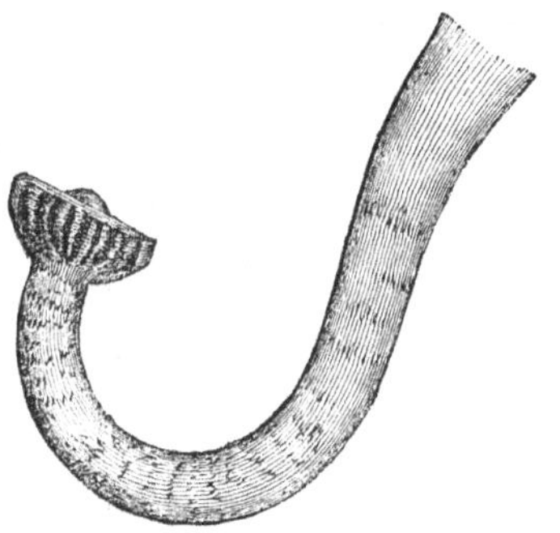

SUCKER OF URCHIN.

its smooth slenderness and translucency for a corrugated semi-opacity. I push the knob aside with a needle's point and thus destroy its adhesion; which done, I take up the severed and shrunken sucker, and lay it in a little sea-water in the live-box.

Under a power of 180 diameters we see that the tube is composed of two series of muscular fibres, the one set running lengthwise, the other transversely or annularly; the former by their contraction diminishing the length of the tube, the latter diminishing its calibre. The muscular walls are covered with a transparent skin, studded with round orange-coloured spots, perhaps glandular, exactly similar to those we saw on the exterior of the spines and *Pedicellariæ*.

Now, to illustrate the action of these tubular feet, I must again have recourse to the denuded shell of a preserved *Echinus*. Taking this globose empty box into your hand, hold it up against the light; looking in at the large orifice, which was once occupied by the mouth;—you see that the whole shell is pierced with minute holes—pores, which are arranged in ten longitudinal or meridional lines, associated so as to make five pairs of lines. Now with a lens scrutinize more minutely a portion of any one of these lines, and you discern that it is composed of a multitude of pores, which have a peculiar order of arrange-

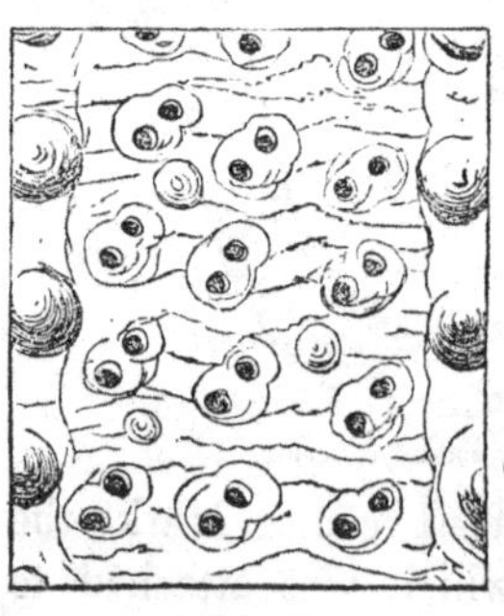

PORES OF URCHIN.

ment among themselves; that is to say, they form minor rows which cross, obliquely or diagonally, the course of the meridional line. These rows are themselves double, the pores running in pairs, not however with mathematical symmetry. In this species, there are three pairs of pores in each row, and so there are in the one which I have here alive, but in other of our native species the rows consist of five pairs.

These pores are intimately connected with the tubular feet, each of which springs from a portion of the shell that is perforated with a pair of pores; so that the cavity of every tube communicates with the interior of the shelly box by two orifices.

Now on the interior side of these two pores,—that is, within the cavity of the shell,—there is placed a little membranous, or rather muscular, bladder, filled with a fluid which is not materially different from sea-water. There is a free communication between the bladder within and the tube without the shell, by means of the pair of pores, through which the fluid passes. By means of the muscular fibres, which are under the control of the Urchin's will, any portion of this double vessel can be contracted to a certain extent. Suppose it is the interior bladder; the effect of the contraction of its walls is to diminish its capacity, and the contained fluid is forced through the pores into the tube without. The longitudinal fibres of this part being at the same moment relaxed, the tube is lengthened, because of the injected water. Suppose, now, in turn, the fibres of the tube contract, while those of the bladder relax; the fluid is driven back, the bladder dilates, and the tube shortens, until, if the animal so please, its swollen tip is brought close up to the pores. By mechanism so beautiful and simple is the prolongation or abbreviation of these organs effected.

We noticed, however, that the extremities of the tubes had an adhesive power, which faculty it is that constitutes them feet. They are prehensile, and thus they afford, as we observed in the living Urchin, the means by which it takes hold of even a smooth and vertical surface, as the side of a glass tank, and drags up its body thereby.

Putting, now, the extremity of this cut-off tube under graduated pressure, having first applied to it a drop of caustic potash, we see that it carries a beautiful glassy plate of extreme thinness, which lies free in the swollen cavity of the termination of the tube. This plate is circular in form, apparently notched at the margin, and cut with four or five (for the number varies) incisions, which reach almost to the centre. The substance is formed of the common clear brittle calcareous matter of the skeleton, hollowed into numberless cavities, according to the general plan. The centre is perforated with a larger round orifice. The appearance of marginal notching is deceptive; and indicates a structure analogous to what we see in the spine. The notched line indicates the extent of the spongy structure; but beyond this the plate extends into a perfectly circular smooth edge, but is constituted of a layer of calcareous substance so thin that there is no room for the ordinary cavities within it.

The round aperture in the centre plays an important part in the function of the organ. The foot adheres on the same principle as that by which children take up large flat stones with a piece of wetted leather, to the middle of which a string is attached. The boy drops his sucker on the stone, and treads firmly on it, to bring it into close contact with the surface; then he pulls at the string perpendicularly, by which the central part of the leather is lifted a little way from the stone, leaving a vacuum there; since the contact of the edges with the stone is so perfect that no air can find entrance between them. Now the pressure of the atmosphere upon the leather is so great that a considerable weight, perhaps half-a-dozen pounds, may be lifted by the string before the union yields.

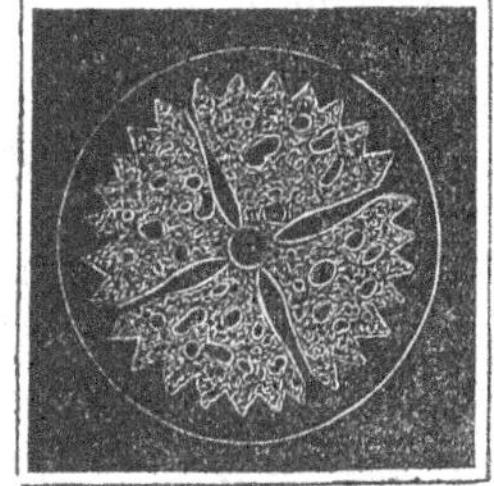

SUCKER-PLATE OF URCHIN.

Well, the very counterpart of this amusing operation is repeated by the clever "Urchin" whose performances we are considering. The tube is his string; the dilated end with the plate in it his leather; his muscular power acts like the other urchin's tread, to press the bottom of the sucker against the surface of the rock. Then he pulls the string; in other words, he drags inwards the centre of the muscular bottom of the sucker, which is, as it were, sucked up into the central orifice of the plate. Thus a vacuum is formed beneath the middle of the sucker, on which the weight of the incumbent water and atmosphere united presses with a force far more than sufficient to resist the weight of his body, when he drags upon it, and, as it were, *warps himself up* to the adhering point.

Here is in my cabinet a specimen of a Sea-urchin of a less regular form: it is the Heart-Urchin (*Amphidotus cordatus*). Essentially, its structure agrees with that of the more globular forms, but it is heart-shaped, and the two orifices, instead of being at opposite poles, are separated only by about one-third of the circumference. It shows also singular impressed marks on its shell, as if made by a seal on a plastic substance.

But what I chiefly wish to direct your attention to are the spines. These differ much from the kindred organs in *Echinus*, being far more numerous, very slender, curved, thickening towards the tip, and lying down upon the shell in the manner of hair, whence the species is sometimes called the Hairy Sea-egg. The array of spines has a glittering silky appearance in this dried state.

We will now put a few of them under a low power of the microscope, using reflected light and a dark background. They thus present a very beautiful appearance; elegantly-formed curved clubs, made of a substance which seems to be between glass and ivory, having the whiteness of the latter and the glittering brilliance of the former. The entire surface appears to be exquisitely carved, with excessively minute oval pits, arranged in close-set lines, with the most charming regularity. It is the light reflected from the polished bottoms of these pits that imparts to the surface its sparkling brilliancy. At the bottom of the spine there is a little depression, which fits a tiny nipple on a wart-like prominence of the shell, as we saw in *Echinus;* but a little way above this point there is a singular projection or shoulder of the calcareous substance, which is set on at a very oblique angle with the axis of the spine, reminding one, as we look at the spine laterally, of the budding tines on the horn of a young deer.

At first, perhaps, you are at a loss to know what purpose this shoulder can serve; but by turning to the shell, and carefully observing the spines in their natural connexion with it, you will observe that the obliquity of its position accurately corresponds with the angle which the individual spines form with the surface of the shell from which they spring; and that the shoulder has its plane exactly parallel with the latter, but raised a little way above it. Now the entire shell, during life, was clothed with a living vascular flesh, having a thickness exactly corresponding to the distance of the shoulder from the shell. This shoulder, then, was an attachment for the muscular bands, whose office it was to move the spine to and fro; the projection affording the muscles a much better *purchase*, or power, than they could have had if they had been inserted into the slender stem itself.

The tubercles on the shell show a structure which corresponds with this. They are very minute; but each of them is regularly formed, and is crowned with its little polished nipple, on which, as I have said, the spine works, as by a ball-and-socket joint. These are arranged with perfect regularity in *quincunx*, and by close examination you will see that each is inclosed in a little area formed by a very low and narrow ridge of the shell, which makes a network. On the lateral portions of the under surface the meshes of this net are particularly conspicuous, and we see that they constitute shallow hexagonal cells, in the midst of which is seated the tubercle; yet not in the exact centre either, but nearer the front than the back of the area inclosed.

Now this elevated ridge affords, doubtless, the insertion of the other end of the muscles that move the spine; the ridge giving a better purchase than a flat surface, as the keel on the breastbone of birds is deep in proportion to the vigour of the muscles used for flight. And, surely, the apparently trivial fact that the space behind the tubercle is greater than that in front, is not without significance, since it implies a thicker muscle at that part, which accords with the circumstance that such would be the insertion of the muscle-band whose contraction produces the *outward* stroke by which the sand is forced away from the bed.

But what is the need of so much care being bestowed upon the separate motion of these thousands of hair-like spines, that each individual one should have a special structure with special muscles, for its individual movement? The hairs of our head we cannot move individually: why should the Heart-Urchin move his? Truly, these hairs are the feet with which he moves. The animal inhabits the sand at the bottom of the sea in our shallow bays, and burrows in it. By going carefully, with the lens at your eye, over the shell, you perceive that the spines, though all formed on a common model, differ considerably in the detail of their form. I have shown you what may be considered the average shape; but in some, especially the finer ones that clothe the sides, the club is slender and pointed; in others, as in those behind the mouth, which are the largest and coarsest of all, the club is dilated into a long flat spoon; while in the long, much-bowed spines which densely crowd upon the back, the form is almost uniformly taper throughout and pointed.

The animal sinks into the sand mouth downwards. The broad spoons behind the mouth come first into requisition, and scoop away the sand, each acting individually, and throwing it outwards. Observe how beautifully they are arranged for this purpose! diverging from the median line, with the curve backwards and outwards. Similar is the arrangement of the slender side-spines; their curve is still more backwards, the tips arching uniformly outwards. They take, indeed, exactly the curve which the fore-paws of a mole possess,—only in a retrograde direction, since the Urchin sinks backwards,—which has been shown to be so effective for the excavating of the soil, and the throwing of it outwards. Finally, the long spines on the back are suited to reach the sand on each side, when the creature has descended to its depth, and by their motions work it inward again, covering and concealing the industrious and effective miner.

Thus we have another instance added to the ten thousand times ten thousand, of the wondrous wisdom of God displayed in the least and most obscure things. "All thy works shall praise thee, O Lord!" (Ps. cxlv. 10.)

There is an order of animals which naturalists put in the same category as the Sea-urchins, but which an unscientific observer would regard as possessing little or no affinity with them. Some are like long, soft, and fleshy worms, and others, which come the nearest to the creatures we have been looking at, have still the lengthened form, which, however, so closely resembles that of a warty angled cucumber that the animals I allude to are familiarly called Sea-cucumbers (*Holothuriadæ*). The marine zoologist frequently finds them beneath stones at extreme low water, and larger forms—as big in every direction as a marketable cucumber—are occasionally scraped from the bottom of the deep sea by means of that useful instrument, the dredge. If you drop one of them into sea-water you will presently see from one extremity an exquisite array unfold, like a beautifully cut flower of many petals, or, rather, a star of ramifying plumes. Soon it begins to climb the walls of your aquarium, and then you catch the first glimpse of its affinity to the Urchins; for the short warts which run in longitudinal lines down the body, corresponding to the angles, gradually lengthen themselves, and are soon perceived to be sucking-feet, analogous in structure and in function to those with which the Star-fish and Sea-urchin creep along.

But the relationship becomes more apparent still when we find that the Cucumber has a skeleton of calcareous substance deposited on exactly the same plan as in the Urchin, viz., around insulated rounded cavities. It is true you may cut open the animal and find nothing at all more solid than the somewhat tough and leathery skin; but a calcareous skeleton is there notwithstanding, though in truth only a rudimentary one. If we were to cut off a considerable fragment of the skin, and spread it out to dry upon a plate of glass, and then cover it with Canada balsam, we should find—assisted by the translucency which is communicated to the tissues by the balsam—that the skin is filled with scattered atoms of the calcareous structure, perfectly agreeing with that with which the solid framework of the Urchin is built up, but minute and isolated in the flesh, instead of being united into one or more masses of definite organic form.

But the atoms I speak of are still more perfectly seen by dissolving the piece of skin in boiling potash, and washing the sediment twice or thrice in pure water; this may then be spread upon a glass slide, and covered with a plate of thin glass, when it forms an interesting and permanent object for study. I have here a slide which is the result of such treatment to the naked eye it appears sprinkled with the finest dust, but under magnifying power it is seen to consist of numberless calcareous bodies, of great beauty, and very free from extraneous matter.

The elegance of the forms is remarkable, and also their uniformity; for though there do occur here and there among them plates of no regular shape, perforated with large or small roundish orifices, yet the overwhelming majority are of one form, subject to slight modifications, in shape and size.

Neglecting, then, the irregular pieces, we perceive

that the normal form is an oval of open work, built up by the repetition of a single element. That element is a piece of clear glassy material, highly refractive, of the shape of a dumb-bell—two globes united by a thick, short column. The oval is constructed thus:—suppose two dumb-bells to be placed in contact, side by side, and soldered together, there would be of course an oval aperture between their columns. Then two other dumb-bells are united to these in a similar manner, but one on each side, so that the globes of each shall rest in the valley between the former globes now united. These then are soldered fast in like manner; and the result is that there are three oval apertures. The next step is that on the top of the four united globes two other dumb-bells stand erect, and lean over towards each other till their upper globes come into contact, their lower ones remaining remote; these are soldered to the mass and to each other, at the points of contact, leaving a fourth aperture. The same is repeated at the opposite end by two other dumb-bells;—and the structure is complete as you see it. In almost all cases the two united globes of these terminal elements are fused into one globe, and in not a few instances the appearance is as if these two dumb-bells were but one, bent over in a semi-circular form; but still a good many specimens occur in which the two dumb-bells can

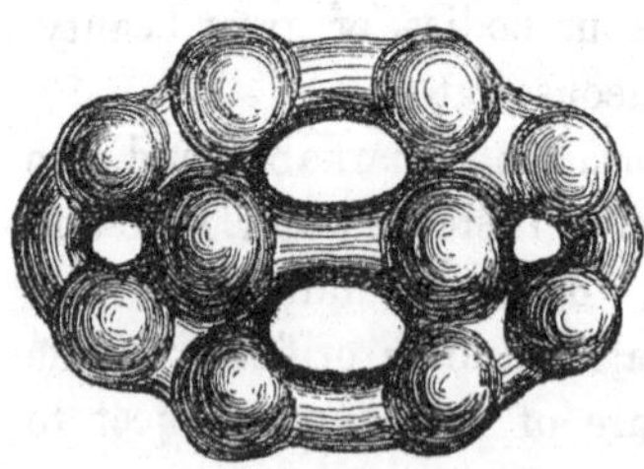

DUMB-BELLS IN HOLOTHURIA.

be quite distinguished from each other. The calcareous matter that solders the elements together seems abundant, and has the appearance that would be presented if they had been made of solid glass, and united by glass in a state of fusion; the latter having apparently run together, so as to smooth and round angles and fill up chinks, even where, as is often the case, the globes themselves have only mutually approximated, and not come into actual contact.

The average dimensions of these oval aggregations may be ·004 inch in length, and a little more than ·002 in width; but some specimens occur which are a little larger, and others a little smaller than this; while the irregular plates are sometimes three times the length.

Some of the more worm-like members of this class have, however, a skeleton composed of pieces imbedded in their skin, of even more remarkable shapes than these. One of these is the *Chirodota violacea*—a native of the southern coasts of Europe. We have indeed a British species of the same genus, a specimen of which is in my possession, but I have vainly examined the skin for any structure analogous to this.* In the Medi-

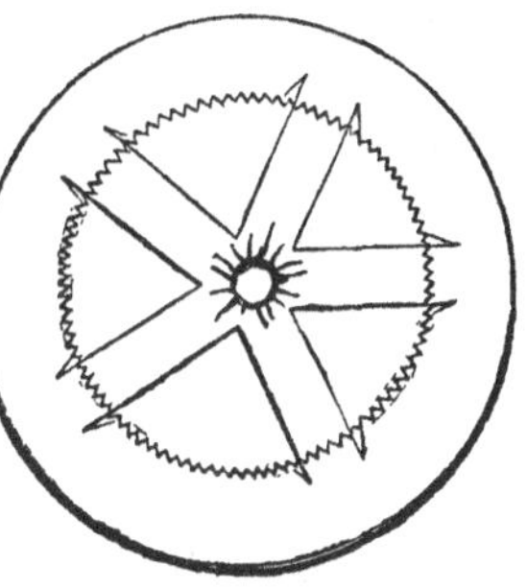
WHEEL IN CHIRODOTA.

* The most careful and repeated search has not availed me to find in the skin the least trace of calcareous atoms; but this may possibly be because I had unfortunately preserved my specimen in

terranean species the skin, especially of the belly-side, is described as filled with plates exactly resembling broad and thin wheels of glass, supported by four, five, or six radiating spokes, and having the inner edge of the hoop cut into teeth of excessive delicacy.

Another animal remarkable for its cuticular furniture is the genus *Synapta*, which is very similar in form, and closely allied to the *Chirodota*. It is very common in the Adriatic and Mediterranean seas, but has not yet been taken on the British coasts. I would counsel you, however, to have your eyes open if you have the opportunity of searching our coasts; for, as Müller found one species, the *Synapta inhærens*, on the shores of Denmark, it is not at all unlikely that we may possess either it or some other. Should it ever come into your hands, slit open the skin of the belly, where you will find, embedded in little papillæ or warts, some highly curious *spicula* or calcareous forms. Each consists of an oblong plate, perforated with large holes in a regular manner, and having a projection on its surface near one extremity, to which is jointed a second piece, having the most singularly true resemblance to an anchor. The flukes of this anchor project from the skin, the shank standing obliquely upward from the plate, to which it is articulated by a dilatation, where the ring would be, which is cut into teeth.

ANCHOR-PLATE IN SYNAPTA.

acetate of alumina, and the acetic acid has perhaps dissolved the lime.

Among the multitude of transparent creatures that swim in the open sea, few are more interesting than those which constitute the infant state of the very animals that we have lately been examining—the Sea-Urchins and their allies. It is a productive way of obtaining subjects for microscopic research, to go out in a boat on a quiet summer's day, especially in the afternoon, when the sun has been shining, or when evening is waning into night, and with a fine muslin net stretched over a brass ring at the end of a pole, skim the surface of the smooth sea. At intervals you take in your net, and having a wide-mouthed glass jar ready, nearly filled with sea-water, invert the muslin in it, when your captures, small and great, float off into the receiver. After a few such essays, unless you have very bad success indeed, you will see even with the naked eye, but much more with a lens, that the water in your jar is teeming with microscopic life; and though many of your captives will not long survive the loss of their freedom, still meanwhile you may secure many an interesting object, and examine it while yet the beauty and freshness of life remain. And, moreover, with care and prudence, some selected subjects may be maintained in vigour, at least long enough to afford you valuable information on the habits, economy, metamorphosis, and development of animals, of which even the scientific world knows next to nothing.

I have just been so fortunate as to obtain in this way the larval stage of one of our Sea-Urchins, and have it now in the thin glass trough which is on the stage of the microscope. It is just visible to the unassisted sight as a slowly moving point in the clear water, when the vessel is held up to the light; but with the low power which I am now using, it is distinctly made out in all its parts, and is an object of singular elegance and beauty.

It is, as you see, somewhat of the figure of a helmet; the crest rising to a perpendicular point, which is rounded, the vizor or mask descending far down and ending in two points, and a long ear hanging down on each side, so as to reach the shoulders of the wearer. Of course such comparisons are fanciful, but they assist one in intelligible description.

Now, the entire helmet is composed of a gelatinous flesh, of the most perfect transparency, so that we can see with absolute clearness everything that is within it. And the first thing that strikes us is, that a framework or skeleton of extreme delicacy, composed of glassy rods, supports the whole structure. Look carefully at this, and mark its symmetry and elegance. There is, then, first, a rod which passes through the crest perpendicularly, and carries at its lower extremity a horizontal ring. To the opposite sides of this ring are soldered two other very slender rods, passing down nearly in a perpendicular direction, but a little diverging; and two other shorter rods pass down from the front of the ring, parallel to these. After a while each lateral pair of rods is united by a short cross-piece, and the result is four lengthened rods, two of which go down through the vizor into the chin-points, and two larger and stouter ones through the ears into the shoulder-points. This, then, is the solid skeleton, the interest of which is much enhanced, when we observe that it is formed, on the common plan, out of perforated lime-glass, the two ear-rods and the crest-rod being pierced with a regular series of oval holes, and bearing on their edges corresponding projecting points.

Now, to turn again to the gelatinous flesh. The inner surface of the vizor, or that which would be in contact with the face of the wearer, supposing it to be a real helmet, has a great squarish orifice with a thickened margin, which we see by its movements to be highly sensitive and contractile. This square orifice is the mouth of the larva, and it leads into a cavity in the upper part of the vizor, which is the gullet; and this in its turn terminates in a narrowed extremity, which passes into the orifice of a greater and higher cavity, the lip of which embraces it just as the bung-hole of a barrel receives and embraces the tube of a funnel. The latter cavity occupies the chief part of the volume of the helmet, the four rods diverging to inclose it. It is the stomach.

It adds to the beauty of the little helmet-shaped creature, that while the greater portion of the substance is of the most colourless transparency, the summit of the crest and the tips of the shoulder-points are tinged with a lovely rose-red. The whole exterior surface is, moreover, studded with those minute and glandular specks, with which every part of the adult Urchin is covered; and the light is reflected from the various prominences with sparkling brilliancy.

The little creature moves through the water with much grace, and with a dignified deliberation; the crest being always uppermost, and the perpendicular position invariably maintained. It does not appear capable of resting, its movements depending on incessantly vibrating cilia. These organs we perceive densely clothing the long ear-pieces, but more especially accumulated and more vigorous in a thickened, fleshy band, which passes partly round the whole helmet, at the origin of these pieces.

You do not discern the slightest resemblance of form between this little slowly-swimming dome and the spined and boxed Urchin which crawls over the rocks; and you wonder by what steps the tiny atom of one-fortieth of an inch in length is led to its adult stage. Fortunately I can satisfy your curiosity on this point, not indeed from my own observations, but by those of Professor Johann Müller, whose discoveries of the developments of these and kindred animals are among the most interesting, because the most startling, of the marvels which modern zoology has revealed to us. The whole process is full of surprising details, to which the change of the caterpillar to a chrysalis, and that of the chrysalis to a butterfly, present no parallel, wonderful as those changes of form appear and are. There we have but modifications of outward form, produced by the successive moults or castings of the external skin, and the gradual growth of the animal, which has from the first been present, though veiled. But the construction of the Sea-Urchin is by no means a process of skin-casting, nor

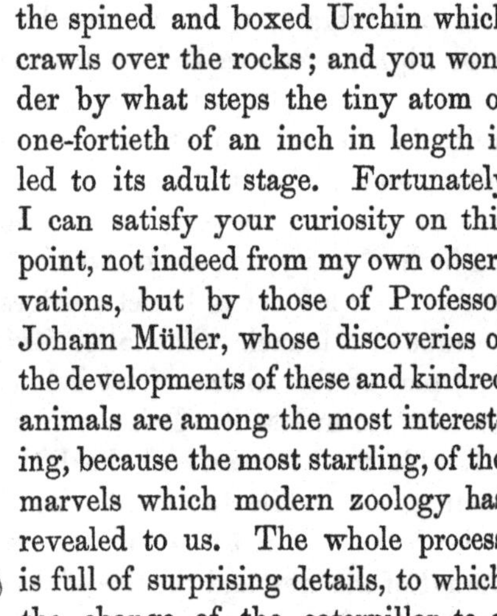

LARVA OF SEA-URCHIN.

has it any recognised parallel in the whole economy of natural history. It is a development perfectly unique. I will endeavour to make you acquainted with the results arrived at from the researches of the eminent German zoologist to whom we are indebted for almost all we know on the matter.

Let me first premise that this beautiful helmet-shaped creature is not the future Urchin; and, strange to say, that only a very small portion of the present structure, namely, the stomach and gullet, will enter into its composition. The helmet is a kind of temporary nurse, within which the future Urchin is to be formed, and by which it is to be carried from place to place by its ciliary action, while the young animal is gradually acquiring the power of independent life, when the whole constitution of the nurse wastes away and vanishes!

The first trace of the young Urchin is a filmy circular plate, which is not symmetrical with the helmet, nor formed even on the same plane, but appears *obliquely* fixed on the exterior of the stomach, *on one side*, close to the arch of transparent flesh which stretches from one of the points of the vizor to one of the ear-points. Herr Müller compares the larva (which is not helmet-shaped in every species) to a clock-case, of which the vizor, with its hanging gullet and mouth, form the pendulum, and then the newly-formed disk represents the face of the clock, only it is put on the side instead of the front. Now this tiny disk gradually grows into the form and assumes all the organs of the Urchin, while the enveloping nurse, flesh, rods, and all, waste away to nothing.

The disk, soon after its appearance, is seen to bear prominences on its surface, in which is traced the figure of a cinque-foil, the elements being five warts set symmetrically. These lengthen and grow into suckers, essentially identical with those of the adult, but most disproportionately large. In the five triangular interspaces between these, little points and needles of solid calcareous glass begin to form, very much like the crystals that shoot across a drying drop of a solution of some salt; these catch and unite, first into T-, and then into H-forms, and then into irregular networks. Meanwhile, fleshy cylindrical columns spring up from the surface, one in each of these interspaces, and presently develop within their substance a similar framework of porous glass; these soon manifest themselves to be the spines, and each is seated on a little nucleus of network, on which it possesses the power of rotating.

At the same time pedicellariæ begin to be formed; and, what is specially marvellous, they are first seen, not on the disk, which alone is to be the future Urchin, but on the interior wall of the helmet, which is even now in process of being dissipated, and even on the opposite side to that which carries the disk. They commonly appear four in number, arranged in two pairs; and one can see in them—they being, like the suckers, large out of all proportion to the disk—the stem, and the three-leaved heads, which already exercise their characteristic snapping movements.

The disk is meanwhile enlarging its area; and the spines and suckers, gradually lengthening, at length push themselves through the walls of the helmet; the hanging points and crest of which are fast diminishing by a kind of insensible absorption; the ciliary movements become less vigorous, and the mouth closes up. But, correspondently, the Urchin is beginning to acquire its own independent power of locomotion; for the suckers, now ever sprawling about, are capable of adhering to any foreign body with which they come into contact, and of dragging the whole structure about, by their proper contractions. The cilia that cover the thickened fringing band still exercise their powers, and are the last to disappear.

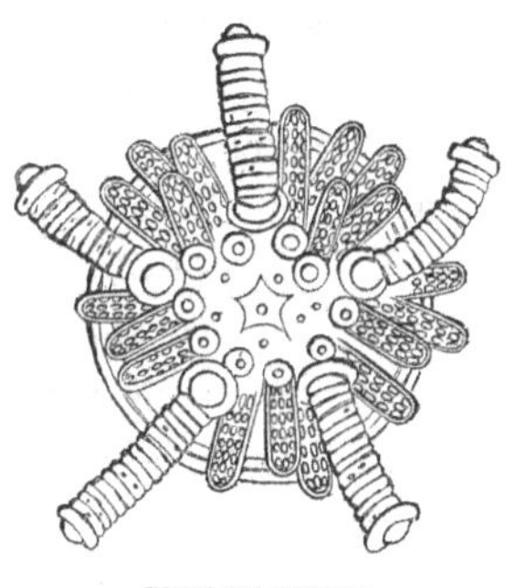
YOUNG SEA-URCHIN: *Development of Disk.*

When the disk has grown to such an extent as to spread over about half of the larval stomach, very little remains of the helmet, except the middle portions of the glassy rods and the ciliary bands; all the rest of this exquisitely modelled framework having vanished by insensible degrees, no one knows how or where. The stomach and gullet, indeed, are gradually sucked into the ever-growing disk; but all the rest, flesh and rods, fringes, bands, and cilia, waste away to nothing.

The mouth of the larva has no connexion with the mouth of the Urchin. The little isolated patches of glassy network continue to spread through the flesh of the disk, until the whole forms one uniform structure, and constitutes a series of plates. The mouth is that spot in the centre, over which the calcareous frame is last extended; and it is first distinguishable by the appearance of five glassy points, which soon develop themselves into the five converging jaws, which we see forming such a curious apparatus on the inferior side of the Sea-Urchin.

Actual observation has not traced the infant animal beyond this stage of the development; but specimens have been taken by Professor Müller, swimming in the sea, in which scarcely a rudiment of the larva remained. They had the form of round flattened disks, which freely moved their spines, and crawled about the sides of the vessel in which they were kept, by means of their suckers, exactly in the manner of the adult Urchin.

Thus "ends this strange, eventful history;" and in reviewing it, one can scarcely avoid being impressed with a sense of the majesty of God in these His humbler works. By what wonderful, what unexpected roads does He arrive at the completion of His designs! and if such things as these are only now bursting upon our knowledge, after six thousand years of man's familiar contact with the inferior creatures, how many more wonders may yet remain to be unfolded, as science pursues her investigations into the Divine handiwork! And yet, how do this and all similar manifestations of power and wisdom, sink into insignificance before the grand marvel, the wonder of wonders, the great mystery of godliness—that GOD WAS MANIFEST IN THE FLESH! We are surprised and delighted when we see one creature change, as it were, into another, but too often the story of that greater wonder, that God should have become man, falls upon listless ears and cold hearts; and yet the former, which we scarcely weary of tracing, concerns only the well-being of a poor dull creature scarcely raised above the life of a plant; whereas the latter had for its object the lifting of creatures from a state of ruin and wrath to immortal life and everlasting glory; and the creatures are—*ourselves!*

CHAPTER XVII.

JELLY-FISHES.

As this afternoon was delightfully calm and warm —the very model of an autumnal day—I took my muslin ring-net and walked down to the rocks at the margin of the quiet sea. Nor was I disappointed; for the still water, scarcely disturbed by an undulation, and clear as crystal, was alive with those brilliant little globes of animated jelly, the Ciliograde and Naked-eyed Medusæ, apparently little more substantial than the clear water itself. Multitudes of them were floating on the surface, and others were discerned by the practised eye, at various depths, shooting hither and thither, now ascending, now descending, now hanging lightly on their oars, and now, as if to make up for sloth, darting along obliquely with quickly-repeated vigorous strokes, or rolling and revolving along, in the very wantonness of humble happiness.

After gazing awhile with admiration at the undisturbed jollity of the hosts, I made a dip with my net, the interior of which, on lifting it from the water, was lined with sparkling balls of translucent jelly. They were far too numerous to allow me to transfer them all to captivity; they would soon have choked up and destroyed one another; I therefore selected the finest and most interesting, shaking an example or two of each kind into my glass jar of sea-water, where they immediately began to frolic and revel as if still in the enjoyment of unrestricted liberty. And here they are.

Among these bright and agile beings which are shooting their wayward traverses across each other, and intertwining their long thread-like tentacles, we will select one or two for examination, as samples of their kindred. And first let me isolate this active little Beroë (*Cydippe pomiformis*), which I dip out with a tea-spoon and transfer to this other glass jar, that we may watch its form and movements unaffected by the presence of its companions.

We see, then, a little ball, almost perfectly globular, except that a tiny wart marks one pole, of the size of a small marble, and apparently turned out of pure glass, or ice, or jelly—according to your fancy,—perfect transparency and colourlessness being its characteristics, so much that it is not always easy to catch sight of the little creature, except we allow the light to fall on the jar in a particular direction. From two opposite sides of the globe proceed two threads of great length and extreme tenuity, which display the most lively and varied movements.

These filaments shall occupy us for a few moments. We trace them to their origin, and find that they proceed each from the interior of a lengthened chamber, on each of two opposite sides of the animal. Suddenly, on the slightest touch of some foreign object, one of the threads is contracted to a point and concealed within its chamber, but is presently darted forth again. When the lovely globe chooses to remain still, the threads hang downward, gradually lengthening more and more, till their extremities lie along the bottom of the jar, extended to a length of six inches from the chamber. Then we see that this delicate thread is not simple, but is furnished along one side, throughout its length, at regular distances, with a row of secondary filaments, which project at right angles from the main thread.

These secondary filaments constitute an important element in the charm which invests this brilliant little creature. They are about fifty in number on each thread, and some of them are half an inch long, when fully extended, but it is seldom that we see them thus straightened; for they are ever assuming the most elegant spiral coils, which open and close, extend and contract, with an ever-changing vivacity. The animal has a very perfect control over the threads, as well as over the secondary filaments in their individuality. One, or both, are frequently projected from their chambers to their full extent by one impulse; sometimes the extension is arrested at any stage, and then proceeded with, or the thread is partially or entirely retracted. Sometimes the secondary filaments are coiled up into minute balls, scarcely perceptible, or only so as to give to the main thread the appearance of small beads remotely strung on a fine hair; then a few uncoil and spread divergently; contract again, and again unfold; or many, or all, interchange these actions together, with beautiful regularity and rhythmical uniformity, repeating the alternation for many times in rapid succession.

The beauty and diversity of the forms assumed by

these elegant organs beguile us to watch them with unwearied interest, and we wonder what is their function. For, with all our watching, this is by no means clear. They are certainly not organs of motion. At times it seems as if they were cables intended to moor the animal, while it floats at a given depth; for we see them with their extremities spread upon the bottom, to which they appear to have a power of adhering, thus forming fixed points, from which the little globe rises and falls at pleasure, shortening or lengthening its delicate and novel cables, maintaining all the while its erect position.

When the *Cydippe* swims, however, which it does with great energy, the threads seem unemployed, streaming loosely behind, and evidently taking no part in the progression, though still adding beauty and grace to the *tout ensemble*. The organs by which the sprightly motions of the whole animal are effected are of quite another character, and shall now engage our attention.

You have doubtless observed, while gazing on the animal, a peculiar glittering appearance along its sides, mingled in certain lights with brilliant rainbow-reflections. Now let us take an opportunity, when it approaches the side of the glass, to examine this appearance with a lens. The globe, you see, is marked by longitudinal bands, eight in number, set at equal distances, and ranging like meridians, except that they do not quite reach to either pole. These bands are the seats of the motile organs, which are highly curious, and in some sort peculiar.

Each band is of considerable width in the middle,

but becomes narrower towards the extremities. It carries a number—usually from twenty to thirty—of flat thin membranous fins, set at regular distances, one above the other, which may be considered as single horizontal rows of cilia, agglutinated together into flat plates. Each plate has a rapid movement up and down, from the line of its insertion into the band, as from a hinge, and thus striking the water downwards, like a paddle. The whole band may be likened to the paddle-wheel of a steamer, except that the paddles are set in a fixed line of curvature instead of a revolving circle. The effect, however, is exactly the same: that of paddling the beautiful little globe

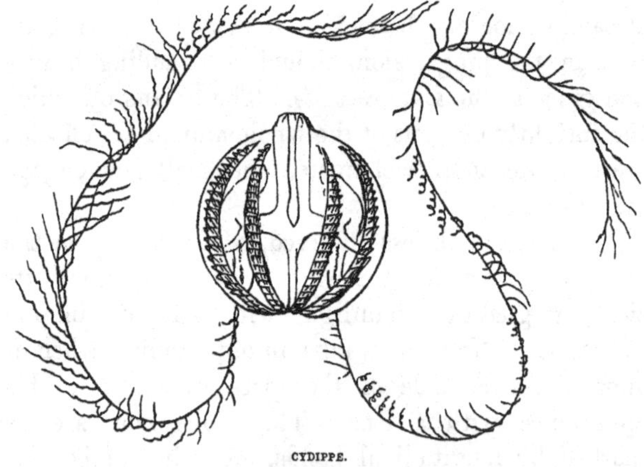

CYDIPPE.

vigorously through the water. The prismatic colours are produced by the play of light on their glittering surfaces, which are ever presented to the eye of the beholder at changing angles.

We rarely see these rows of paddle-fins wholly at

rest, but occasionally one or two bands will be alone in a state of vibration; or one or more will suspend their action while the rest are paddling. Sometimes in a band that is at rest, a minute and momentary wave will be seen to run rapidly along its length. All these circumstances show that the ciliary motion is perfectly under the control of the animal's will, not only in the aggregate, but in every part.

In an excellent memoir on this animal by Mr. R. Patterson, of Belfast,* there are some interesting observations on the power of its tissues to become tinged with extraneous colours, a fact which may be useful to you in your researches, as enabling you with more ease and precision to demonstrate the internal structure.

"From the inconsiderable quantity of solid material," remarks this observer, "which enters into the body of the Beroës, and the rapid circulation of water, which is apparent throughout their frame, we would naturally suppose that any tinge which the body might accidentally acquire would be extremely fugitive. It was found, however, to be much less so than *à priori* would have been expected. My attention was drawn to this peculiarity by the circumstance of all my glass vessels being one evening occupied by Beroës and Crustacea, so as to compel me to place a small Medusa in a tin vessel, which chanced to be rusted at the seams. Next morning the colourless appearance of the animal was changed into a bright yellow, which appeared to pervade every part, and doubtless arose from the oxide of iron,

* Trans. Roy. Irish Academy, vol. xix. pt. 1.

diffused through the sea-water. This tint remained during the entire day, although the animal was transferred to pure sea-water. Wishing to try if the vessels of the Beroë would become distinct, if filled with some coloured fluid from which the animal could suddenly be withdrawn, and viewed through the usual transparent medium of sea-water, I placed a Beroë in a weak infusion of saffron. At the end of twenty minutes its colour had undergone a perceptible change. I allowed it, however, to remain immersed for about six or seven hours, when it had assumed a bright yellow hue. It was then placed in pure sea-water, but retained its yellow colour for twenty-four hours afterwards; and though it gradually became fainter, it was very perceptible even at the expiration of forty-eight hours."

I am sure you will pardon my interrupting your microscopic gazings for a moment by quoting the following charming lines by the Rev. Dr. Drummond, which were elicited by his having watched with pleasure the elegant form and motions of this little creature.

"Now o'er the stern the fine-meshed net-bag fling,
And from the deep the little Beroë bring:
Beneath the sun-lit wave she swims concealed
By her own brightness;—only now revealed
To sage's eye, that gazes with delight
On things invisible to vulgar sight.
When first extracted from her native brine,
Behold a small round mass of gelatine,
Or frozen dew-drop, void of life or limb:
But round the crystal goblet let her swim
'Midst her own element—and lo! a sphere
Banded from pole to pole—a diamond clear,

Shaped as bard's fancy shapes the small balloon
To bear some sylph or fay beyond the moon.
From all her bands see lucid fringes play,
That glance and sparkle in the solar ray
With iridescent hues. Now round and round
She wheels and twirls—now mounts—then sinks profound.
Now see her, like the belted star of Jove,
Spin on her axis smooth—as if she strove
To win applause—a thing of conscious sense,
Quivering and thrilling with delight intense.
Long silvery cords she treasures in her sides,
By which, uncoiled at times, she moors and rides;
From these, as hook-hairs on a fisher's line,
See feathery fibrils hang, in graceful twine,
Graceful as tendrils of the mantling vine,
These, swift as angler, by the fishy lake,
Projects his fly, the keen-eyed trout to take,
She shoots with rapid jerk to seize her food,
The small green creatures of crustaceous brood;
Soon doomed herself a ruthless foe to find,
When in th' Actinia's arms she lies entwin'd.
Here prison'd by the vase's crystal bound,
Impassable as Styx's nine-fold round,
Quick she projects, as quick retracts again,
Her flexile toils, and tries her arts in vain;
Till languid grown, her fine machinery worn
By rapid friction, and her fringes torn,
Her full round orb wanes lank, and swift decay
Pervades her frame till all dissolves away.
So wanes the dew, conglobed on rose's bud,
So melts the ice-drop in the tepid flood:
Thus too shall many a shining orb on high
That studs the broad pavilion of the sky,
Suns and their systems fade, dissolve, and die."

While we have been admiring our lovely little *Cydippe*, and comparing notes with other observers and admirers, other species as small, as transparent, as sprightly, and scarcely less elegant, have been impatiently waiting for their share of admiration; shooting to and fro, tossing their little bells of ductile glass about, and alternately lengthening and snatching-in their sensitive tentacles, in astonishment at our stoical indifference to their charms, and saying, *suo more,* with the little urchin whose feelings were hurt by the neglect of his papa's visitor,—"You don't notice how beautiful I be!"

A thousand pardons, sweet little *Sarsia!* We will now give you our undivided attention; and for this end we must take the liberty of catching you, and of transferring your translucency to isolated grandeur in this other glass. Ha! but you don't want to be caught, eh? And so you pump and shoot round and round the jar as the spoon approaches! Truly you are a supple little subject, difficult to catch as a flea, and difficult to hold (in a spoon) as an eel. But here you are at last, lying as motionless and as helpless in the silver as a half-melted atom of calf's-foot jelly, to which, indeed, you possess no small resemblance.

Look at the pretty little Medusa in his new abode, at once recovering all his jelly-hood as he feels the water laving him, and dashing about his new domain with a vigour which makes up for lost time.

It is a tall bell of glass, a little contracted at the mouth—its outline forming an ellipse, from which about a third has been cut off. The margin of this bell carries four tiny knobs, set at equal distances, and thus quartering the periphery; and these are the more conspicuous because each one is marked with a bright orange-coloured speck. Physiologists are pretty well agreed to consider such specks as these, on the margins of the smaller *Medusæ*, as eyes,—rudimentary organs of vision, capable, probably, of appreciating the presence and the stimulus of light, without the power of forming any visual image of external objects. You will not gain much information about their function from microscopic examination; for all you can discern is an aggregation of coloured specks (pigment-granules) in the midst of the common jelly.

The knobs, however, are connected with other organs; for from each of them depends a highly sensitive and very contractile tentacle. Sometimes one, or more, or all, of these organs hang down in the water motionless, lengthening more and more, especially when the bell is still, until they reach a length some twelve or fifteen times that of the bell, or *umbrella.* Then suddenly one will be contracted, and, as it were, shrivelled, to mere fragments of a quarter of an inch long; then lengthened again to an inch or two; then shortened again. Now the little bell resumes its energetic pumping, and shoots round and round in an oblique direction, the summit always going foremost, and the tentacles streaming behind in long trailing lines. Now it is again arrested; the bell turns over on one side and remains motionless, and the tentacles, as "fine as silkworms' threads," float loosely in the water, become mutually intertangled, instantly free themselves, pucker and shrivel up, slowly lengthen, and hang motionless again, or, as the bell allows itself to sink slowly, are thrown into the most elegant curves and arches.

Though these tentacles look at first like simple threads of extreme tenuity, yet when viewed closely they are seen to be composed of a succession of minute knobs separated by intervals—like white beads strung on a thread; the beads being more remote from each other in proportion as the tentacle is lengthened.

This structure is worthy of a more minute investigation. We will, therefore, confine our little *Sarsia* in this narrow glass trough, which is sufficiently deep to allow its whole form to be immersed, though somewhat flattened; which is an advantage, as its movements are thereby impeded. Now, with a power of 300 diameters you see that each of the knobs of the tentacle is a thickening or swelling of the common gelatinous flesh, in which are imbedded a score or two of tiny oval vesicles, without any very obvious arrangement; but for the most part so placed that the more pointed end of each is directed toward the circumference of the thickening. The intermediate slender portions of the tentacle—the thread on which the beads are strung—is quite destitute of these vesicles.

These little bodies are called *cnidæ*, and, in the whole of this class of animals, and also in that of Zoophytes, they play an important part in the economy of the creature. I shall probably take occasion to exhibit them to you under conditions more favourable than are presented here, viz. in the Sea-Anemones, where they attain far greater dimensions; and therefore I will merely say here that each one of these tiny vesicles carries a barbed and poisoned arrow, which can be shot forth at the pleasure of the animal with great force, and to an amazing length,—that hundreds are usually shot together,—and that this is the provision which the All-wise God has given to these apparently helpless animals for securing and subduing their prey.

There is, however, another organ still more conspicuous in our little *Sarsia*, of which I have not yet spoken. As the whole animal has the most absolute transparency, we see that the roof of the bell is much thicker than the sides, and that it gradually thins off to the edge. The interior surface is called the sub-umbrella, and it carries within its substance four slender tubes, which, radiating from the centre of the roof, proceed to the margin, where they communicate with another similar canal which runs round the circumference, sending off branches into the tentacles. This is the circulatory system; and you may see, with the magnifying power which you are at present using, that a clear fluid is moving rapidly within all these canals, carrying minute granules; not with an even forward current, but with an irregular jerking vacillating movement, as if several conflicting eddies were in the stream. Yet we

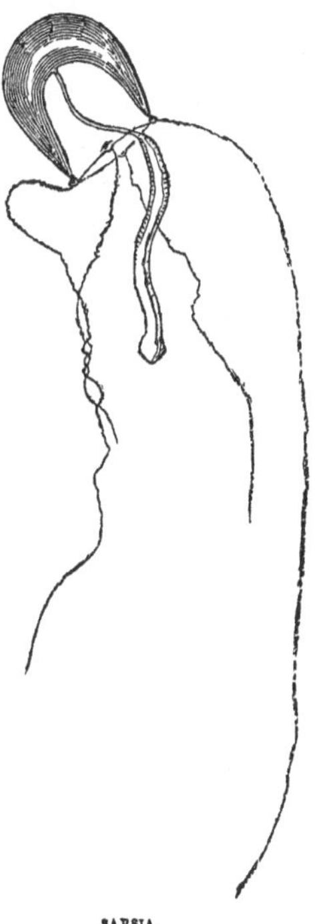

SARSIA.

discern that, on the whole, the granules are moved forward; passing from the centre of radiation towards the margin, when we see them slip into the marginal canal from the several mouths of the radiating canals.

This is a very simple and rudimentary blood-system. There is here no heart with its pulsations, no proper arteries or veins, no lungs for oxygenation; but the products of digestion are themselves thus circulated through the system. And this brings me back to the central point, whence you see depending the curious organ I spoke of. A long cylinder of highly moveable and evidently sensitive flesh hangs down from the middle of the roof exactly like the clapper of a bell; and, as if to add to the resemblance, this same clapper is suspended by a narrow cord, and is terminated by a knob.

Sometimes this whole organ is allowed to hang about as low as the edge of the bell; then it gradually lengthens to twice, thrice, nay to five times that length; the tongue lolling out of the mouth to a most uncouth distance, and even the suspending cord (as I presume to term the attenuated basal portion) reaching far beyond the margin; then, on a sudden, like the tentacles, the tongue is contracted, thrown into wrinkles, curled into curves, and the whole is sheltered within the concavity; presently, however, to loll out again.

This proboscis-like organ is called the peduncle, and its office is that of a stomach, of which the knob at the end is the mouth, having a terminal orifice with four minute lips. The flexible substance and rapid motions of this peduncle are suited to enable it to seize small passing animals that constitute its prey; and I have seen the *Sarsia* in confinement seize with the mouth, and swallow, a newly-hatched fish, notwithstanding the activity of the latter. For hours afterwards, the little green-eyed fry was visible, the engulfment being a very slow process; but the Medusa never let go its hold; and gradually the tiny fish was sucked into the interior, and passed up the cavity of the peduncle, becoming more and more cloudy and indistinct as digestion in the stomach dissolved its tissues.

The greater portion of the food is by-and-by discharged from the mouth, the fluids which have been extracted from it being on the other hand carried up through the base of the peduncle, and distributed along the four radiating vessels, conveying nutrition, supply of waste, and growth to all parts of the system.

We may now liberate our little *Sarsia*, with thanks for the gratification he has afforded us, to resume his active play among his many companions. Meanwhile we will look for one of another kind among the group.

Here is a pretty and interesting species. Active it is, but less vigorously rapid in its movements, than either the *Beroë* or the *Sarsia*. It is, as you see, something less than a hemisphere, or resembling a watch-glass in shape, about three-fourths of an inch in diameter. In general character it resembles the *Sarsia*, but the peduncle is small, never reaching to the level of the margin, and its mouth is terminated by four expanding fleshy lips, which are extremely flexible and versatile.

The four radiating vessels here carry, just before they merge into the marginal canal, a dilatation of the common flesh, which, as you see, bulges out the surface of the umbrella. We will examine one of these dilatations with the microscope.

It is, as you perceive, occupied by a number of clear globes, each of which has another minute globose body in its interior. They are very diverse in size, some being very small, others comparatively large, and it is to the dimensions of these latter that the swelling of the surface of the umbrella is due. These vesicles are the eggs of the animal considerably advanced towards maturity; and the dilatations around the radiating vessels are the ovaries.

The margin, however, presents us with the most obvious, and perhaps the most interesting, points of diversity from the little *Sarsia.* In the little beauty before us,—whose name, by the way, *Thaumantias pilosella,* I have not yet told you,—the outline is fringed with about fifty short and slender tentacles, each of which springs from a fleshy bulb, in which is set a speck of deep purple. These collections of coloured granules, which I have already explained to be rudimentary eyes, have a very charming effect, and give a beautiful appearance to the little creature, as if its translucent crystalline head were encircled with a coronet of gems.

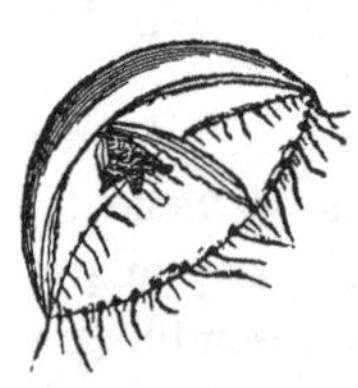
THAUMANTIAS.

You shall see them, however, under circumstances

which will make them appear more lustrously gemmeous still. Come with me, and I will carry the glass containing our little *Thaumantias* into the next room. You need not bring the candle, or what I am going to show you will be quite invisible.

Take hold of this pencil, and having felt for the glass, disturb the water with it. Ha! what a circle of tiny lamps flashes out! You struck the body of the *Thaumantias* with the pencil, and instantly, under the stimulus of alarm, every purple eye became a phosphoric flame. Touch it again; again the crown of light flashes out, but less brilliantly; and each tiny lamp after sparkling tremulously for a moment, wanes, and the whole gradually, but quickly, go out, and all is dark again.

But it is tired of lighting up for nothing,—or its gas is exhausted,—or it is become used to the pencil and is not alarmed,—or,—at all events you may knock it, and push it, but it refuses to shine any more. Back with it then to the microscope, and let us see if it possesses any other points of interest for us there.

Yes: we have not exhausted the organs of the margin yet. Between the tentacles which spring from bulbs there are a good many more, far more minute, without any bulbs;—from four to seven between every two of the primary ones. We won't mind these, but bringing the margin itself into focus, and moving it along the stage horizontally, we presently see one and another singular organs. They are eight in all, two being placed, but irregularly, in each of the four quadratures of the circle formed by the radiating canals.

These are auditory vesicles, or organs of hearing, very closely similar to those which we see imbedded in the bosom of the Snail and other Mollusca. Here they are comparatively large, and unusually well

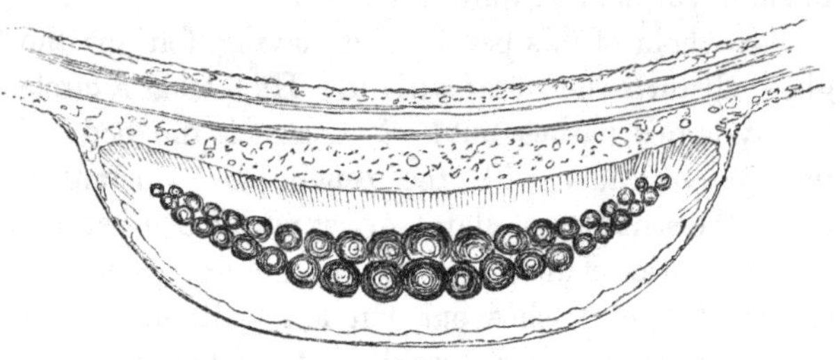

OTOLITHES OF THAUMANTIAS.

furnished. Each is a semi-oval enlargement of the flesh of the margin, in close connexion with the walls of the marginal canal, hollowed so as to inclose a capacious cavity, in which are placed a considerable number,—from thirty to fifty in this individual—of otolithes, or spheres of solid, transparent, highly refractive substance. They are arranged in a double line, forming a crescent, and those which are nearest the centre are longer than those towards the extremities of the line. I believe some observers have seen oscillatory and rotatory movements among these spherules, as in the *Mollusca;* but I have invariably found them motionless in all the species of Medusa that I have examined, as you see them here.

One more little beauty from our stock, and we have done with these. There is one that moves among the rest like a bead of coral, the smallest of all, yet the most brilliant. Here is another, and here another of the same sort; which has been named by Professor Edward Forbes, *Turris neglecta,* because naturalists before him had neglected to notice it, just as we have been doing, engrossed by its larger *confrères.*

Beautiful as is this little gem, it is not so large as a dried pea, scarcely larger than a grain of hemp-seed. It is described as "mitre-shaped;" in other words it is a tall bell, with the margin slightly bent inwards, and the sides a little constricted. The umbrella is thick, and being very muscular is not so translucent as those we have been examining; hence it has a pellucid white appearance. But through this shines its chief beauty; the peduncle is very large, and globose at the upper or basal part, which is usually, as here, of a pale scarlet or rich orange hue.

Imbedded in this orange-coloured flesh are seen many points of a lovely rose-purple, which two colours blending together, and softened by their transmission though the sub-pellucid umbrella, have a peculiar brilliancy. But stay! here I have one more advanced in age, which will exhibit some peculiarities of interest in the economy of these frail but charming creatures.

In this specimen, which is somewhat larger than the former, the margin of the umbrella is a little turned back, displaying more clearly the peduncle with its brilliant ovaries. These, too, are more turgid, and the rosy points are seen projecting from their interior, and some of them even ready to fall. And look! here on the bottom of the glass are lying half-a-dozen or more of similar purple points, whose rich hue renders them plainly discernible, after a slight searching, to the unassisted eye. I will collect one or two with a capillary tube of glass, and submit them to your examination under the microscope.

You now discern that these bodies are perfectly oval in form. One might, indeed, call them eggs, for they perform the part of such organisms; but that these have soft walls, covered on their whole external surface with fine vibratile cilia, by the action of which they are endowed with the power of free locomotion. We see them, in fact, gliding about the water of the live-box under view, with an even and somewhat rapid motion, which *appears* to be guided by a veritable will. Under this power they are seen to be of a soft rich lake-crimson hue, all over.

These little gemmules have a somewhat romantic history of their own. I am afraid that these we see are too recent to afford us any help in tracing it, and therefore I must be satisfied with telling you what I have observed of it on former occasions.

After the beautiful little Coral Jelly has swum about a few days, the umbrella begins to turn outward and backward, and to contract more and more, until at length, it lies in shrivelled folds around the top, leaving the whole peduncle exposed. Long before this, it has lost its power of swimming, and lies helpless on its side upon the bottom. Meanwhile the orange ovaries have swollen; the purple gemmules have become developed, and have gradually worked their way through the ovaries, and fallen one by one upon the bottom. There then they glide about for a little time, perhaps for a day or so, by means of their vibrating cilia.

At length each little gemmule loses its power of wandering, its motion becomes feebler, and more intermitted, and finally ceases altogether. The little being now rests on some solid body,—a stone or a shell,—to which it firmly adheres. Its two extremities grow out, adhering as they extend, and sometimes branching, but still in close and entire contact with the support. At length, after a day or two, from some point of the upper surface of this creeping root, a kind of wart buds forth, and soon grows into an erect slender stem, which presently divides into four straight, slender, slightly divergent tentacles, which grow straight upward to a considerable length. The whole structure retains the rich purple hue of the original gemmule.

Beyond this point I have not pursued the history of the little *Turris* from personal observation; nor am I aware that any naturalist besides has studied the development of this particular genus. But the history of other genera is known; and as the phenomena they exhibit are quite parallel to those which I have been describing, so far as these have been traced, we may fairly conclude that there is the same parallelism in the subsequent stages.

Assuming this, then, the little crimson stem with four rays,—a veritable polype,—buds forth four more tentacles in the interspaces, making the total number eight; these in like manner increase progressively to sixteen, thirty-two, and sixty-four. It now possesses a close resemblance to the Hydra of our ditches, only having more tentacles; and, like it, the Medusa-larva buds forth from its sides young Hydra-like polypes, which take the form of their immediate parent, fall off, attach themselves, bud forth more, and so on. All these catch living prey with their tentacles, swallow them with their mouths, and digest them with their stomachs, exactly like real polypes, and thus produce generation after generation of similar beings.

Years may pass in this stage, during which numberless polypes are formed. At length the original stock, or any one of its descendants, takes on an important change. Its body lengthens, and becomes cut as it were into a number of rings, as if tied tightly round with thread, or like the body of an Annelid. These

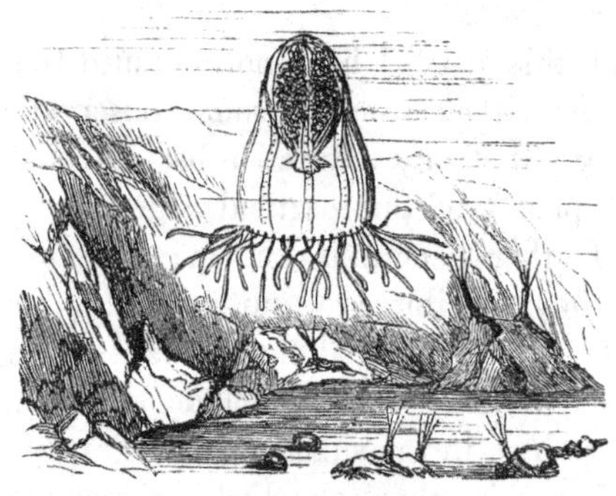

TURRIS AND ITS YOUNG.

segments become increasingly distinct, until at length each is seen to be a shallow cup, notched at its margin, and sitting in the concavity of the one next below it. This structure is developed first in those at the free extremity of the polype, and progressively downwards; and the terminal cups are nearly free, rocking in their successors with every wave, while the lowest segments are scarcely visible as such.

At length the extreme cup rocks and oscillates until the slender thread of connexion is snapped, and it is free. It at once turns itself over, so as to present its concavity downwards, and contracting its margin with the well-known pulmonic spasm, shoots away with the movement as well as the form of a veritable Medusa. The little progeny has at length, after passing through so many changes, returned to the image of its parent.

Such are, in brief, the phenomena of one of the most remarkable series of facts that modern zoology has discovered, and which have been propounded under the title of the Law of Alternation of Generations.

CHAPTER XVIII.

ZOOPHYTES.

It is pleasant to go down to the shore on a bright autumnal morning at low water, when the tide has receded far, exposing great areas of wet sand and wildernesses of rugged rocks draped with black and red weed. It is pleasant to make our way on cautious foot round some frowning point whose base is usually beaten by the billows; to travel among the slippery boulders, now leaping from one to another, now winding between them, now creeping under their beetling roofs; to penetrate where we have never ventured before, and to explore with a feeling of undefined awe the wild solitudes where the hollow sea growls, and the grey gull wails. It is pleasant to get under the shadow of the tall cliffs of limestone, to creep into low arching caves, and there to stoop and peer into the dark pools, which lie filled to the brim with water as clear as crystal, and as unruffled as a well. What microcosms are these rugged basins! How full of life all unsuspected by the rude stonecutter that daily trudges by them to and from his work in the marble quarry of the cliff above! What arts, and wiles, and stratagems are being practised there! what struggles for mastery, for food, for life! what pursuits and flights! what pleasant gambols!

what conjugal and parental affections! what varied enjoyments! what births! what deaths! are every hour going on in these unruffled wells, beneath the brown shadow of the umbrageous oarweed, or over the waving slopes of the bright green *Ulva*, or among the feathery branches of the crimson *Ceramium!*

I have just been examining some of these rock-wells, and have rifled them of not a few of their living treasures, bringing home the *opima spolia*,* that you may share with me in the enjoyment of examining them.

The Zoophytes are here in their glory. Such places as those I speak of are the very metropolis of the zoophytic nation. Look at this great leaf of the fingered Tangle: see how its broad olive-brown expanse is covered with tiny forests of white branching threads, which spread and spread till they run off into the fingers of the much split leaf; and not only on one side, for the under surface is as densely clad with the shaggy burden as the upper; the smooth leathery tissue being covered with a network of creeping roots, branching and anastomosing everywhere, like the railways on Bradshaw's map.

This double forest is wholly composed of a single species, called *Laomedea geniculata;* nay, I believe it is but one single individual. That is to say, the whole of these multitudinous ramified threads and stems, with their innumerable polypes, have all extended by gradual though rapid growth from a single germ, and all are connected even now, so that a common life pervades the whole. But we will look

* Rich spoils.

awhile at it in detail till we have mastered its external features, and then I will tell you something of its history and economy.

With the unassisted eye we can discern plainly enough the outline and plan of this compound organism. Along the smooth and lubricous surface of the olive weed runs a fine thread of a pellucid white appearance, so firmly adherent that if you attempt to remove it with a needle's point, you find that you only tear either the leaf or the thread. The course is generally in a straight line, but does not ordinarily pursue the same direction far, commonly turning off with an abrupt angle at intervals of about an inch, and thus meandering in a zig-zag fashion, very irregularly, branching frequently, and uniting with a thread already formed, when the creeping one has to cross it.

Thus the basal network is formed; but meanwhile, from every angle, and often from intermediate points, a free erect thread has shot up—like the stem of a tiny plant—to the height of an inch, rarely more; not, however, straight, but with frequent zig-zag angles, whence the name *geniculata*, or "kneed." At every angle a slender branch is sent forth, pursuing the same direction as that of the joint from the summit of which it issued, and terminating in a tiny knob. In the angles of some of these branchlets are seated oblong vesicles, twice or thrice as large as the terminal knobs. And this is pretty well all that we can make out with the naked eye.

Cutting carefully off with scissors a narrow strip of the leaf, I drop it into the parallel-sided cell of glass half-filled with sea-water, and examine it first with a low power, and afterwards with a higher. We now see that the creeping thread is a tube of horny substance, flattened on its under side, and that the erect stems and their branches are similar tubes, whose cavities are in free communication with that of the creeping root. The wall is thin, and perfectly transparent and colourless; the whiteness of the whole being dependent on a soft medullary core of living jelly, which permeates the whole structure, on which the horny sheath is as it were moulded.

This medulla is pierced with a canal, through which a fluid circulates, carrying along numerous minute granules with a quivering, jerking motion; this is doubtless the nutrient fluid conveying the products of digestion to every part of the common structure.

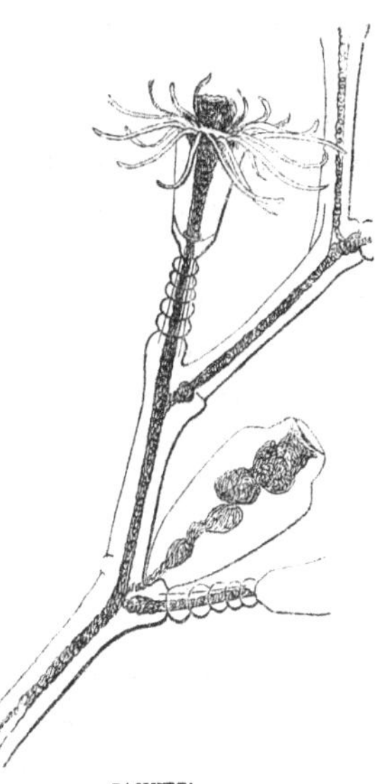

LAOMEDEA.

Where the branches issue from the angles of the stem, the medulla, and consequently the horny sheath, is dilated into a knob; immediately above this there is a joint-like constriction in the tube, and the branch itself is insected by four or five such constrictions, so as to form as many rings. Its extremity then expands into an elegant cup, or vase, of extreme tenuity and transparency, shaped like a wine-glass, with the rim undivided, but so thin and subtle as to be seen with the greatest difficulty.

These cups, or cells, are each the proper habitation of a polype, which is nothing else but the termination (in this direction) of the living, growing, vascular pith. The latter becomes exceedingly attenuated to pass through a very narrow orifice in the centre of a horny diaphragm, or sort of false bottom, which passes across the bottom of each cell. It then dilates into a soft contractile animal, whose body—but look for yourself; for here, full in the field of the microscope, is one expanding in the highest vigour and beauty.

It is a long trumpet-shaped body of granular flesh, the mouth of which just reaches the brim of the cup, over which it spreads on all sides. From its margin spring some eighteen or twenty tentacles—the exact number varying in different individuals—arranged in one or two close-set circles, like a crown. These organs, which, as you see, fall into elegant double curves, like the branches of a chandelier, are roughened with knobbed rings, something like the horns of a goat; this structure we will presently submit to more close examination.

In the midst of the space surrounded by the tentacular crown there is protruded, at the pleasure of the animal, a large, fleshy, funnel-shaped mouth, the lips of which are highly sensitive and versatile, continually changing their form—protruding, contracting, bending in upon themselves, now closing, now opening the mouth, and, as it were, testing the immediate vicinity, like a very delicate organ of some unknown sense.

The whole polype is much too minute for us to attempt, with any probability of success, the amputation of one of the tentacles with scissors. But by cutting off a polype, cell and all, and putting it into the compressorium, we may be able, by means of the graduated pressure, to flatten the whole, and thus discern the gnarled structure of the tentacles. A very high magnifying power is needed for this.

Here, then, we have one of the tentacles flattened between the glass plates, but still retaining its integrity. We find that the thickenings are similar in character to those of the tentacles of *Sarsia*, which we lately observed. They are, in fact, accumulations of *cnidæ*—those peculiar weapons of power, which I shall presently describe in full—but here they are symmetrically arranged in single rows, each pointing upward and outward.

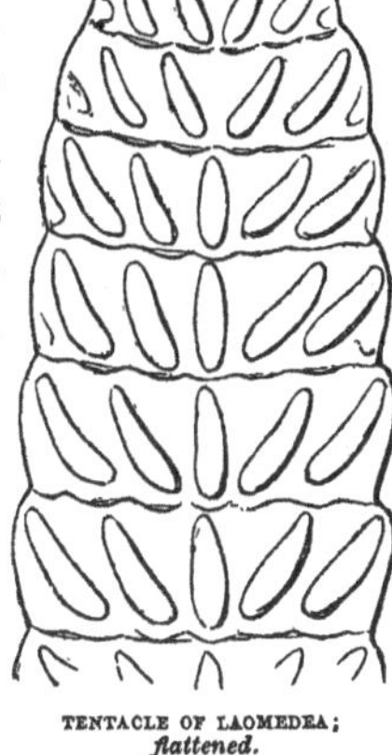

TENTACLE OF LAOMEDEA; *flattened.*

To return to the living specimen on the leaf: you see seated in the angles of the branches here and there elegant urn-shaped cells, larger than the polype-cells, each with a sort of shoulder and a narrow neck. The common pith passes from the joint into the bottom of these, and then extends through the centre till it reaches the mouth. In some of the urns this forms merely a slender column, expanding at the mouth, but in others it enlarges at irregular intervals into large knobs or masses of granular flesh, which are confusedly grouped together, eight or ten in one capsule. This latter is the most interesting condition; let us watch it.

While doing so, let me inform you that these urns are the reproductive organs, and the fleshy masses are embryos of peculiar character, which are developed out of the nutrient medulla. The largest of those now under observation is, as you see, moving, and slowly working its way out of its glassy prison. Two or three flexible finger-like bodies are protruding from the orifice of the urn, and more are joining them; we see they are tentacles, protruded in a loose bundle, just as the polype emerges from the cell.

It is a somewhat slow process; but at length the fleshy mass squeezes itself forth, as if pushed out by some contractile force behind; while we see the fluids, carrying granules, run into the parts of the tentacles which are already free. The embryo is liberated.

For a few seconds it appears helpless, and falls through the water in a collapsed state, so that we cannot discern its proper form. It gives a spasmodic contraction or two, feeble at first, then more vigorous; the tentacles lengthen, the body expands, and—lo! it is not a polype, but a Medusa!

And now take your eye for a moment from the microscope, and glance at this glass jar, in which the oarweed with its colony of Zoophytes has been standing for a few hours. Hold it between your eye and the light; do you not see that the water is alive with tiny dancing atoms? Hundreds are there, playing and pumping through the fluid with a sort of flapping motion, which, when you get one sidewise in clear view, will not fail to remind you of the flagging flight of some heavy-bodied, long-winged bird. These are the Medusa-shaped progeny of the *Laomedea.*

But let us return to the one of which we have just witnessed the birth, and which is still flapping to and fro in the narrow glass trough. You see a pellucid colourless disk or umbrella of considerable thickness, about one-sixtieth of an inch in diameter in its average state of expansion. Its substance has a reticular appearance, probably indicating its cellular texture. Internally, the disk rises to a blunt point in the centre, whence four vessels diverge to opposite points of the margin. These form elevated ribs, the surface being gradually depressed from each to the centre of the interspace. Externally, the centre of the disk is produced into a fleshy peduncle, having a narrow neck, and then expanding into a sort of secondary disk, of a square form, with the angles rounded. This organ, which is capable of varied, precise, and energetic motions, corresponds to the peduncle of a true Medusa, the angles being the lips. These lips, which correspond in their direction to the four internal ridges, are very protrusile, and when the little animal is active are continually being thrust out in various directions, sometimes everted, but more commonly made to approach each other in different degrees; sometimes one being bent-in towards the centre, sometimes all closing-up around a hollow interior. These four lobes, thus perpetually in motion, and changing within certain limits their form and their relation to each other, remind one of the lips or the tongues of more highly organized animals. The substance of this peduncle appears to be delicately granular; but there is a very manifest tendency to a fibrous character in its texture, the fibres being directed from the exterior towards the interior, supposing the lobes to have their points in contact.

Let us now look at the margin of the disk. Here are attached twenty-four slender tentacles, six in each quadrant formed by the divergent ribs, or radiating canals. Each tentacle springs from a thickened bulb, which is imbedded in the margin of the disk; it is evidently tubular, but the tube is not wider in the bulb than in the filament. The general surface is rough with projecting points, which in some assume a very regular muricate appearance, and the tentacle terminates in a blunt point. The discal part of the bulb is fringed with a row of minute bead-like spherules. Around the edge of the circumference of the disk, on the exterior, are arranged eight beautiful and conspicuous auditory vesicles. They are placed in pairs, each pair being approximate, and appropriated to each of the quadratures of the circle. Each of these organs consists of a transparent globe, not enveloped in the substance of the disk, but so free as to appear barely in contact with it: it contains a single otolithe, of high refractive power, placed not in the centre, but towards the outer side. The inexperienced naturalist, on first seeing these organs, would unhesitatingly pronounce them eyes, and the otolithe

the crystalline lens. They are, however, pretty certainly, rudimentary organs of hearing; the crystalline globule or otolithe being capable of vibration within its vesicle. Their exact counterparts are found in many of the smaller Medusæ, as we lately saw in the *Thaumantias*.

The disk is endowed with an energetic power of contraction, by which the margin is diminished, exactly like that of a Medusa swimming; and the tentacles have also the power of individual motion, though in general this is languid, their rapid flapping being the effect of the contraction and expansion of the disk, producing a quick involution and evolution of the margin, and carrying the tentacles with it. Occasionally, however, all the tentacles are strongly brought together at their tips, with a twitching grasping action, like that of fingers, which is certainly independent of the disk, and may be connected with the capture of the prey.

Now every detail of the structure here, as well as the general form, appearance, and habits, agrees with the small naked-eyed Medusæ, so closely that if we had not witnessed the birth of the little creature from the reproductive cell of a *Laomedea*, we should have pronounced it with unhesitating confidence a true Acaleph. The peduncle, it is true, seems out of place, being on the outside of the dome, instead of hanging suspended from its interior; but this difference is only apparent, and arises from the circumstance that the disk is reverted. If you suppose the edge of the disk to be turned in the opposite direction, you will have the peduncle in its normal place: the umbrella in these specimens is carried within, and the sub-umbrella without; an inversion which is probably accidental.

Comparing now this strange production of a Medusa by a Polype, with what I lately told you of the production of Polypes by a Medusa (as in the case of the lovely little *Turris*), you will have some acquaintance with the wondrous phenomena which have of late years been surprising and interesting naturalists,—viz., those of the Alternation of Generations; in which as Chamisso, the first discoverer of the strange facts, observed,—"a mother is not like its daughter, or its own mother, but resembles its sister, its grand-daughter, and its grandmother." The Polype gives birth to a generation of Medusæ which lay eggs, which develop into Polypes. The Medusa on the other hand lays eggs (gemmules), which develop into Polypes, which at length divide themselves into colonies of Medusæ.

At first you will perhaps see nothing remarkable in another object which I collected in my rock-ramble to-day. A Hermit-crab in an old *Natica* shell; both common things enough. Yet look more narrowly. The greater portion of the shell is not smooth, has no such porcelain-like polish as the *Natica* usually has, but is clothed with a sort of downy nap, a coarse sponginess of a greyish hue, splashed with yellowish and pink tints. The shell is invested with *Hydractinia.*

We restore the strange partnership,—shell, fleece, and crab,—to the glass of sea-water; where we soon see the whole tumbling about the bottom in uncouth agility.

Assist your eye with this pocket-lens, and look again. The shaggy nap upon the shell now bristles with tall slender polypes, crowded and erect, like ears of corn in a field.

No high power of magnification is necessary to furnish us with considerable entertainment from this populous colony. The polypes stand individually nearly half an inch in height; each consists of a straight slender column, surmounted by eight straight rod-like tentacles, four of which stand erect, slightly diverging, and the other four, alternating with these at their origin, extend horizontally like the arms of a turnstile.

The rough jolting of the crab over the stones the expanded polypes bear with equanimity; they are used to it; and though their tentacles wave and stream hither and thither, they are not retracted on this account. But just touch with the point of the pencil in your hand any part of the shaggy fleece, and instantly the whole colony retire together, as if by a common impulse, apparently shrinking into the substance of the shell. Yet they soon re-appear, one after another quickly protruding its closed tentacles, which are presently expanded as before.

The explanation of this phenomenon is, that the whole colony of polypes are but the free points, or feeding mouths of a common living film, which invests the shell; just as in *Laomedea,* the polypes that inhabit the vase-like cells are the off-shoots or free points of the common medulla.

The investing film will sometimes in captivity spread upon the glass side of a tank, and then develop all the polypes and organs proper to the complete organism. When this is the case, an admirable opportunity is presented for studying with ease and precision the economy of the creature ; and it is to the skill with which Dr. T. Strethill Wright has availed himself of such an opportunity* that I am indebted for the chief part of the facts that I am going to tell you, connected with the form and appearance, of which you can here judge for yourself.

The spreading film or polypary is a thin coat of transparent jelly, slightly coloured with various tints, which secretes and deposits within its substance a still thinner horny layer of chitine. This rises here and there into numerous spines and points, which are curiously ridged with toothed keels; and these ridges run in various directions over the horny layer also, making a fine network over it. The investing flesh, however, fills up all the cavities and areas so inclosed.

The mode in which the polypary increases is by throwing out from its edge a creeping band, exactly analogous to the root-thread of the *Laomedea*. This "propagative stolon, after leaving the point of its origin, increases rapidly in diameter, and sends out irregular branches. The tips of these branches are covered with a glutinous cement, by which they attach themselves tenaciously to glass, or other surface near them. Having attached themselves, they expand laterally, at the same time throwing out finger-like prolongations, which, as they come in contact with each other, coalesce, until a fleshy plate is found

* See Edin. New Philosophical Journal, for April, 1857.

dherent to the glass. Polypes are developed both from the loose branches and the attached polypary; and the latter is clearly seen to be permeated by a beautiful system of anastomosing canals, connected with the hollow bodies of the polypes. Within these canals may be detected an intermittent flow of fluid, containing particles, the dancing motion of which indicates the presence of ciliary action, and which, having passed in one direction for a short time, are arrested, and after a slight period of oscillation, commence to flow in an opposite direction."

The polypes which are developed from this living carpet are not all of the same form. No fewer than five distinct sorts exist, at one and the same time, and I doubt not we shall be able to find and to identify them all, on this well-grown specimen.

First, there are the alimentary polypes, which we have already cursorily glanced at. Within the space inclosed by the two circles of tentacles, there is a mouth with soft protrusile lips, which can be pushed out and folded back so as to hide tentacles, column, and all.

Scattered amongst these we see numerous polypes, which agree in general form with these, but with some remarkable abstractions and additions. They have no mouth nor stomach, and the tentacles are reduced to the smallest possible warts or protuberances denticulating the dilated tip. But the additions are still more peculiar. From the middle part of the column a number, from four to nine, of great oval sacs project, each attached by one end, while the other stretches out horizontally, thus surrounding the slender column. Each of these sacs is an ovarian capsule, and contains several ova of a brilliant yellow or crimson hue. Thus we have the second form,—that of the reproductive polypes.

In some places single ovarian capsules stand up alone from the fleshy carpet, agreeing in every respect with those which we have just examined, except that they are sessile, instead of being carried by a polype.

The fourth form is that of the tentacular polype. Here and there, from amidst the forest of shorter polypes,—alimentary and reproductive,—white threads are seen protruding, which extend to a length four or five times as great as theirs, and hang down or loosely float in the water. They are found on the outskirts of the whole compound structure, and at each extremity of the long diameter of the mouth of the supporting shell, so that they must, in their natural condition, reach to the ground, along which the crab-tenanted shell is carried, enabling the Zoophyte to seize and appropriate the atoms scattered by the crab whenever he takes his meals. The tips of these organs are covered with a dense pavement of large thread-cells; and they must doubtless perform the office of general purveyors to the composite animal.

But still more remarkable, more extraordinary than all we have been considering, are the objects which are now in view in the field of the microscope. You see a number of bodies, which Dr. Wright calls ophidian or spiral polypes, and which, as he truly observes, are "like small white snakes, closely coiled in one, two, or three spirals, and grouped immediately round the mouth of the shell." The habits of these polypes are still stranger than their forms. "When touched, they only draw their folds more closely together. But if any part of the polypary, however distant from them, be irritated, the spiral polypes uncoil, extend, and lash themselves violently backwards and forwards, and then quickly roll themselves up again; and that not irregularly or independently of each other, but all together, and in the same direction, as if moved by a single spring. A violent laceration of the polypary causes these polypes to remain extended and stretched like a waving and tremulous fringe across the mouth of the shell, for several minutes. The ophidian polypes (evidently a barren modification of the reproductive polype) are never found in any other situation on the polypary than in that before described, or round the margins of accidental holes in the shell. They have no mouth, and the tentacles are rudimentary. The walls of the body are very transparent, from the extreme vacuolation of the inner tissue. The muscular coat, as might be expected from the active movements of the polypes, is highly developed, and forms a beautiful object on the dark polarized field of the microscope, each spiral coil shining out as a bright double ring, divided by four dark sectors. The outer tissue of the whole body and tentacles is crowded with the larger thread-cells. The ophidian polypes are, doubtless, organs of defence or offence, like the motile spines and bird's head processes of the *Polyzoa*, or the pedicellariæ of the *Echinodermata;* but it is difficult to assign a reason for their peculiar situation. They vary much in number and size in different specimens of *Hydractinia*, but are rarely altogether absent." *

The reflections of the able zoologist who first called attention to these varied developments, and his comparisons of them with those of another polype-form which we have lately been observing, are so interesting and instructive that you will not deem it needful that I should apologize for citing them. "In our consideration of the *Hydractinia*," he observes, "our attention is arrested by the multitude of objects grouped together to constitute a single animal, their variety in form, and the sympathy which subsists between the different parts. The singular spinous skeleton; the expanded membrane of the polypary, with its beautiful internal network of tubes and delicate peripheric prolongations; the alimentary polyps, some white and filiform, others thick, fleshy, crimson, or yellow sacs, obligingly everted, to expose their interior to our microscopic eye; the reproductive polyps, with their richly coloured generative sacs; the sessile generative organs of the polypary; the ophidian polyps, coiled in neat spirals when at rest, but starting into furious action, like a row of well-drilled soldiers, when injury is inflicted on the body to which they are attached; and, lastly, the tentacle polyps, floating in the water like long and slender threads of gossamer, or dragging up heavy loads of food for the common good;—these, together with the intimate relation and sympathy subsisting between the polypary and its associated organs, all combine to form an object of the highest interest, and indicate

* Dr. Wright, op. cit.

that, in this fixed yet travelling zoophyte, we have a type of structure transitional between the dendritic *Hydroidæ* and the more highly organized *Acaleph*. In the simplest acalephoid form, such as the medusoid of *Companularia* [or *Laomedea*] (which is nothing more than an extension of the polypary specially organized for independent and motile life), we have (as in *Hydractinia*) an expanded polypary, represented by the umbrella, and permeated by vascular tubes from the confluence of which last spring, at the centre, the tentacular polyps, various in number; and between them the reproductive polyps, represented by the sessile generative sacs."*

You see here a jar, on the glass side of which are traced a number of very fine white lines, barely discernible by the unassisted eye. But by the aid of the lens you see that each line is a long and slender thread, which creeps along the glass, and at length starts out from it free for a short distance, and is then terminated by a long club-shaped body, which carries at its extremity four horizontally divergent organs, like the arms of a turnstile. Tracing down the threads to their lower extremities, you see that they are branches of one thread, which creeps irregularly over a filamentous sea-weed growing from a stone in the jar. The sea-weed had been in the vessel for several weeks, and the water having been undisturbed, the knobbed thread, which was originally confined to the plant, continued to grow, and coming into contact with the glass spread upon it. Many other threads have extended from the creeping

* Ibid., op. cit.

root, some of which stand up freely in the water, with their knobbed extremities floating in the wave.

This is one of the Polype tribe, named *Stauridia producta*, and as its form and structure are interesting, we will devote a few moments to its examination. We can easily sever one or two of the freely floating threads, and transfer the amputated portions to one of the live-boxes of the microscope. The motions and appearance of the club with its organs will be, for a while, little affected by the violence.

The long cylindrical thread is enveloped in a transparent horny tube, which, however, so closely invests it, that it is with difficulty distinguished. The club-shaped head, or individual polype, is an enlargement of the thread, which protrudes from the investing tube. It is swollen in the middle and rounded at the end, and many of the heads, which are more ventricose than the rest, contain a bubble of air in the centre. This air is doubtless taken in at the mouth, which is situated at the extremity; for, though you can discern no perforation, yet there is an aperture capable of being opened widely at the pleasure of the animal, and surrounded by protrusile, contractile, and expansile fleshy lips. I have several times seen this mouth opened, and partly everted, in kindred species; and once I had an opportunity of witnessing a quite unexpected use to which it was applied, viz. that of a great sucking disk. I had put the animal in such a live-box as this—the two glass surfaces being just sufficiently wide apart to allow it free liberty to turn about in all directions as far as it wished. On my looking at it after a momentary interval, I saw that the extremity had suddenly become a large circular disk of thrice the diameter of the body: its substance was gelatinous, full of oblong granules, arranged concentrically. I neither saw this disk evolved nor retracted; but after some time, on looking at it, the same phenomenon was repeated. In order to obtain a better sight of it, but without suspicion of what I was about to effect, I slightly turned the tube of the box, carrying with it the alga to which the polype was attached, my eye upon it attentively observing all the time. The base of the polype moved away from its position, but the broad disk was immoveable. I continued to turn the upper glass, until at length the body was dragged out so as to be considerably attenuated; still the disk maintained its hold on the lower glass, with no other change than that of being elongated in the direction in which it was dragged. At length it slowly gave way, and resumed its original shape by gradual and almost imperceptible diminution of the circumference.

Around this expansile, but now fast-closed mouth, you observe four tentacles, radiating in a plane at right angles to the axis of the thread, towards the four cardinal points; they are long, slender, straight, and each is terminated by a globose head of considerable size, resembling the arms of certain screw-presses, which are loaded with terminal globes of metal to increase their impetus when turned.

The structure of these tentacles is very interesting. The stem contains a core or central chain of large cells, which take a somewhat square outline from mutual pressure. The surface is roughened with small swellings, from each of which projects a long and excessively attenuated hair (*palpocil*), which is probably a very delicate organ of touch. The terminal globe is filled with proportionally large oval vesicles, each with a central cavity, which are arranged in a divergent manner around the centre, so that their tips shall reach the surface of the globe; these are those potent weapons of offence called thread-cells (*cnidæ*). The surface of the globe is covered with short thick palpocils, which Dr. T. S. Wright considers as prehensile organs. "These palpocils arise, each as a somewhat rigid process, from the side of one of the large thread-cells, buried in the head of the tentacle; and they probably convey an impression from bodies coming into contact with them, to the thread-cell, causing the extrusion of its duct."

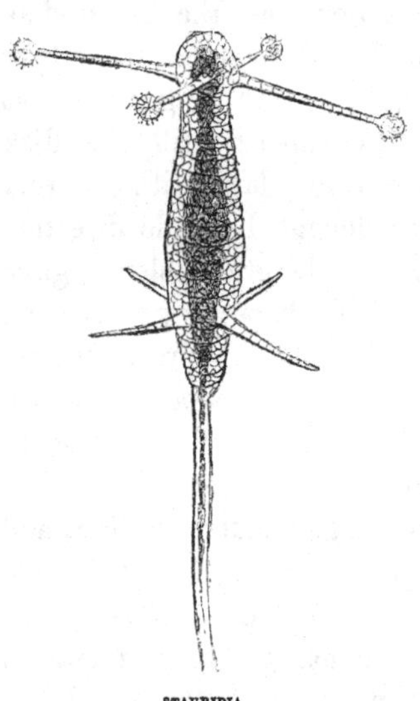

STAURIDIA.

Besides these globe-headed tentacles, there are, on the lower part of the club-foot, four other organs similar in every respect, except that they are not furnished with heads, nor any terminal dilatation whatever. They project horizontally as the knobbed ones, but their origin, and the respective lines of their radiation, are intermediate or alternate; in other words, if we consider the globe-heads as pointing N.-E. S. and W., the simple ones point N.-E., S.-E., S.-W., and N.-W.

From the carefully made observations of several excellent naturalists,—as Dujardin, Steenstrup, Dalyell, Lovén, and others—it appears that this beautiful and elegant little Polype gives birth to medusa-shaped young. Contrary, however, to the rule in *Laomedea*, the Medusa is in this case pushed forth as a bud from the side of the club, without any protecting capsule. The process is exceedingly like a plant developing a flower; for the bud grows until it at length expands blossom-like, and a beautiful little umbrella-form Medusa is seen adhering to the Polype. At length the brilliant little living flower becomes detached; and, after swimming freely for a time, discharges ova or gemmules from its ovaries, which develop into a creeping-root-thread, and finally into the club-headed threads of the *Stauridia*.

Some objects which I have to exhibit to you are altogether unique as to their appearance; and, if you are not as imperturbable as a philosopher of the Στοὰ, or a Mohawk Indian, will certainly excite both your risibility and your wonder. For some little time I have been keeping in this tank a specimen of that rather rare and very interesting *Sabella*, the *Amphitrite vesiculosa* of Montagu.* You see it is a worm, inhabiting a sort of skinny tube, much begrimed with mud, about two inches of its length being

* Linn. Trans. xi. 19.

exposed; the remainder, or about as much more, being concealed among the sand and sediment of the bottom.

A beautiful object is presented by the gill-fans of this worm. These organs are always elegant, whatever species of the genus is before us; but here, in addition to the charm of the slender filaments, so delicately fringed with their double comb-like rows of cirri, the tip of each bears a dark purple spherule. That of the anterior filament on each side is much larger than the rest, and forms a stout, globose, nearly black ball; the others diminish to about the twelfth on each side, where they disappear. These balls are placed on the inner or upper face of the filament-stem, at the point where the pectination ceases, the stem itself being continued to a slender point beyond it, and constituting the "short hyaline appendage" of Montagu. From their great resemblance to the tentacle-eyes of the Gasteropod Mollusca, I have little doubt that these are organs of vision. If so, the profusion with which the *Sabella* is furnished in this respect may account for its excessive vigilance; which is so great, that not only will the intervention of any substance between it and the light cause it to retire, but very frequently it will dart back into its tube almost as soon as I enter the room, even while I am ten feet distant.

It is not, however, to the tube, nor to the worm, that I wish specially to direct your attention: yet it is necessary that I say a preliminary word about the former. Ordinarily the tubes of these worms are formed of the fine impalpable earthy matters (clay, mud, &c.) held in suspension in the sea, incorporated with a chitinous secretion from the body of the animal; and therefore the surface of the tube is always rough and opaque. But in this individual case, probably owing to the habitual stillness of the water in the vessel not holding in suspension the particles of mud that ordinarily enter into the composition of the tube, the latest-formed portion is composed of pure transparent *chitine*, without any perceptible earthy element. This clear terminal portion of the tube you may perceive to be occupied by a curious parasite. About twenty bodies, having a most ludicrously-close resemblance to the human figure, and as closely imitating certain human motions, are seen standing erect around the mouth of the tube, now that the *Sabella* has retired into the interior, and are incessantly bowing and tossing about their arms in the most energetic manner.

LARES.

As soon as you have a little recovered from your surprise at this strange display, we will begin to examine the performers more in detail. A slender creeping thread, irregularly crossing and anastomosing, so as to form a loose network of about three meshes in width, surrounds the margin of the *Sabella's* tube, adhering firmly to its exterior surface, in the chitinous substance of which it seems imbedded. Here and there free buds are given off, especially from the lower edge; while from the upper threads spring the strange forms that have attracted our notice. These are spindle-shaped bodies, about $\frac{1}{40}$th of an inch in height, whose lower extremities are of no greater thickness than the thread from which they spring; with a head-like lobe at the summit, separated from the body by a constriction, immediately below which two lengthened arms project in a direction towards the axis of the tube.

Such is the external form of these animals, and their movements are still more extraordinary. The head-lobe of each one moves to and fro freely on the neck, the body sways from side to side, but still more vigorously backward and forward, frequently bending into an arch in either direction; while the long arms are widely expanded, tossed wildly upward, and then waved downward, as if to mimic the actions of the most tumultuous human passion.

Whenever the *Sabella* protrudes from its tube, these guardian forms are pushed out, and remain nearly in contact with the Annelid's body, moving but slightly; but no sooner does it retire than they begin instantly to bow forward and gesticulate as before. These movements are continued, so far as I have observed, all the time that the *Sabella* is retracted, and are not in any degree dependent on currents in the surrounding water, whether those currents be produced by the action of the Annelid or by other causes. They are not rhythmical; each individual appears to be animated by a distinct volition.

Applying a higher magnifying power than we have yet used to the animals, we find that the head-lobe encloses a central cavity; that the arms are also hollow, with thick walls, marked with transverse lines, indicating flattened cells, and muricated on the exterior; and that the body contains an undefined, sub-opaque nucleus, doubtless a stomachal cavity.

I cut out, with fine scissors, a segment of the tube, including two of the parasites, with the portion of the network of threads that carried them. They have become immediately paralysed by the division of the threads, but those that remain on the tube are unaffected by the violence. Subjecting one of the animals so cut out to the action of the compressorium, with a power of 560 diameters, the arms are seen to be formed of globose cells, made slightly polyhedral by mutual pressure, set in single series. The interior of these organs is divided by partitions, placed at intervals of about the diameter. Some at least of the cells contain a small bright excentric nucleus.

When the tissues were quite crushed down by the pressure of the compressorium, a quivering motion was visible among the disjointed granules, but it was very slight. No trace of cilia, nor any appearance of ciliary motion, was perceptible during life.

When I first discovered these strange beings, I was as much astonished by what I saw as you are; nor could I imagine to what class of animals they were to be referred. Neither did I know whether their presence on the tube of the worm was a mere accident, or whether it indicated a predominant instinct. On both these points, however, light has been shed.

This larger *Sabella* tube was not the only one infested with the parasites. I observed them on at least two smaller specimens of the same species, in the same situation, and with precisely the same movements. The extremity of one of these smaller tubes I cut wholly off, and placed in the live-box of the microscope. Two of the parasites only were on it, which were active at first, but in about an hour—probably from the exhaustion of the oxygen in the small quantity of water inclosed—they decomposed, or rather disintegrated, the outline dissolving, and the external cells becoming loose and ragged, and the whole animal losing its definite form.

One of these specimens, however, while yet alive and active, afforded me an observation of value. I had already associated the form conjecturally with the Hydroid Polypes, and was inclined to place it in the family *Corynidæ*, considering the arms to be tentacles, and the head-lobe to be homologous with them in character, but abnormal in form. It appeared to be a three-tentacled *Coryne*, with the tentacles simple instead of capitate. But while I was observing the individual in question, I saw it suddenly open the head-lobe, and unfold it into the form of a broad shovel-shaped expanded disk, not however flat, but with the two halves inclining towards each other, like two leaves of a half-opened book. This immediately reminded me of the great sucking-disk which, as I lately told you, I had seen evolved from the obtuse summit of *Stauridia producta,* and confirmed my suggestion of the natural affinities of the form.

Altogether unlike, in their shape, and in the unwonted vivacity and peculiarly human character of their movements, all the other members of their natural family that I had ever seen or heard of, these curious creatures have afforded much entertainment, not only to myself, but to those scientific friends to whom I have had opportunities of exhibiting them. When I see them surrounding the mansion of the *Sabella,* gazing, as it were, after him as he retreats into his castle, flinging their wild arms over its entrance, and keeping watch with untiring vigilance until he reappears, it seems to require no very vivid fancy to imagine them so many guardian demons; and the Lares of the old Roman mythology occurring to memory, I described the form under the scientific appellation of *Lar Sabellarum.* You may, however, if it pleases you better, call them "witches dancing round the charmed pot."

The Polypes that we have as yet been looking at are all of simple structure individually, though some of them we have seen united into a very populous community of compound life. We will now look at some whose organization is of a higher, that is, more complex character.

On this old worm-eaten oyster shell which has been

dredged up from the bottom of the sea, you observe several rounded lumps. They are of a cream-white hue, of somewhat solid texture, tough and hard to the touch, and studded all over with shallow depressions or pittings. The largest of these is not more than an inch and a half in height, by two-thirds of an inch in thickness; but specimens often occur of twice or thrice these dimensions, and much more divided than this; sometimes forming a rude resemblance to a hand of stumpy round fingers of sodden flesh,—whence the fishermen call the object, "Dead men's fingers," or sometimes, by a comparison equally apt, "Cows' paps." To zoologists it is known as *Alcyonium digitatum*.

Certainly there is nothing very attractive in these white lumps, as they now appear; but then they are now in undress; they do not expect to see company out of water. Their drawing-room is beneath the waves, in some submerged cave of ocean, where the sun's ray never penetrated, and where the only light is that dim green haze reflected from the sand and shingle of the sea-floor, save when, on gala occasions perchance, the *Laomedeæ* that fringe the walls light up their myriads of fairy lamps, and the tiny *Medusæ* crowd in to the watery festivities with their elfish circlets and spangles of living flame. It is then that the Cows' paps "take their hair out of paper," and display their loveliness to advantage.

Unfortunately, we have no card of invitation to these submarine routs, but perhaps we may induce one of the more juvenile of these beauties to indulge us, as a special favour, with a sample of the effect;

particularly if we can improvise a ball-room suited to the occasion. Let us try.

Selecting the very smallest specimen—a tiny thing no larger than a pea—I try to detach it without injury, by inserting the tip of my pocket-knife under the frilled lamina of oyster-shell on which it rests, and working off the fragment. I have succeeded: here it is; its attachment unbroken: it is still firmly adherent to the severed slice of shell, which is so small that I can drop it with its burden into this narrow trough of glass. The whole concern—trough, shell, and polype—is now to be dropped into this capacious jar of freshly dipped sea-water, and put away for an hour into a dark closet.

. . . .

Now let us see the result. Yes, it is as I expected. The united stimulus of the darkness and the sea-water has acted on the Cow's pap, just as would the rising and covering tide in its native cavern, after it had been left exposed for some hours by the recess of the sea. It is fully expanded, and is now as lovely as just now it was unpleasing.

In the first place it is swollen to twice its former dimensions, and has acquired at the same time a semi-pellucidity, and a more delicate hue. But in place of the pits on the surface (there were not more than half a dozen in this little specimen, which makes it more suitable for examination), it is covered with tall polypes, standing out on all sides, of crystalline clearness and starry forms, each eminently beautiful in itself, and surrounding the whole mass with a sort of atmosphere of almost invisible and impalpable lustre peculiarly charming.

Coy as these deep-water strangers are of displaying their beauties in our glaring aquariums, they will bear, when once they *are* expanded, a good deal of shaking with equanimity. Hence I may be able to transfer the trough with its contents from the jar to the stage of the microscope, and thus enable you to gaze on its details for a little while, before the dull sensorium of the creature is sufficiently warned of its ungenial position to cause it to shut itself up and resume its ugliness.

As the protruded polypes are exactly alike, it will be enough to confine our attention to one. It is an elevated tubular column of translucent substance, terminating in an expanded flower of eight slender pointed petals, which spring outward with a graceful swell, so as to give the form of a shallow bell to the general outline. The base springs, like the foot of a tree, from the margin of a cell, which penetrates the substance of the mass, into which we can see far down, and into which the whole of the now extended and expanded blossom was withdrawn when we first saw it, leaving only the shallow depression to mark its situation.

The form of the column is in general that of the trunk of a tree, or that of a long cone; but there is a sudden constriction just above the base, and another below the point, where what may be called the flower expands. It is the petals of this latter which constitute the principal charm of this creature. They are, properly speaking, the tentacles of the polype, answering in function and position to those on the *Laomedea*, but differing considerably from them in form. Each of the eight is thick and broad at its origin, and quickly tapers to a point: on each of two opposite sides—viz. those which look towards the two adjoining tentacles, runs a row of delicately slender filaments, which at the middle part of the

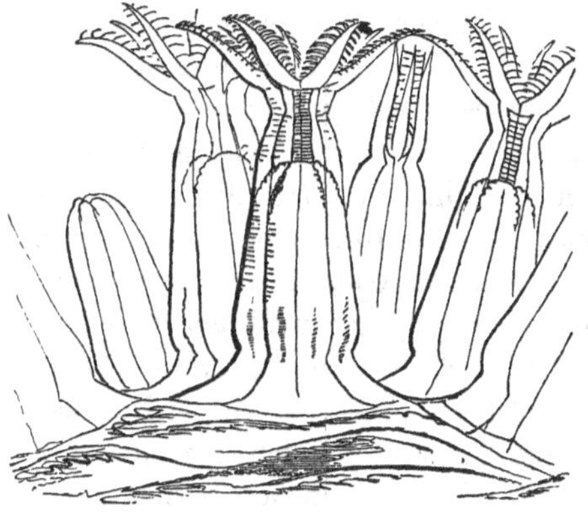

POLYPES OF COW'S PAP.

tentacle are moderately long, but diminish regularly as they approach either end. Starting from the side of the tentacle, in the plane of its transverse diameter, these elegant pinnæ presently arch downwards, but with perfect uniformity and symmetry. By means of the high magnifying power which I have now applied, each of these pinnæ is seen to be roughened with whorls of knobs, which are accumulations of cnidæ, analogous to those which we lately demonstrated in the tentacle of *Laomedea.*

In the midst of the area surrounded by the petal-like tentacles, a narrow slit opens into the stomach. This organ is a flat sac, resembling an empty pillow-case hanging down in the centre of the column, and open at the lower end. From this end, which does not extend to more than one-sixth of the depth of the cavity, three threads, much convoluted and irregularly thickened, spring off at each side and arch downwards, for a short distance. These are the reproductive organs, which fringe the free edges of as many delicate membranes which run up as perpendicular partitions between the stomach and outer wall, uniting with both, and thus dividing the space surrounding the stomach into chambers open at the bottom. There are eight of these septa, but one on each side is destitute of the fringing convoluted thread.

The whole surface of the interior—the walls, the stomach, and the septa—is clothed with fine vibratile cilia, by the action of which constant currents are maintained in the water, which bathes every part of the cavity, freely entering at the mouth. We can distinctly trace these ciliary currents hurling along with irregular energy the products of digestion, in the form of translucent granules, especially along the edges of the septa.

Though the substance of the polype be soft and flexible, it contains solid elements. Just below the expansion of the tentacular blossom, we see imbedded in the skin a vast aggregation of calcareous spicula. Individually, these are very minute, and their form is swollen in the middle, and taper at each extremity, the whole roughened with projecting knots. Collectively, they are grouped in regular forms, crowded into dense masses at the foot of each tentacle; the mass having a three-pointed outline, of which the central and largest point runs up into the tentacle.

Towards the lower region of the column, spicula again occur, scattered throughout the skin, and crowded into groups, one on each interseptal space. These spicula are of a very different shape from the upper ones; for they form short thick cylinders, with each end dilated into a star of five or six short branches, which are again starred at their truncate ends.

If we now sacrifice our little Cow's pap to our scientific curiosity, we shall see something of its internal structure. When removed from the water, the flower-like polypes soon retract. I now cut open the mass lengthwise with a keen knife, and you see that it is permeated by canals running from the base towards every part of the circumference, dilating here and there to form the cells which protrude and retract the polypes. This is a complete system of water-supply: the surrounding sea-water entering at the mouths of the several polypes, bathes the whole interior, and conveys oxygen and the products of digestion together to every part of the compound organism.

The fleshy substance which surrounds these canals is of a loose spongy character, and grates beneath the knife; a circumstance which is owing to the predominance of the calcareous element here, as you will see when I extract a small portion of it, and, laying it on a slip of glass, treat it with caustic potass. The microscope now reveals a large number of spicula, far larger than those we have hitherto observed, and different from either sort in form. These resemble very gnarled branches of oak, with the branchlets broken off close to their origin, leaving ragged and starred ends.

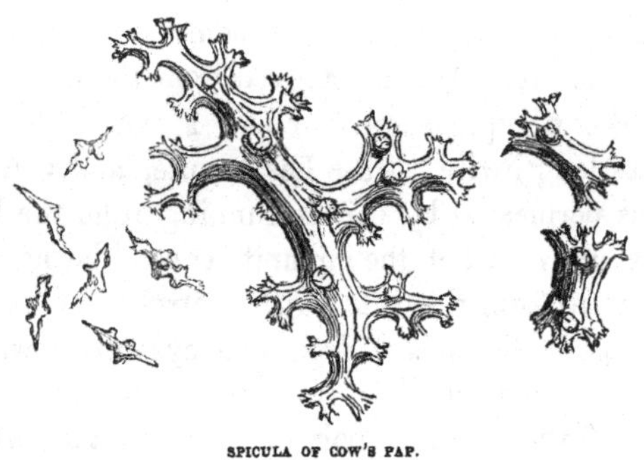

SPICULA OF COW'S PAP.

CHAPTER XIX.

SEA-ANEMONES: THEIR WEAPONS.

A VERY vast amount of the energy of animal life is spent either in making war, or in resisting or evading it. Offence and defence are sciences which the inferior creatures can in nowise neglect, since all are interested in one or other, and many in both; and various are the arts and devices, the tricks and stratagems, the instincts and faculties, employed in that earnest strife which never knows a suspension of hostilities. All classes of animals, invertebrate as well as vertebrate, are warriors by profession: the Spider is as carnivorous as the Lion, and more strategic; and the invisible Brachion is as ruthless and insatiable as either.

An enumeration and description of the diverse *weapons*, by means of which this truceless warfare is carried on, would make a volume: nor would the subject be then exhausted; for since it enters so largely into the very existence of animal life, the discoveries of advancing science are ever bringing to light new forms and modifications, strange and unexpected contrivances, all calculated to enhance our view of the inexhaustible resources of the Lord God Omnipotent, "who is wonderful in counsel, and excellent in working."

I am going to bring under your notice this evening some highly curious examples of animal weapons, of which the very existence was until lately altogether unsuspected; yet so profusely distributed that they are eminently characteristic of the two great classes of animals we have been recently considering—viz. the Medusæ and the Zoophytes. They have repeatedly fallen under our observation in examining the specimens of these creatures which we had selected, but I had reserved the fuller elucidation of them for an occasion in which they should come before us under circumstances of such unusual development as greatly to facilitate our researches. The weapons I speak of are the *cnidæ* or nettling-cells.

Look at this beautiful Scarlet-fringed Anemone (*Sagarta miniata*), expanding to the utmost its disk and tentacles in the clear water of the tank. I touch its body; instantly the blossom-like display is withdrawn; the column closing over it in the form of a hemispherical button, which goes on contracting spasmodically. At the same time see these white threads which shoot out from various points of the surface; new ones appearing at every fresh contraction, and streaming out to a length of several inches—resembling in appearance fine sewing cotton, twisted and tangled irregularly.

Now the animal has attained its utmost contraction, and the threads lengthen no more. But already they are disappearing; each is returning into the body by the orifice at which it issued. It is, as you may see by examining it carefully with a lens, gradually contracting into small irregular coils, at that end which is attached to the animal; and these little coils are, one after another, sucked in, as it were, through an imperceptible orifice.

Before the whole have disappeared, we will secure a portion for examination. For this end I cut off with a sharp scissors about one-sixth of an inch of the extremity of one of the threads, which now I transfer to a drop of sea-water in the compressorium. These threads are called *acontia.*

Examining this fragment under a low power of the microscope, we readily see that, though at first it seems a solid cylinder, it is really a flat narrow

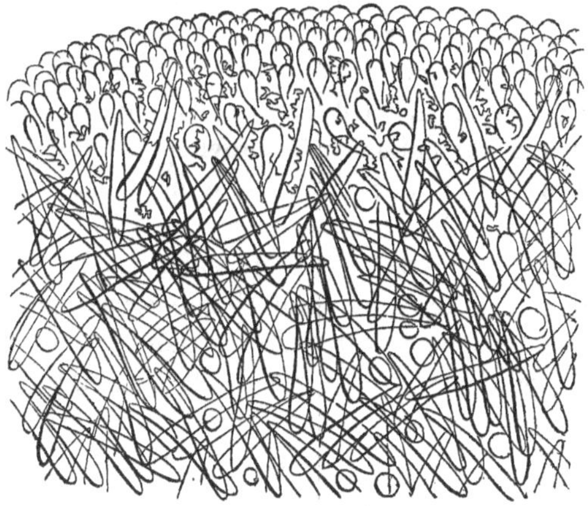

PORTION OF ACONTIUM (*flattened*).

ribbon with the edges curved in, which can at pleasure be brought into contact, and thus constitute a tube. Like all other internal organs in these animals, its surface is richly ciliated, and the ciliary currents not only hurl along whatever floating atoms chance to approach the surface, but cause the detached fragments themselves to wheel round and round, and to swim away through the water. Though there is not the slightest trace of fibre in the structure of the *acontium*, when scrutinized even with a power of eight hundred diameters, the clear jelly, or sarcode, of which its basis is composed, is endowed with a very evident contractility; the filament can contract or elongate; can extend itself in a straight line, or throw its length into spiral curves and contorted coils; can bring its margins together, or separate them in various degrees; can perform the one operation in one part, and the other at another, and thus can enlarge or attenuate the general diameter of the cord, apparently at will; and some of these changes can be effected even in the fragment detached from the animal, thus proving that the motile power, whatever it is, is localized in the constituent tissue itself.

Under pressure the edges of the flattened *acontium* appear to be thronged with clear viscous globules, overlapping one another, and protruding; indicating one or more layers of superficial cells, doubtless forming the epithelium. As the pressure is increased, these ooze out as long pear-shaped drops, and immediately assume a perfectly globular form, with a high refractive power. Below these are packed a dense crowd of *cnidæ*, arranged transversely.

Before we proceed to the examination of these curious organs in detail, it may be well to devote a moment's attention to the mechanism by which the *acontia* themselves are projected from the body. As this was first described (so far as I am aware) by myself,* I will take the liberty of citing some of my observations on the matter.

"The emission of the *acontia* is provided for by the existence of special orifices, which I term *Cinclides.* The integument of the body, in the *Sagartiæ*, is perforated by minute foramina, having a resemblance in appearance to the *spiracula* of Insects. They occur in the interseptal spaces; opening a communication between these (and therefore the general visceral cavity) and the external water. It follows that they are placed in perpendicular rows, but I have not been able to trace any other regularity in their arrangement. So far as I have seen, they are so scattered, that one, two, or even more contiguous intersepts may be quite destitute of a *cinclis.* I would not, however, attach too much weight to this negative evidence, since the animal has the power of closing them individually at will, and that so completely that the most careful scrutiny does not detect their presence.

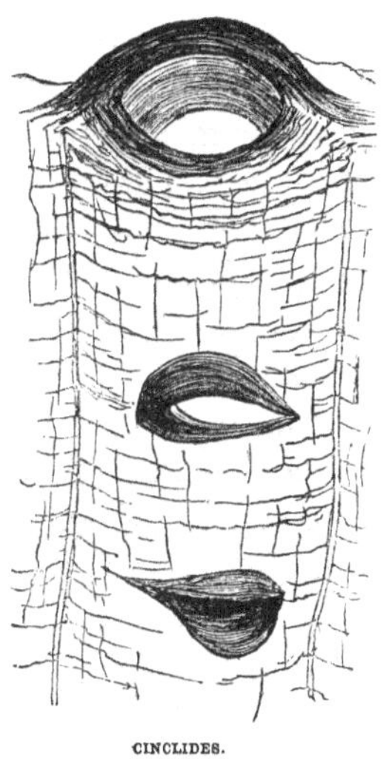
CINCLIDES.

Perhaps the best mode of examining them is to put a small specimen of *S. dianthus* or *S. bellis* into

* In a memoir entitled, "Researches on the Poison Apparatus in the Actiniadæ," read before the Royal Society, Feb. 4th, 1858.

a narrow parallel-sided glass-cell, filled with sea-water. After a while the animal will be much distended; the exhaustion of the oxygen impelling the Anemone to bathe its organs with as large a quantity of the fluid as it can inhale. The pellucidity of all the integuments will be thus greatly increased. A strong lamp-light being now reflected, by means of the mirror through the animal on the stage of the microscope, an inch or a half-inch object-glass will probably reveal the orifices in question with much distinctness.

"The appearance of the *cinclides* may be compared to that which would be presented by the lids of the human eye, supposing these to be reversed; the convexity being inwards. Each is an oval depression, with a transverse slit across the middle. When closed, this slit may sometimes be discerned merely as a dark line—the optical expression of the contact of the two edges; but, when slightly opened, a brilliant line of light allows the passage of the rays from the lamp to the beholder. From this condition the lids may separate in various degrees, until they are retracted to the margin of the oval pit, and the whole orifice is open.

The dimensions of the *cinclides* vary not only with the species, and probably also with the size of the individual, but with the state of the muscular contraction of the integument, as, also, I think, with the pleasure of the animal. In a small specimen of *S. dianthus*, I found the width of a *cinclis*, measured transversely, $\frac{1}{275}$th of an inch; but that of another, in the same animal, was more than twice as great, viz.

$\frac{1}{130}$th of an inch. This was on the thickened marginal ring, or parapet, which in this species surrounds the tentacles, where the *cinclides* are larger than elsewhere. Watching a specimen of *S. nivea* under the microscope, I saw a *cinclis* begin to open, and gradually expand till it was almost circular in outline, and $\frac{1}{275}$th of an inch in diameter. I slightly touched the animal, and it in an instant enlarged the aperture to $\frac{1}{200}$th of an inch. In a specimen of *S. bellis,* less than half-grown, I found the *cinclides* numerous, and sufficently easy of detection, but rather less defined than in *dianthus* or *nivea.* They occurred at about every fourth intersept, three intersepts being blind for each perforate one, and about three or four in linear series, but not quite regularly in either of these respects. In this case they were about $\frac{1}{66}$th of an inch in transverse diameter—a large size; and I measured one which was even $\frac{1}{55}$th of an inch. By bringing the animal before the window, I could discern the light through the tiny orifices with my naked eye.

From several good observations, and especially from one on a *cinclis,* widely opened, that happened to be close to the edge of the parapet of a *dianthus,* I perceived that the passage is not absolutely open, at least in ordinary, but that an excessively thin film lies across it. By delicate focussing, I have detected repeatedly, in different degrees of expansion, and even at the widest, the granulations of a membrane of excessive tenuity, and one or two scattered *cnidæ,* across the bright interval. On another occasion, in the case of a *cinclis* at the edge of the parapet, a position singularly favourable for observation, I saw that this subtle film was gradually pushed out until it assumed the form of a hemispherical bladder, in which state it remained as long as I looked at it. At the same time the outline of the *cinclis* itself was sharp and clear, when brought into focus further in. The film, whatever it be, is superficial, and does not appear to be a portion of the integument proper. I take it to be a film of mucus (composed of de-organized epithelial cells), which is constantly in process of being sloughed from all the superficial tissues in this tribe of animals, and which continues tenaciously to invest their bodies, until, corrugated by the successive contractions of the animal, it is washed away by the motions of the waves. As, however, one film is no sooner removed than another commences to form, one would always expect external pores so minute as these to be veiled by a mucus-film in seasons of rest.

The pressure of this film is sufficient evidence that the *cinclides* are not excretory orifices for the outflow of the respired water, in the manner of the discharging siphon in the Bivalve MOLLUSCA:—at least that no current constantly, or even ordinarily, passes through them. I have watched them continuously for periods sufficient to detect such discharge if it were periodic. On one occasion (viz. that in which the film was protruded like a blown bladder) a minute Infusorial animalcule chanced to pass across, close to the surface of the film: this would have been a decisive test of the existence of a ciliary current; but not the slightest deviation in the little atom's course could be detected.

That the *cinclides* are the special orifices through which those missile weapons—the *acontia*—are shot and recovered, rests not merely on the probability that arises from the co-existence of the two series of facts I have above recorded, but upon actual observation. In a rather large *S. dianthus*, somewhat distended, placed in a glass vessel between my eye and the sun, I saw, with great distinctness, by the aid of a pocket-lens, many *acontia* protruded from the *cinclides*, and many more of the latter widely open. The *acontia*, in some cases, did not so accurately fill the orifice but that a line of bright light (or of darkness, according as the sun was exactly opposite or not) was seen partially bordering the issue of the thread, while the thickened rim of the *cinclis* surrounded all.

The appearance of the orifices whence the *acontia* issued was that of a tubercle or wart, and the same appearance I have repeatedly marked in examples observed on the stage of the microscope; namely, that of a perforate pimple, or short columnar tube. This was clearly manifest when the animal, slowly swaying to and fro, brought the sides of the *cinclis* into partial perspective.

On another occasion I witnessed the actual issue of the *acontia* from the *cinclides*. I was watching, under a low power of the microscope, a specimen of a *S. nivea*, while, by touching its body rudely, I provoked it to emit its missile filaments. Presently they burst out with force, not all at once, but some here and there, then more, and yet more, on the repeated contractions of the corrugating walls of the body. Occasionally, the free extremity of a filament would appear, but more frequently *the bight of a bent one,* and very often I saw two, and even three, issue from the same *cinclis.* The successive contractions of the animal under irritation, caused the *acontia* already protruded to lengthen with each fresh impetus, the bights still streaming out in long loops, till perhaps the free end would be liberated, and it would be a loop no longer; and sometimes a new thread would shoot from a *cinclis,* whence one or two long ones were stretching already; while, as often, the new-comers would force open new *cinclides* for themselves. The suddenness and explosive force with which they burst out, appeared to indicate a resistance which was at length overcome:—perhaps (in part at least) due to the epithelial film above mentioned, or to an actual epiderm, which, though often ruptured, has ever, with the aptitude to heal common to these lowly structures, the power of quickly uniting again.

It appeared to me manifest from this and other similar observations, that no such arrangement exists as that which I had fancied:—that a definite *cinclis* is assigned to a definite *acontium,* or pair of *acontia;* and that the extremity of the latter is guided to the former, with unerring accuracy, by some internal mechanism, whenever the exercise of the defensive faculty is desired. What I judge to be the true state of the case is as follows: The *acontia,* fastened by one end to the *septa* or their mesenteries, lie, while at rest, irregularly coiled up along the narrow interseptal fossæ. The outer walls of these fossæ are pierced with the *cinclides*. When the animal is irritated, it immediately contracts: the water contained in the visceral cavity finds vent at these natural orifices, and the forcible currents carry with them the *acontia*, each through that *cinclis* which happens to lie nearest to it. The frequency with which *a loop* is forced out shows that the issue is the result of a merely mechanical action; which is, however, not the less worthy of admiration because of the simplicity of the contrivance, nor the less manifestly the result of Divine wisdom working to a given end by perfectly adequate means. The ejected *acontia*, loaded with their deadly *cnidæ* in every part of their length, carry abroad their fatal powers not the less surely than if each had been provided with a proper tube leading from its free extremity to the nearest *cinclis*."

Curious as these contrivances are, there is yet much more to be told: these are preparatory and ancillary, as it were, to the elaborate mechanism by which the ultimate object of the whole provision is to be attained. The *acontium* is but a reservoir for the weapons,—a kind of quiver for the arrows; and the *cinclis* is a provision for getting them ready for action: we have not yet looked at the arrows themselves.

They occur under three principal forms; and for the investigation of these we shall find it convenient to have recourse to different species.

The first and most generally distributed form is the Chambered Cnida, as it is also the most elaborately organized. I know of no species in which it can be examined under so favourable circumstances as the pretty Madrepore (*Cyathina Smithii*) of our south-western coasts; and as I have several specimens of that species in my aquarium, subjects are at hand for our investigation. The clear tentacles are, as you perceive, crowned with opaque globular heads; if I should nip off one of these heads and flatten it by means of the compressorium, you would

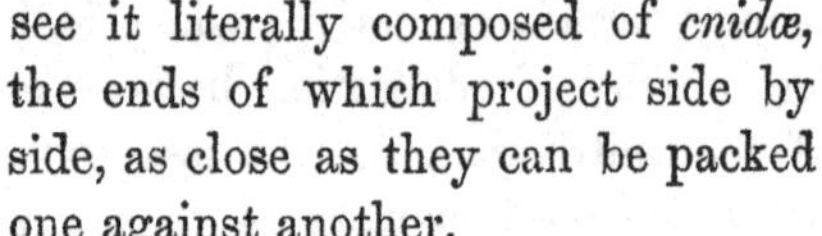

see it literally composed of *cnidæ*, the ends of which project side by side, as close as they can be packed one against another.

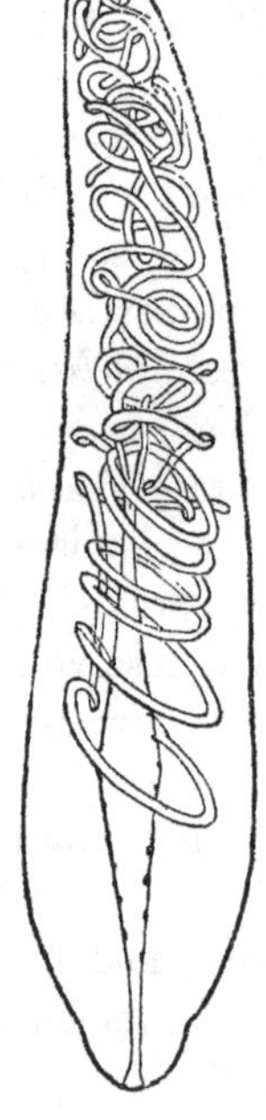

CNIDA OF MADREPORE.

But still larger examples may be obtained from the *craspeda*. With a smart sudden blow I break the stony skeleton of the Madrepore in sunder—the flesh tearing apart also; and thus I expose the interior of the living animal. A great number of pellucid ribbons are now seen, very much convoluted, which are named *craspeda*. These are almost composed of large *cnidæ*.

I remove with fine pliers a small fragment of one of these ribbons, and placing it between the plates of the compressorium, flatten it gradually till the plates are brought into as close contact as they can be. A high power now being put on, examine the organs in question.

You see a multitude of perfectly transparent, colourless vesicles, of a lengthened ovate figure, considerably larger at one end than at the other; one of average dimensions measures in length $\frac{1}{280}$th of an inch, and in greatest diameter $\frac{1}{2000}$th. "In the larger (the *anterior*) moiety, passing longitudinally through its centre, is seen a slender chamber, fusiform or lozenge-form, about $\frac{1}{6500}$th of an inch in its greatest transverse diameter, and tapering to a point at each extremity. The anterior point merges into the walls of the *cnidæ* at its extremity, while the posterior end, after having become attenuated like the anterior, dilates with a funnel-shaped mouth, in which the eye can clearly see a double infolding of the chamber-wall. After this double fold the structure proceeds as a very slender cord, which passing back towards the anterior end of the capsule, winds loosely round and round the chamber, with some regularity at first, but becoming involved in contortions more and more intricate, as it fills up the posterior moiety of the cavity. The fusiform chamber appears to be marked on its inner surface with regularly recurring serrations, which are the optical expression of that peculiar armature to be described presently.

"Under the stimulus of pressure when subjected to microscopical examination, and doubtless under nervous stimulus, subject to the control of the will, during the natural exercise of the animal's functions,—the *cnidæ* suddenly emit their contents with great force, in a regular and prescribed manner. It must not be supposed, however, that the pressure spoken of is the immediate mechanical cause of the emission: the contact of the glass-plates of the compressorium is never so absolute as to exert the least direct force upon the walls of the capsule itself; but the disturbance produced by the compression of the surrounding tissues excites an irritability, which evidently resides in a very high degree in the interior of the *cnidæ;* and the projection of the contents is the result of a vital force.

"In general, the eye can scarcely, or not at all, follow the lightning-like rapidity with which the chamber and its twining thread are shot forth from the larger end of the *cnidæ.* But sometimes impediments delay the emission, or allow it to proceed only in a fitful manner—a minute portion at a time; and sometimes, from the resistance of friction (as against the glass-plate of the compressorium), the elongation of the thread proceeds evenly, but so slowly as to be watched with the utmost ease; and sometimes the process which has reached a certain point normally, becomes from some cause arrested, and the contents of the cell remain permanently fixed in a transition state. Thus, a long continued course of patient observation is pretty sure to present some fortuitous combinations, and abnormal conditions, which greatly elucidate phenomena, that normally seemed to defy investigation.

"In watching any particular *cnida,* the moment of its emission may be predicted with tolerable accuracy, by the protrusion of a nipple-shaped wart from the anterior extremity. This is the base of the thread. The process of its protrusion is often slow and gradual, until it has attained a length about equal to twice its own diameter, when it suddenly yields, and the contents of the *cnida* dart forth. At this instant

I have, in many instances, heard a distinct crack or crepitation, both in the examination of this species and of *Sagartia parasitica*.

"When fully expelled, the thread or wire, which is distinguished by the term *ecthoræum*, is often twenty, thirty, or even forty times the length of the *cnida*; though in some species, as in most of the *Sagartiæ*, it frequently will not exceed one-and-a-half, or two times the length of the *cnida*.

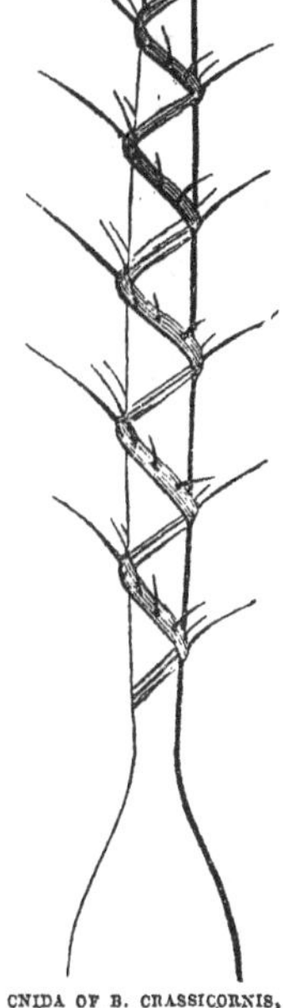
CNIDA OF B. CRASSICORNIS, *discharged.*

"The *ecthoræa* which are discharged by *chambered cnidæ*, are invariably furnished with a peculiar armature. The basal portion, for a length equal to that of the *cnida*, or a little more, is distinctly swollen, but at the point indicated it becomes (often abruptly) attenuated, and runs on for the remainder of its length as an excessively slender wire of equal diameter throughout. In the short *ecthoræa* of *Sagartia*, the attenuated portion is obsolete.

"It is chiefly upon this ventricose basal portion that the elaborate armature is seen, which is so characteristic of these remarkable organs. For around its exterior wind one or more spiral thickened bands, varying in different species as to their number, the number of volutions made by each, and the angle which the spiral forms with the axis of the *ecthoræum*. The whole spiral formed of these thickened bands, is termed the screw, or *strebla.*

"In the *ecthoræa* emitted by chambered *cnidæ* from the *craspeda* of *Tealia crassicornis,* the screw is formed of a single band, having an inclination of 45° to the axis, and becoming invisible when it has made seven volutions. In those from the same organ in *S. parasitica,* we find a screw of two equidistant bands, each of which makes about six turns—twelve in all,—having an inclination of 70° from the common axis. In those similarly placed in *Cyathina Smithii,* [now under your observation,] the *strebla* is composed, as you may perceive, of three equidistant bands, each of which makes about ten volutions—thirty in all—with an inclination of about 40° from the axis. In every case the spiral runs from the east towards the north, supposing the axis to point perpendicularly upwards.

"Sometimes, especially after having been expelled for some time, the wall of the *ecthoræum* becomes so attenuated as to be evanescent, while the *strebla* is still distinctly visible. An inexperienced observer would be liable, under such circumstances, to suppose that the screw, when formed of a single band, as in *T. crassicornis,* is itself the wire; an error into which I had myself formerly fallen. An error of another kind I fell into, in supposing that the triple screw of the wire in *C. Smithii* was a series of overlapping plates: the structure of the armature is the same in all cases (with the variations in detail that I have just indicated); and the structure is, I am now well assured, a spiral thickened band running round the wall of the *ecthoræum* on its exterior surface. I have been able when examining such large forms as those of *Corynactis viridis* and *Cyathina Smithii*, with a power of 750 diameters, to follow the course of the screw, as it alternately approached and receded from the eye, by altering the focus of the object-glass, so as to bring each part successively into the sphere of vision.

" These thickened spiral bands afford an insertion for a series of firm bristles, which appear to have a broad base and to taper to a point. Their length I cannot determinately indicate, but I have traced it to an extent which considerably exceeds the diameter of the *ecthoræum*. These barbed bristles are denominated *pterygia*.

" The number of *pterygia* appears to vary within slight limits. As well as I have been able to make out, there are but eight in a single volution of the one-banded *strebla* in *T. crassicornis*; while in the more complex screws of *S. parasitica*, *Cor. viridis*, and *Cy. Smithii*, there appear to be twelve in each volution.

" The barbs, when they first appear, invariably *project* in a diagonal direction from the *ecthoræum*; and sometimes they maintain this posture. But more commonly, either in an instant, or slowly and gradually, they assume a reverted direction.

" From some delicate observations made with a very good light, I have reason to conclude that the *strebla*, and even the *pterygia*, are continued on the attenuated portion of the *ecthoræum*, perhaps throughout its length. In *Cor. viridis* and *Cyathina Smithii*, I have succeeded in tracing them up a considerable distance. In the latter I saw the continuation of all these bands, with their bristles, but what was strange, the angle of inclination had become nearly twice as acute as before, being only 22° from the axis. The appearance of the attenuate portion, as also of the base of the ventricose part, is exactly that of a three-sided wire, twisted on itself; the barbs projecting from the angles.

"The next form of these organs is the Tangled Cnida. This form is very generally distributed, and is mingled with the former in the various tissues. In the genus *Sagartia*, however, it is by far the rarer form, while in *Actinia* and *Anthea* it seems to be the only one.

"The pretty little *Corynactis viridis* is the best species that I am acquainted with for studying this kind of *cnidæ*. [A fragment of its *craspeda* I have here ready for your observation, prepared exactly like that of *C. Smithii*.] Their figure is near that of a perfect oval, but a little flattened in one aspect, about $\frac{1}{250}$th of an inch in the longer, and $\frac{1}{700}$th in the shorter diameter. Their size, therefore, makes them peculiarly suitable for observations on the structure and functions of these curious organs. Within the cavity lies a thread (*ecthoræum*) of great length and tenuity, coiled up in some instances with an approach to regularity, but much more commonly in loose contortions, like an end of thread rudely rolled into a bundle with the fingers.

"The armature of this kind does not differ essentially from that already described. It is true, I have detected it only in *Corynactis*, where the short *ecthoræum* of the Tangled Cnida is surrounded throughout its length by a barbed *strebla* of three bands. The barbs are visible, under very favourable conditions for observation, even while the tangled wire remains enclosed in the *cnida*, but their optical expression is that of serratures of the walls, without the least appearance of a screw. This, I say, is the only species in which I have actually seen the armature of the *ecthoræum* in this kind of *cnidæ*, but I infer its existence from analogy in other species, where the conditions that *can* be recognised agree with those in this, though the excessive attenuation of the parts precludes actual observation of the structure in question.

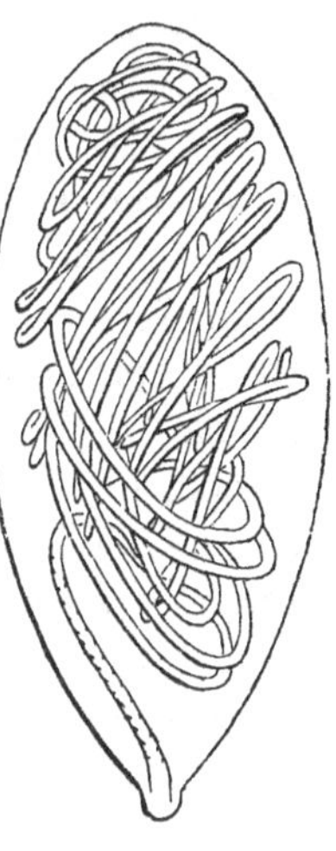
CNIDA OF CORYNACTIS.

"Spiral Cnidæ constitute the third form. In a few species, as *Sagartia parasitica*, *Tealia crassicornis*, and *Cerianthus membranaceus*, I have found very elongated fusiform *cnidæ*, which seem composed of a slender cylindrical thread, coiled into a very close and regular spiral. In some cases the extremities are obtuse, but in others, as in *T. crassicornis* [an example of which I now show you] the posterior extremity runs off to a finely attenuated point, the whole of the spire visible even to the last, the whole bearing no small resemblance to a multispiral shell; as one of the *Cerithiadæ* or *Turritelladæ*. The *ecthoræum* is discharged reluctantly from this form, and I have never seen an example in which the whole had been run off. So excessively subtile are the walls of the *cnidæ*, that it was not until after many observations that I detected them; in an example from *T. crassicornis*, which had discharged about half of the wire, I have not seen the slightest sign of armature on the *ecthoræum*. So far as my investigations go, these Spiral Cnidæ are confined to the walls of the tentacles, in which, however, they are the dominant form."

Such, then, is the form and armature of these organs. But I have not yet done with them. The emission of the wire, strange to say, is a process of distinct eversion from beginning to end. The *ecthoræum* is not a solid, but a tubular, prolongation of the walls of the *cnida*, turned-in, during its primal condition, like the finger of a glove drawn into the cavity. Of this fact you may convince yourself by a careful watching of the phenomena before you. Many of the *ecthoræa* from the *tangled cnidæ* now under your eye run out, not in a direct line, but in a spiral direction. Select one of these, and you will perceive that each bend of the spire is made, and stereotyped, so to speak, in succession, while the tips go on lengthening; the tip only progresses, the whole of the portion actually discharged remaining perfectly fixed; which could not be on any other supposition than that of evolution.

In the discharge of the *chambered* kind — to revert to those which we were just now examining—we saw the ventricose basal part first appear; the lower barbs flew out before the upper ones, and all were fully expanded before the attenuated portion began to lengthen. "This, again, is consistent only with the fact of the evolution of the whole. On several occasions of observation on the chambered *cnidæ* of *Cyathina Smithii*, I have actually seen the unevolved portion of the *ecthoræum* running out through the centre of the evolved ventricose portion. But perhaps the most instructive and convincing example of all was the following. One of the large tangled *cnidæ* of *Corynactis viridis* had shot about half of its wire with rapidity, when a kind of twist, or 'kink,' occurred against the nipple of the *cnida*, whereby the process was suddenly arrested. The projectile force, however, continuing, caused the impediment to yield, and minute portions of the thread flew out piecemeal, by fits and starts. By turning the stage-screw I brought the extremity of the discharged portion into view, and saw it slowly evolving, a little at a time. Turning back to the *cnida*, I saw the kink gradually give way, and the whole of the tangled wire quickly flew out through the nipple. I once more moved the stage, following up the *ecthoræum*, and presently found the true extremity, and a large portion of the wire still inverted; slowly evolving, indeed, but very distinct throughout its whole course, within the walls of the evolved portion.

" From all these observations there cannot remain a doubt of the successive eversion of the entire *ecthoræum*."

You ask, What is the nature of the force by which

the contained thread is expelled? "That it is a potent force is obvious to any one who marks the sudden explosive violence with which the nipple-like end of the *cnida* gives way, and the contents burst forth; as also the extreme rapidity with which, ordinarily, the whole length is evolved. A curious example of this force once excited my admiration. The *ecthoræum* from a *cnida* of *Corynactis viridis* was in course of rapid evolution, when the tip came full against the side of another *cnida* already emptied. The evolution was momentarily arrested, but the wall of the empty capsule presently was seen to bend inward, and suddenly to give way, the *ecthoræum* forcing itself in, and *shooting round and round the interior* of the *cnida*.

"The most careful observations have failed to reveal a lining membrane to the *cnida*. I have repeatedly discerned a double outline to the walls themselves, the optical expression of their diameter; but have never detected any, even the least, appearance of any tissue starting from the walls, as the *ecthoræum* bursts out. My first supposition, reluctantly resigned, was, that some such lining membrane, of high contractile power, lessened, on irritation, the volume of the cavity, and forced out the wire.

"The *cnida* is filled, however, with a fluid. This is very distinctly seen occupying the cavity when, from any impediment, such as above described, the wire flies out fitfully; waves, and similar motions, passing from wall to wall. Sometimes, even before any portion of the wire has escaped, the whole mass of tangled coils is seen to move irregularly from side to side, within the capsule, from the operation of some intestine cause. *The emission itself is a process of injection;* for I have many times seen floating atoms driven forcibly along the interior of the *ecthoræum*, sometimes swiftly, and sometimes more deliberately. Nothing that I have seen would lead me to conclude that the wall of the *cnida* is ciliated.

"I consider, then, that this fluid, holding organic corpuscles in suspension, is endowed with a high degree of expansibility; that, in the state of repose, it is in a condition of compression, by the inversion of the *ecthoræum;* and that, on the excitement of a suitable stimulus, it forcibly exerts its expansile power, distending and, *consequently*, projecting, the tubular *ecthoræum*—the only part of the wall that will yield without actual rupture."

It has been proved that the execution of these weapons is as effectual as their mechanism is elaborate. The wire shot with such force penetrates to its base the tissues of the living animals which the Anemone attacks, when its barbs preclude the withdrawal of the dart. But the entrance of bodies so excessively slender would of itself inflict little injury; there is evidently the infusion at the same time of a highly subtile poison into the wound; some venomous fluid escaping with the discharge of the *ecthoræum*, which has the power, at least when augmented by the simultaneous intromission of scores, or hundreds, of the weapons, of suddenly arresting animal vigour and speedily destroying life, even in creatures—fishes for example—far higher than the zoophyte in the scale of organization. I have seen a little fish in perfect health come in accidental contact with one of the *acontia* of an irritated *Sagartia*, when all the evidences of distress and agony were instantly manifested; the little creature darted wildly to and fro, turned over, sank upon the bottom, struggled, flurried, and was dead.

"Admitting the existence of a venomous fluid, it is difficult to imagine where it is lodged, and how it is injected. The first thought that occurs to one's mind is, that it is the organic fluid which we have seen to fill the interior of the *cnida*, and to be forced through the everting tubular *ecthorœum*. But if so, it cannot be ejected through the extremity of the *ecthorœum*, because if this were an *open* tube, I do not see how the contraction of the fluid in the *cnida* could force it to evolve; the fluid would escape through the still inverted tube. It is just possible that the barbs may be tubes open at the tips, and that the poison-fluid may be ejected through these. But I rather incline to the hypothesis, that the cavity of the *ecthorœum*, *in its primal inverted condition, while it yet remains coiled up in the cnida*, is occupied with the potent fluid in question, and that it is poured out gradually within the tissues of the victim, as the evolving tip of the wire penetrates farther and farther into the wound."

I do not think that the whole range of organic existence affords a more wonderful example than this of the minute workmanship and elaboration of the parts; the extraordinary modes in which certain prescribed ends are attained, and the perfect adaptation of the contrivance to the work which it has to do.

We must remember that all this complexity is found in an animal which it is customary to consider as of excessively simple structure. But the ways of God are past finding out. These are but parts of His ways.

CHAPTER XX.

PROTOZOA AND SPONGES.

We are so accustomed to see certain of the vital functions of animals performed by special organs or tissues, that we wonder when we find creatures which move without limbs, contract without muscles, respire without lungs or gills, and digest without a stomach or intestines. But thus we are taught that the function is independent of the organ, and, as it were, prior to it; though in nine hundred and ninety-nine cases out of a thousand it be associated with it. In truth, the simplest forms of animal life display very little of that division of labour, the minuteness of which increases as we ascend the organic scale; the common tissue is not yet differentiated (to use the awkward term which is becoming fashionable among physiologists) into organs, but is endowed with the power of fulfilling various offices, and performing many functions. In all probability, the function is but imperfectly performed; the specialization of certain tissues, and their union into organs, and the complexity of such combinations, no doubt, perform the given function in a far more complete degree; and it is the number and elaborateness of these that constitute one animal higher in the scale than another. The human lung is no doubt a more complete breathing apparatus than the entire ciliated surface of an Infusory, and the human eye sees more perfectly than the loose aggregation of pigment granules on the edge of a Medusa. But this diversity is essential to creation, as the great and wondrous plan which we see it to be; and, meanwhile, we may rest satisfied that the humble requirements of the lowest organism are met adequately by its humble endowments.

This evening I propose to show you some of these humble conditions of animal life—the lowest of the lowly. I have here two or three phials of very rich water dipped from the fresh-water ponds in the neighbourhood. All collections of water are not equally productive; and very far indeed is the popular notion from correctness, that every drop of water which we drink contains millions of animalcules. You may find many collections of clear water, springs, streams, and pools, from which you may examine drop after drop in succession, with the highest powers of the microscope, and scarcely discover a solitary animalcule. Again, it is not stagnant and fetid pools that are the richest in vitality; though no doubt you will always obtain some forms abundantly enough in such conditions. According to my own experience—an experience of many years—the paucity or profusion of animal life in any given collection of water can never be determined beforehand; the season, the situation, the aspect, the character of the country, and many other unsuspected conditions, may influence the result; which yet one may often give a shrewd guess at. Generally speaking, small ponds, in which a good deal of sub-aquatic vegetation grows—and particularly if this be of a minutely-divided character, such as *Myriophyllum*, *Chara*, &c., and whose surface is well covered with duckweed (*Lemna*), yields well; and, in collecting, it is desirable so to dip as that some of the fine loose sediment of the bottom may flow into your phial, and then to pluck up one or more of the filamentous water-plants, and introduce these into your vessel.

Now, to examine such a collection, proceed as I am about to show you. I hastily glance with the pocket-lens over the foliage, and selecting such filaments as seem the most loaded with dirty floccose matter, I pluck off with pliers one or two, together with one or two of the cleaner ones that are higher up on the plant, nearer the growing point. Having laid these on the lower glass of the live-box, I take up with the tip of a fine capillary tube, or a pipette, a minute quantity of the water at the bottom, which flows in as you see, carrying a few granules of the sediment. This drop I discharge upon the glass of the live-box, put on the cover, and place the whole on the stage of the microscope.

First let us use a low power—one hundred diameters or so—in order to take a general glance at what we have got. Here is an array of life, indeed! Motion arrests the eye everywhere. "The glittering swift and the flabby slow" are alike here; clear crystal globules revolve giddily on their axes; tiny points leap hither and thither like nimble fleas; long forms are twisting to and fro; busy little creatures are regularly quartering the hunting-ground, grubbing with an earnest devotedness among the sediment, as they march up the stems; here are vases with translucent bodies protruding from the mouths; here are beauteous bells, set at the end of tall threads, ever lengthening and shortening; here are maelströms in miniature, and tempests in far less than a teapot; rival and interfering currents are whirling round and round, and making series of concentric circles among the granules. Surely here is material for our study.

I see an object slowly creeping along the glass, which will be just the thing for our purpose. It is the Proteus (*Amœba diffluens*). Let me put on a higher power, and submit it to your observation.

You see a flat area of clear jelly, of very irregular form, with sinuosities and jutting points, like the outline of some island in a map. A great number of minute blackish granules and vesicles occupy the central part, but the edges are clear and colourless. A large bladder is seen near one side, which appears filled with a subtile fluid.

But while you gaze on it, you perceive that its form is changing; that it is not at two successive moments of exactly the same shape. This individual, which when you first looked at it was not unlike England in outline, is now, though only a few minutes have passed, something totally different; the projecting angle that represented Cornwall is become rounded and more perpendicular; the broken corner that we might have called Kent has formed two little points, up in the position of Lincolnshire; the large bladder which was in the place of the Eastern counties is moved up to the Durham coast, and is, moreover, greatly diminished; and other like changes have taken place in other parts.

Lo! even while speaking of these alterations, they have been proceeding, so that another and a totally diverse outline is now presented. A great excavation takes the place of Dorset; Kent is immensely prolonged; the bladder has quite disappeared, &c.; but it is impossible to follow these changes, which are ever going on without a moment's intermission, and without the slightest recognisable rule or order.

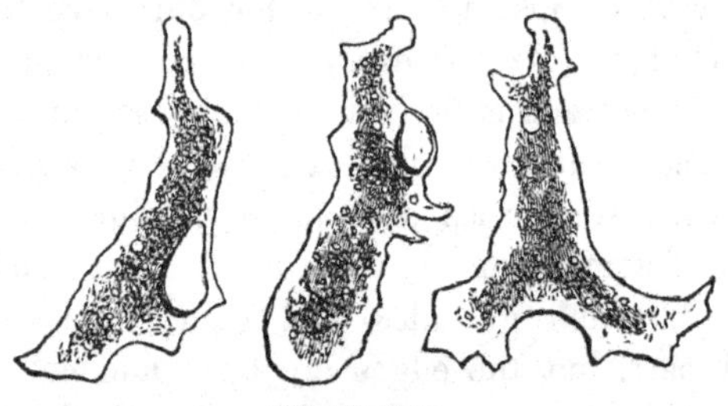

FORMS OF AMŒBA.
Successively drawn from one individual.

The projections are obliterated or exaggerated; the sinuosities are smoothed, or deepened into gulfs, or protruded into promontories; firths form here, capes there; not by starts, but evenly, and with sufficient rapidity to be appreciable to the eye while under actual observation; though the alterations are more striking if you take your eye off the object for a few seconds, and then look again; and still more so, if you try to sketch the outline. Individuals vary greatly in dimensions; this specimen is about one hundred and twentieth of an inch in long diameter, but others I have seen not more than one-tenth as large as this, and some twice as large.

Disregarding now this peculiarity of change of

form, which has procured for it the name of the old versatile sea-god that was so difficult to bind, we will concentrate our attention on some other points not less interesting. That great bladder undergoes changes besides those gradual alterations of place which are dependent on the general form. It slowly but manifestly increases in size up to a certain extent, when it rather suddenly diminishes to a point, and immediately begins to fill again, as slowly as before. These alternations go on with some regularity, and we cannot observe them without becoming convinced that it is a process of filling and emptying; that the bladder gradually fills with a fluid which is either secreted by its walls or percolates into it from the surrounding tissue; which fluid, when full, the bladder discharges by a sudden contraction of its outline. But whither the fluid goes it is difficult to determine; I have never been able, in this or in any other instance of its occurrence—though this contractile bladder is characteristic of the extensive classes *Infusoria* and *Rotifera*—to see any issue of fluid from the body at the moment of contraction, and therefore conclude that it is discharged into the body, perhaps back again into the tissues whence it was taken up, and whence it is about to be collected again. Hence, it is probably the first obscure rudiment of a circulation; the fluids impregnated with the products of digestion being thus collected and then diffused throughout the soft and yielding tissues.

The smaller bladder-like spaces that you see in considerable numbers in the substance of the animal, are collections of fluid contained in excavations of that substance, which are called *vacuoles*, differing from vesicles, inasmuch as they seem to have no proper wall or inclosing membrane, but to be merely casual separations of the common substance, such as would be made by drops of water in oil. These vacuoles appear to be connected with the digestive function; for very many of them are not clear, but are occupied with granules more or less opaque, and of exceedingly various dimensions. That these collections of granules are food, you will see by this experiment.

I mingle a little carmine with the water, just enough to impart a visible tinge to it, and close the live-box again. Already you perceive that some of the tiny globules are become turbid and red, and that their opacity and colour are deepening perceptibly. We see by this that the particles of carmine have been taken into the jelly-like sarcode, and are accumulating in little pellets surrounded by fluid, in these casual hollows of its substance. The process is rendered still more obvious when, as is often the case, some *Diatomacean*, with a hard siliceous shell, becomes the food of the *Amœba*. The apparently helpless jelly spreads itself over the organism, so as soon to envelope it; the flesh, which having no skin can unite with itself wherever the parts come into contact, closes over the Diatom, which is thus brought into the midst of the sarcode, a vacuole being new-made for its reception. This, then, performs the part of a temporary stomach, the digestible portions of the prey are extracted, and then the insoluble shell of flint is, as it were, gradually squeezed to some part of the exterior, and gradually forced out, the vacuole disappearing with it, or perhaps retaining a minute portion of the fluid, and thus perpetuating itself for a while. This is the earliest condition in which the process of digestion can be recognised.

Another genus somewhat similar is *Arcella*, but it differs in being furnished with a more or less rounded shell (*lorica*), like a little box. In examining the matters that adhere to the stems of Duckweed, and other water plants, we frequently observe little circular bodies of a yellowish or reddish brown colour, some much darker than others, but all having a central round spot paler than the rest. On first examination they seem inert and dead, but if we closely watch one, we perceive that it is endowed with the power of motion; and we directly discern thrust out from its edge, variable processes, in the form of arms, of clear, perfectly colourless, and most delicate jelly, sometimes pointed, sometimes blunt, which slowly change their form and position. By the aid of these, a feeble and irregular motion is given to the box, which is sometimes turned partly over; when we perceive that its under-side is flat or probably concave, and that its outline is cut into facets. The *lorica* is somewhat flexible, for the edges at two opposite points are sometimes bent down towards each other, so as to give the creature the form of a crescent. The internal viscera are dimly discernible through the coloured *lorica*, and resemble those of *Amœba*. A dark oval ring is commonly seen at one side, which is probably the outline of the contractile bladder. It may, in fact, be considered as an *Amœba*, whose external surface has the power of secreting a symmetrical shell of horny, or chitinous substance. The *lorica* is about $\frac{1}{500}$th of an inch in diameter. This species is named *Arcella vulgaris.*

Laying aside our live-box with its contents for the present, we will have recourse to the tank of sea-water for one or two other objects of intermediate interest. On the green and brown mossy sea-weed which covers the rocks on the bottom, you see many white specks clinging to the filaments ; and there are several adhering to the sides of the tank. These are little living shelled animals of the class *Foraminifera,* and these which you see include several species. By bringing your eye assisted by the lens to bear upon one of these latter, you perceive that it is a little discoid spiral shell, of very elegant form, marked with curved diverging grooves. This is the pretty little *Polystomella crispa,* a fair sample of its class, and though not more than $\frac{1}{30}$th of an inch in diameter, it is a giant compared with the *Arcella.*

There is more however than the shell to be seen ; though so filmy and shadowy that I wonder not at your overlooking it. Extending from two opposite sides of the shell to a distance each way considerably exceeding its diameter, you discern fine threads of clear jelly, running out in long points. The power you employ is not sufficient to enable you to resolve their detail : and for this, I will try to secure a specimen for the microscope.

In this other live-box, then, I inclose one of the white specks from the moss-like clothing of the stones. It is, I see, of another species, namely,

Polymorphina oblonga, but it will answer our purpose equally well.

At present we see only the shell, the removal of the animal having induced it in alarm to withdraw the whole of its softer parts within the protection of its castle. We must have a few minutes' patience.

Now look again. From the sides of the opaque shell we see protruding tiny points of the clear sarcode; these gradually and slowly,—so gradually and slowly that the eye cannot recognise the process of extension—stretch and extend their lines and films of delicate jelly, till at length they have stretched right across the field of view. The extension is principally in two opposite directions corresponding to the long axis of the shell; though the branched and variously connected films often diverge considerably to either side of these lines, giving to the whole a more or less fan-shaped figure.

These films are as irregular in their forms and sizes as the expansion of the sarcode of *Amœba*, with which they have the closest affinity. Their only peculiarity is their tendency to run out into long ribbons or attenuated threads, which however coalesce and unite whenever they come into mutual contact, and thus we see the threads branching and anastomosing with the utmost irregularity, usually with broad triangular films at the points of divergence and union.

There can be no doubt that the object of these lengthened films, which are termed *pseudopodia*, is the capture of prey or food of some kind; perhaps the more sluggish forms of minute animalcules, or the simpler plants. These the films of sarcode probably entangle, surround, and drag into the chambers of the shell, digesting their softer parts in temporary vacuoles, and then casting out the more solid remains, just as the *Amœba* does.

Though this beautiful array was so very deliberately put forth, it is, as you perceive, very rapidly withdrawn on any disturbance to the animal, as when we agitate the water, by slightly moving or turning the cover of the live-box. Another fact, of which you may convince yourself, by watching manifest though small changes of position in the shell while under observation, is, that it is by means of the adhesion and contraction of the *pseudopodia* that the animal drags itself along a fixed surface. This it can effect so assiduously, that I frequently find them in the morning adhering to the tank-sides three or four inches from the bottom, though on the previous evening none were visible on the glass. Thus they must crawl, on occasion, from a hundred to a hundred and fifty times their own diameter in a night.

The structure of a Sponge is much the same as that of these animals, with the exception that its solid part or skeleton is not a continuous covering by which the sarcode is invested, but consists of fibres or points or rods of varying form, which are clothed with the sarcode. This loose sort of skeleton may be of horny or chitinous matter, like that of *Arcella*, or calcareous, like that of the *Foraminifera*, or it may be siliceous,—that is, composed of flint (*silex*).

In some cases, as in the common Turkey Sponge, the horny skeleton consists of a network of solid but slender fibres, very tough and elastic, which branch and anastomose in every direction, at very short intervals, as you may see by looking at this atom, which I cut off from a dressing sponge.

In the lime and flint Sponges, however, the continuity and cohesion of the skeleton does not depend upon the organic union of the constituent parts, as it does in the loose and open network of the Turkey sponge. For it is made up of an immense multitude of glassy needles, all separate and independent, between themselves, yet so contrived that they do hold together very firmly, and in a great number of cases are arranged on a prescribed plan, so as to give a certain form and outline to the aggregate.

If you have ever shaken up a box of dressing-pins, and have then endeavoured to take one out, you know how by their mere interlacement they adhere together in a mass, so that by taking hold of one you may lift a bristling group of scores. Somewhat on the same principle are the calcareous and siliceous pins (*spicula*) of a Sponge held together by mutual interlacement. Yet their cohesion is aided by the tenacity of the living sarcode which invests them; for I have found that specimens of *Grantia* (calcareous Sponges with needles of three rays), when long macerated in water, so that the sarcode is dissolved, have very slight power of cohesion among their spicula.

To understand the structure of a Sponge we will shave a thin sectional slice from this *Halichondria suberea.* This when alive is of an orange colour; and is always found closely investing turbinate shells which are inhabited by Hermit-crabs. We will macerate the slice in tepid water for a quarter of an hour, and then examine it in the live-box.

The surface is a thin layer of greater density than any other part, and is composed of coloured fleshy granules,—omitting for the present, the skeleton. Of the same substance is the whole slice composed, but looser and more open as it recedes from the surface. It is separated by blank spaces which are larger towards the centre, smaller and more numerous as they approach the exterior.

These openings are sections of so many canals, by which the whole substance of a sponge is permeated.

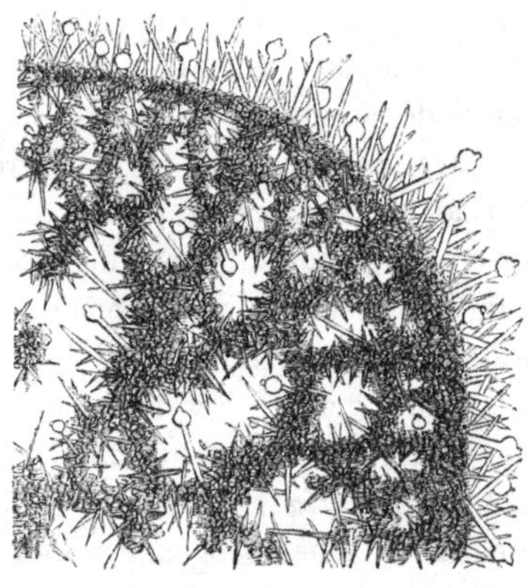

SECTION OF SPONGE.

The surface is perforated with minute pores, at which the surrounding water enters on all sides. These presently unite into slender pipes, which, irregularly meandering, are continually uniting into larger and yet larger canals; of which the greater open spaces that you see are the oblique divisions. These have certain outlets, called *oscula*, on the surface, from which the stream is poured that has thus made the grand tour of the whole interior. Such *oscula*, as you perceive on the remainder of the *Halichondria*, are usually raised on slight eminences; and resemble, especially when in living action, miniature volcanoes, vomiting torrents of water and granules of effete matter, instead of fire and ashes.

During life these granules were much more diffused, and formed a considerable portion of the living flesh, the remainder being composed of a glairy sarcode, almost fluid. The whole was maintained in position by the solid spicula of flint, which you see abundantly in this slice. These take a curious form, exactly that of the pins which we use on our dressing tables; each consisting of a cylindrical slender rod, pointed at one end, and at the other surmounted by a globular head, the whole formed of glass,—*flint glass* literally. You see them bristling all round the edge of the section, being stuck into the surface of the sponge, exactly as pins are loosely stuck into a pin-cushion. The heads and points, too, project into the cavities; more, however, than they did during life, for you must make allowance for the shrinking of the soft parts; and thus you perceive how the whole structure is permeated by these glassy pins, which seem to be entangled together quite at random without rule or arrangement. And yet there is an arrangement discernible here; for the canals are formed by the manner in which these are grouped; and this is seen much more clearly, in the case of the three-rayed needles of lime in the *Granticæ.* Mr. Bowerbank has shown that in *G. compressa* the substance is divided into very regular chambers in a double series, separated by a diaphragm, whose axis is at right angles to the axis of the sponge; and that these chambers are defined by walls made up of the three-rayed needles in their mutual interlacement.

CHAPTER XXI.

INFUSORIA.

Revenons à nos moutons. We will resume our examination of the drop of pond-water, and the fragments of *Myriophyllum*, which have been waiting for us in the live-box.

Our attention then shall first be given to some elegant creatures of a brilliant translucent green hue, which are gracefully gliding about. They are of the genus *Euglena*, so called because each is furnished with a very conspicuous spot of a clear red hue, situated near the head, which Ehrenberg, on account of its resemblance to the lowest forms of eyes in the *Rotifera*, that are somewhat similar in colour and appearance, pronounced to be an organ of vision. More recent physiologists, however, doubt the correctness of the conclusion.

The animals are of several kinds. The most numerous is an active little thing of about $\frac{1}{250}$th of an inch in length when extended, though from its extreme versatility it is as difficult to assign to it a definite size, as a definite shape. It seems to be the *E. sanguinea*, so called because it is said to occur sometimes of a deep red hue, and in such vast profusion, as to give the waters the appearance of blood. I have never seen it, however, other than as it now appears, rich emerald green in the body, with the two extremities perfectly clear and colourless. I might, perhaps, describe its ordinary form as spindle-shaped, with a pointed tail, and a blunt, rounded head; but it is remarkable for the variableness of its shape. It is capable of assuming an appearance very diverse from what it had half a minute before, so that you would hardly identify it, if you were not watching its evolutions. Whether this ability to prove an *alias* be at all dependent on the remarkable *clear-headedness* of the subject, I leave for you who are skilled in metaphysics to determine. Away they go, tumbling over and over, revolving on the long axis as they proceed, which they do not very rapidly, with the blunt extremity forward.

Here is another form, a little larger than the former, but much more slender; yet from the slowness and steadiness of its movement more easy of observation. It is named *E. acus,* or "the Needle Euglena." This is an animalcule of great elegance and brilliance; its sparkling green hue, with colourless extremities, and its rich pale crimson eye, are very beautiful. It commonly swims extended, with a slow gliding motion, turning round on its long axis as it proceeds, as may be distinctly seen by the rotation of certain clear oblong substances in its body. These then are seen not in the interior, but near the surface, as they would appear if imbedded in the flesh around a hollow centre. The interior is probably not hollow, but occupied with pellucid sarcode. These were assumed by Ehrenberg, but on no adequate grounds, to be organs connected with reproduction. They vary in number in different individuals, and those which contain the greatest number are thereby more swollen. They appear to be separated into two series, one anterior, the other posterior. The animal is capable of bending its head and body in various directions, but is most beautiful when straight. The front is furnished with a slender thread-like proboscis. This species affords us a good opportunity of observing the red spot which, for convenience sake, we may still term an eye. It seems to be an irregular oblong vacuole, or excavation in the sarcode, filled with a clear ruby-red fluid. The red spot in the *Rotifera* is connected with a well-defined crystalline lens, whose definite form, and high refractive power, may in many cases be distinctly marked; but here nothing of the kind is seen; the spot itself has no certain shape, and does not appear to be bounded by a proper wall. Some forms, which are by general consent admitted to be plants, have similar spots; and hence it has been, rather too hastily, I venture to think, concluded, that they can have no connexion with vision. I think it still possible, that a sensibility to the difference between light and darkness may be the function of the organ.

I have found that this animal, when allowed to dry on a plate of glass, retains its form and colour perfectly; but in about two days the eye-spot, which at first becomes much larger in the drying, gradually loses all traces of its brilliant colour, probably by the evaporation of the contained fluid.

Another pretty species you see gliding along amongst the rest, called *E. triquetra*, or the Three-sided. It bears a resemblance to a broad rounded leaf, with the footstalk forming a short transparent point, and the mid-rib elevated into a sharp ridge. The under side seems slightly concave. This is equally attractive with the others. It is persistent in form, and appears not to be even flexible. Its motion is slow, and as it goes, it rolls irregularly over and over in all directions, not revolving on its long axis, and thus giving you very satisfactory views—though only momentary—of the keel with which the back is furnished. It is in the turnings of such minute creatures that the microscopist often gets a glimpse of peculiarities of form, which a view of the animal when in repose, however long continued, fails to reveal. Longitudinal interrupted lines are seen running down the body of this pretty leaf, which do not appear to mark irregularities of the surface, and therefore are probably internal. Ehrenberg calls these and similar collections of granules "ova," or eggs; but this is to cut the knot instead of untying it. There is no sufficient reason to believe that these animals increase by ova.

THREE-SIDED EUGLENA.

About the front of all these *Euglenæ*, you may discern now and then a slight flickering or quivering in the water. The power we are using, though best for the general display of the form, is insufficient to resolve this appearance: I will put on a higher objective. You now see that there proceeds from the frontal part of the body, a long and very slender filament, which is whisked about in the manner of a whip-lash. This is considered to be the organ of locomotion; but I rather doubt that such is the function; the smooth and even gliding, often rotating, of the creature, seems more like that produced by minute and generally distributed ciliæ, than that caused by the lashings of a single long thread.

Yet two more species of this extensive genus we discern in this well-stocked drop of water. They have received the appellations of the Pear (*E. pyrum*), and the Sloth (*E. deses*). The former is the most minute we have yet seen, and seems to be scarce; but it is highly curious and interesting in appearance. It much resembles, in outline, a fish of the genus *Balistes;* the muzzle being somewhat protruded and truncate, and the form rhomboidal; it terminates in a slender pointed tail. The body is obliquely fluted, which gives a very singular effect; for from the transparency of the tissues the lines of the opposite side can be discerned crossing those next the eye, and dividing the animal into lozenge-shaped areas. The colour is sparkling green, but the tail and the edges of the body are clear and colourless; and there is a bright red eye. At other times this *Euglena* takes the form of a claret-bottle, or an oil-flask; the muzzle being broadly truncate, or even indented.

Its motion is rapid; a swift gliding in the direction

of its long axis; it turns continually on the same axis, which gives a waving irregularity to its course; and has a pretty effect from the continual crossing of the flutings in the revolving. This specimen is about $\frac{1}{700}$th of an inch in length, including the tail.

Euglena deses is much larger, being about $\frac{1}{250}$th of an inch in length, though the tail is very short. It has a thick body; with a round blunt head; it tapers suddenly to the tail. Its colour is bright green with a red eye; but the presence of an infinite number of irregular oblong granules and lines, with several globular vesicles, gives an opacity and a blackness to its appearance. In its manners it is sluggish; it never swims or glides gracefully and swiftly among its playful congeners, but contents itself with twining slowly among the floccose stems and filaments of the water-plants, or crawls upon the surface of the live-box. It does not appear to change its form, otherwise than its soft and flexible body necessitates, as it twines about.

But enough of the *Euglenas*. For I have just caught sight of a still more curious creature, the Swan Animalcule (*Trachelocerca olor*). It is reposing on one of the leaves of the *Myriophyllum*, its long and flexible neck lengthening and contracting at pleasure, the tip thrown about in quick jerks, in every direction, somewhat like a caterpillar when it touches several points impatiently with its head.

If we admire the graceful sailing of a swan upon a lake, the swelling of its rounded bosom, the elegant curves of its long neck, we shall be struck with the form and motion of this animal. The form has much resemblance to that of a swan, or still more to that of a snake-bird (*Plotus*); the body, swelling in the middle, tapers gradually into a slender pointed tail, at one extremity, and at the other, into a very long and equally slender neck, which is terminated by a slight dilatation. The whole is perfectly transparent, but the body is filled with numerous minute globular vessels, or temporary stomachs. The grace of its motion as it glides along with a free and moderately swift progression through the clear water, or winds through the intricate passages of the green conferva, throwing its long neck into elegant curves, is very remarkable. There are, I see, two of them, which however take no notice of each other, even when passing close to each other; the neck of one is much longer than that of the other. Now and then, when gliding along, the neck is suddenly contracted, but not wholly, as if something had alarmed or displeased the animal: the body also can be swollen or lengthened at pleasure; it can move in either direction, but the neck usually goes foremost, extended in the direction of the motion, and seems to be used to explore the way.

I had once an opportunity of seeing the process of increase by spontaneous self-division in this creature. It was an unusually large specimen, found in an old infusion of sage leaves. When I discovered it, it was darting about its long neck in the most beautiful contortions. As it was partly hidden by the vegetable fibres present, I partly turned the glass cover to alter the position of the contents. On again looking, the Swan was in a clear part of the field, but in the form of a dark globose mass, the neck being entirely contracted. It was quite still, except a continual slight alteration of the form by the protrusion or contraction of parts of the outline. The body seemed full of minute globules, set in a granular mass of a blackish hue, and the outline was not a continuous line, but formed a multitude of rounded elevations. Presently it protruded the clear neck, but only for a short distance, and then retracted it as before; when the only indication of the presence of

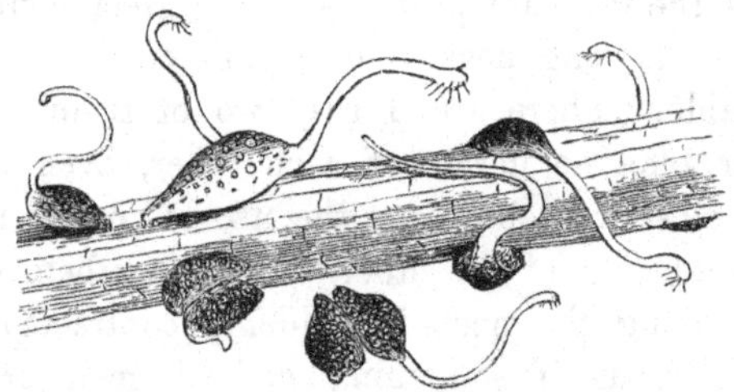

SWAN-NECK AND ITS DIVISIONS.

this organ was a depression in one part of the surface, somewhat like the mouth of a closed Actinia, where there was a slight but incessant working, very much like the irregular motion on the surface of boiling water, in miniature; there was also an indistinct ciliary action at this part, not of rotation, nor of vibration, but a sort of waving. At this point I had occasion to get up from the table, and though I was not away more than a minute, on my return I observed a strong constriction around the middle of the body. It was transverse, for the depressed and ciliated mouth was at a point exactly at right angles to the constriction. From the depth to which this latter extended in so few minutes, I supposed the process of separation would be very rapid; for I could very soon see a line of light all across at intervals, and the two halves seemed to slide freely on each other. Yet they remained long without much apparent progress, or even change, except that the anterior half at one time threw forth its neck a short distance; at this time it looked extremely like a bird, bridling up its lithe neck and swelling bosom; while to make the resemblance perfect, it began to imitate the action of a fowl picking up grain, bobbing its head hither and thither: so curious are the analogies of nature! Along the dividing line, there had appeared very early in the posterior half, a distinct ciliary action; after a while (how, I do not exactly know) without the general relation of position being changed, the mouth of the anterior (which must now be called the old) animal appeared on the side, and at the point correspondent in the other, a similar ciliary wreath appeared, while the action along the dividing line was no longer seen. So that the division which was at first transverse now appeared longitudinal. I believe, however, the animals were really separated before this, though they remained in contact, for as they slid over each other, it was manifest that each had an independent action.

At length, about an hour and a half after the first appearance of the constriction, the new animal threw out its clear neck to a great length, writhing it about with rapid agility, and forming the most elegant curves, like those of a serpent, often completely encircling its own body with it. It still remained, however, in contact with its parent, which, after a time, also protruded its neck in the same manner. Both then retracted and remained still for a while; and again, almost simultaneously, threw out their long necks, and again retired to sluggish repose.

Among the sediment, the grains of which are driven hither and thither by their spasmodic jerking movements, you see several individuals of another sort of creature,—the Chrysalis Animalcule (*Paramœcium aurelia*). This is a "whale among minnows;" for it is greatly larger than any of those we have yet observed; and is just visible to the naked eye, when we hold up the live-box obliquely against the light; for then the animals appear as the smallest possible white specks.

Bringing them again under the microscope, each presents a pellucid appearance, and an oblong figure of which the fore part is somewhat narrowed. The back rises in a rounded elevation; and the mouth is situated as far back as the middle of the body upon the under surface, where its position is marked by a sort of long fold, the sides of which are fringed with long cilia, whose vibrations are very marked. The whole surface, on both sides, is covered with minute cilia, arranged in longitudinal rows, of which, according to the great Prussian professor, there are from thirty to sixty on each surface, each row bearing sixty or seventy cilia. This must be considered as an approximation; for we may well doubt the accuracy of the counting, when the objects are so very evanescent as these vibrating cilia.

The vacuoles, and the temporary stomachs, more

or less completely filled with the brown and green food, which the animals are collecting from the decayed vegetable matters, are sufficiently numerous and conspicuous; but they may be rendered still more so by the device of mixing a little carmine with the water. The ciliary currents are thus instantaneously rendered strikingly visible. The crimson atoms are attracted from all quarters towards the tail of the animal, whence they are urged in a rapid stream along one side towards the head, around which they are hurled, and then down the other side to the tail, pouring off in a dense cloud in a direction contrary to that in which they originally approached.

But now the gathered currents have produced their expected result; for many of the globular vacuoles are already become of a beautiful rosy hue, from the minute particles of the pigment which have been whirled to the mouth, and swallowed.

The feature of greatest interest, however, in this animal is the contractile bladder. Two of these organs are usually seen co-existent in each individual; placed, the one on the front, the other in the rear of the mouth, but near the opposite,—*i. e.* the *dorsal*, surface of the body; for as the creature slowly revolves on its longitudinal axis, the line of

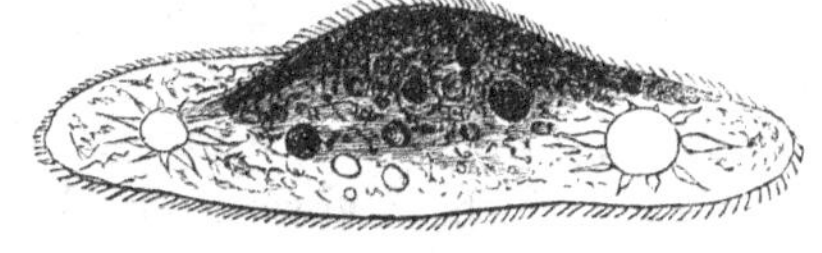

PARAMŒCIUM.

the vesicles alternately approaches and recedes from that of the mouth. They are remarkable for their structure. Far from the simplicity in which the organ is usually presented to us in the animals of this class, the contractile bladders are here very complex. Each when distended is globular; and it is surrounded by a number of others of much smaller dimensions, and of a drop-like form, so set as to radiate around the principal vesicle as a centre, the rounded portion of each in apparent contact with the vesicle; and the slender extremity running off as an attenuated point till lost to sight in the sarcode. The main vesicles alternately become distended, and suddenly contract to a point; while the radiating cells are continually varying in size, though in a less degree. It is customary to describe the secondary vesicles as coming into view at the instant of the contraction of the primary one, and to suppose that the emptying of the one is the filling of the other, but I have not been able to observe this mutual relation satisfactorily made out. The smaller as well as the larger vesicles are conspicuous from their colourless transparency; for the general sarcode of the body, though pellucid, is only so in the same degree as glass, slightly smoked; besides that its clearness is often impaired by crowds of granules and minute globules.

You ask what is that comparatively large oval body attached by its side to one of the leaves of the plant. It is the egg of some considerable Rotifer, probably *Euchlanis*, which is always glued to some filament or stem of a water-plant. It may interest you to watch the progress of the contained embryo, which you can readily do, since the egg-shell is as transparent as glass, and the infant animal already displays the movements of independent life. Meanwhile I will tell you the tragical and lamentable history of just such an embryo as this, that was eaten up before it was born, under my own eye. One of the depredators was a very amusing animalcule, which is sufficiently scarce to make its occurrence a thing of interest, especially to a young microscopist, as I was at the time.

A large egg of (as I believe) *Euchlanis dilatata* had been laid during the night on a leaf of *Nitella*, in the live-box. When I observed it, the transparency of the shell allowed the enclosed animal to be seen with its viscera; which occasionally contracted and expanded; the place of the *mastax* I could distinctly make out. The cilia were vibrating, not very rapidly, but constantly, on the front, where there was a vacant space between the animal and the shell. From 7 A.M. when I first saw it, I watched it for about eight hours, without perceiving any change; but at that hour having withdrawn for a short time, I perceived on my return that a portion of the animal was outside the shell. The appearance was that of a small colourless bladder oozing out, at an imperceptible aperture; and this oval vesicle quickly but gradually increased, until it was half as large as the egg itself. A little

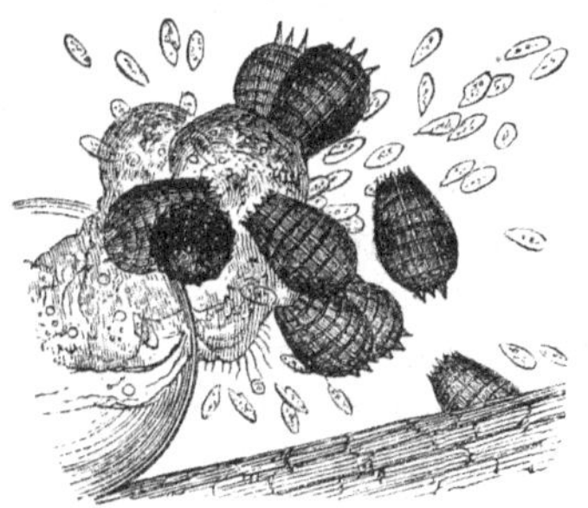
COLEPS AND CHILOMONAS.

earlier than this point, the cilia were seen on the front or lower side ot the excluded portion, and these began to wave languidly in a hooked form. They thus seemed much longer and more substantial than when rotating in the perfect animal. When excluded to the extent just named, some little creatures that were flitting about found it, and began to assemble round it. These were far too rapid in their movements to allow me to identify them before, or to perceive any thing else than their swift motion and oval form, but this attraction causing them to become still, allowed me to perceive their singular and beautiful structure. Each consists of an oval vase open at the top, the margin of which is cut into a number of little points; the sides are marked by a series of ribs, which run down longitudinally, and are crossed by other transverse ones; the rounded bottom is furnished with three short points; so that the whole reminded me of a barrel with its staves and hoops, set on a three-legged stool. Within the body, which is colourless, are seen small dark spots, which are probably the stomach-vacuoles. Thus I identified these little barrels with *Coleps hirtus* of Ehrenberg, but I found no record of their carnivorous propensities. One after another whirled into the field, and after a few gyrations became stationary at the head of the half-born *Euchlanis*, just as I have seen vultures gather one by one to a carcase. Very soon there were a dozen or fifteen of them, some of which were ever shifting their places, and some were playing around, or revolving on their longitudinal axis. I found that their object really was to prey on the soft parts of the creature just excluded from the egg; for by carefully watching one, I distinctly perceived particles of the flesh fly off, as it were, and disappear in the body of the *Coleps.* The appearance was that of steel-filings drawn to a magnet, for the mouth of the *Coleps* was not in actual contact with the flesh; and therefore, I suppose, the surface having been in some way ruptured (which I could see it was), the loose gelatinous atoms were sucked off by a strong ciliary current. They did not attack any other part, and after having continued their murderous occupation about ten minutes, they one by one departed. The ciliary motion of the *Euchlanis* ceased immediately after it was first attacked, and I suppose it was soon killed, for it did not increase in size in the least afterwards. When the *Colepes* left it, a great portion, perhaps a third, of the excluded parts was devoured.

As soon as the depredators were gone, or even before, others more diminutive, but more numerous, were ready to take their place. The drop of water under review had been found amazingly full of a small *Monas*, perfectly transparent, of an oval form, with some granules visible in the interior. They were about $\frac{1}{2000}$th of an inch in length. They filled the whole field, gliding about very nimbly, but so close as but just to allow space for motion, and that in several strata. By the morning these were collected in masses, which to the naked eye looked like little undefined white clouds, but which under the microscope showed the Monads in incalculable multitudes, but for the most part in still repose. Some were still moving to and fro, however, and, in the course of the day, most of them became again active. As soon as the *Colepes* had forsaken their prey, the Monads began to gather around it, cleaving to the same parts, and apparently imbibing the juices, for the extruded parts still slowly decreased, until at length these were reduced to about one-third of their original dimensions.

A close examination of these latter when they had settled to rest, showed me that they were of the species *Chilomonas paramœcium.* There is an indentation on one side of the front, where the mouth is situated; here there is a ciliary action; the projecting part, called the *lip,* is said to be furnished with two slender flexible proboscides, but my power was not sufficient to discern any trace of these. A sort of a ridge, or keel, runs down the length of the body, perceptible by a slight line; numbers of stomach cells also are perceptible. The motion of these lip-monads was not very rapid when unexcited; it is performed by a sort of lateral half-roll, the two sides alternately being turned up, like a boat broadside to a swell, and the line of progression is undulating.

And now having pretty well exhausted the contents of this live-box, let us try a dip from this other phial from another locality, equally productive, if I am not mistaken. Yes; for, to begin, the stalks of *Nitella* here are fringed with populous colonies of the most attractive of all the Infusoria, the beautiful *Vorticellæ.* The species is not the common bell-shaped one, but the smaller with pursed mouth, the little *V. microstoma.*

Look at this active group, consisting of a dozen or so of glassy vases, shaped something like pears, or elegant antique urns, elevated on the extremities of long and very slender stalks, as slender as threads, and about six times as long as the vases. The stalks grow from the midst of the floccose rubbish attached to the plant, and diverge as they ascend, thus carrying their lovely bells clear of one another.

Each vase is elegantly ventricose in the middle, terminating below in a kind of nipple to which the stalk is attached, and above in a short wide neck

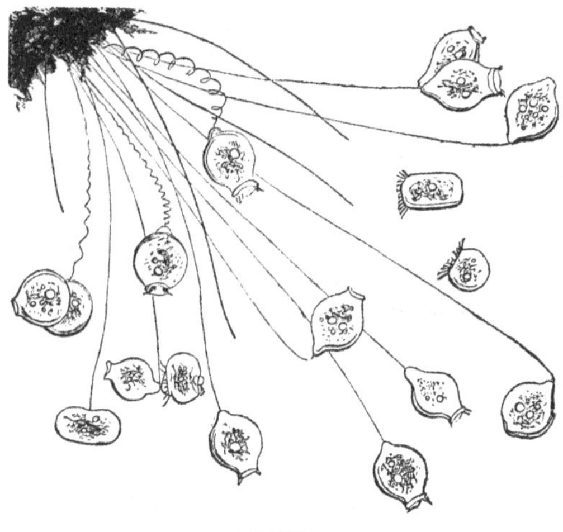

VORTICELLÆ.

with a thickened rim. This last is highly sensitive and contractile; its inner edge is set round with a circle of vibratile cilia, which, when in full play, produce a pair of small circular vortices over two opposite points of the brim. The cilia themselves cannot be distinguished, but their optical expression is curious. At the two opposite points of the circular margin, as seen in perspective when slightly inclined towards the observer, viz. at those points where the cilia, from their position with regard to the eye, would be crowded together, there are seen two dark dashes, representing, doubtless, two ciliary waves, but which have all the appearance of tangible objects, sometimes withdrawn, sometimes protruded, and often vibrating with a rapid snatching movement.

These vases are of the usual appearance in Infusoria. Their substance is the clear transparent colourless sarcode, but it contains within it more or less of the cloudy nebulous matter which we have been lately familiar with. There are several globular vesicles or vacuoles, some ready to imbibe colour from pigment, and others already occupied with brown food, while in each case we see, near the centre of the vase, a longish body of clear granular texture, which is called the nucleus, and which seems to play an essential part in the vital economy of the animal.

The movements of a group such as that we are looking at are very sprightly and pleasing. The vases turned in all directions, some presenting their mouths, some their sides, some their bases to the eye; inclined at various angles from the perpendicular, and bending in diverse degrees upon the extremity of their stalks; swaying slowly and gracefully to and fro, as driven hither and thither by the ciliary currents, and, above all, ever flying up and down within the length of their radius, as a bird when confined by a string:—all these circumstances impart a charm to this elegant animalcule, which enables us to look long at it without weariness.

This last movement is peculiar, and worthy of a moment's closer examination. The stalk, when extended to the utmost, is an elastic glassy thread, nearly straight, like a wire, but never so absolutely straight as not to show slight undulations. The stalk when thus rendered tense by extension, is highly sensitive to vibrations in the surrounding medium; and as in the circumstances in which we observe the animals, such vibrations must be every instant communicated to the vessel in which they are confined, the stalks are no sooner tense than they contract with alarm. This depends on a contractile cord which passes throughout the entire length of the stalk, and which is distinctly visible in the larger species as a narrow band. We can scarcely err in considering this ribbon as a rudimentary condition of muscle, though we do not recognise in it some of the characteristic conditions in which we are accustomed to see it in higher animals.

The contraction of the muscle is very sudden, energetic, and complete. With a rapidity which the eye cannot follow, the vase is brought down almost to the very base of the stalk. Then it slowly rises again, and now we see, what we could not discern in the act of contraction itself, that in that act the stalk was thrown into an elegant spiral of many turns, which at the utmost point of contraction were packed close on each other, but which in the extending act gradually separate, and at length straighten their curves.

In any stage of the extension, the sudden contact of the vase with any floating or fixed object apparently causes alarm, and induces the vigorous contraction; but vibrations, even when so violent as those produced by tapping the stage of the microscope with the finger-nail, have no effect unless the stalk be tense, its own power of vibration being then only developed, just as a cord becomes musical in the ratio of its tension.

It is not until we view these creatures with a good microscope that we acquire an adequate idea of their beauty: for myself, at least, it was so. I had seen engravings of many of the invisible animalcules, and had read technical descriptions; but of their brilliant transparency, their sudden and sprightly motions, their general elegance and delicacy, and the apparent intelligence with which they are endowed, neither books nor engravings had given me any conception.

Some of the individuals under our present examination are exhibiting phenomena of no less interest than their form and motions. Some of the stalks are terminated by *two* vases instead of one, which appear to spring from a common point. These, however, are the result of the spontaneous splitting of one; and in other examples you may see the process in different stages, or, if your patience endure a couple of hours' watching, you may trace the whole phenomena, as I have done, from the moment when it first becomes recognisable, to its completion in the freedom of one of the newly formed animalcules.

For instance, you perceive that one of the bells instead of being vase-shaped has assumed a globular form. By keeping your eye on this for only a few moments, you detect a depression forming in the midst of its front outline, which momentarily deepens, until it is manifestly a cleft. The division proceeds downwards, the two halves healing simultaneously, so that they are at all times perfectly smooth and rounded; at length two vases appear, side by side, where a few minutes before there had been but one.

One of these is destined to be ultimately thrown off, while the other retains sole possession of the stalk. You soon see which it is that is going to emigrate: for though the two are alike in size, the roving one early closes the mouth of the vase, becoming smooth and globular there, never to open again. The cilia, now therefore become useless, disappear by absorption; but meanwhile a new circle of these organs are developed around the basal extremity of the vase, and these, every instant becoming more vigorous in their motions, sway the little globe about on its point of attachment. At length the connexion yields, breaks, and the animalcule shoots away, rowed by its hundred oars, to find a new abode, and to found a new colony.

Here and there you see shooting through the group, with a rapid gliding movement, an oblong clear body. This is one of the vases, formed by self-division, and exercising its newly found power of locomotion. It is giddily roving hither and thither, until the instinct of wandering ceases, when it will soberly settle down, affix itself by the point which was formerly its mouth, whence a new stalk will gradually grow, and opening a new mouth in the midst of the new crown of cilia.

I believe that the division is sometimes transverse instead of longitudinal, the cleft occurring by constriction across the middle of the vase; but this I have not seen. In whatever direction it takes place, it is essential that the oblong granular body, called the nucleus, which you see in each vase, be divided, the cleft passing through the middle of this substance, a portion of which is therefore appropriated to each new made animal.

That the essential vitality of the creature resides in this nucleus is shown by another and highly curious mode of increase, namely, that which is effected by *encystion.* Let us search the live-box carefully, for amidst so great a profusion of *Vorticellæ* as we have on this *Nitella,* it will go hard if we do not find some individuals in the encysted stage.

Look at this elegant object. It resembles a trumpet of the clearest glass, with a rounded extremity, and with the base affixed to the weed, from which it stands up erect. Within the expanded part of the trumpet there is a turbid mass, with a perfectly defined outline, from several points of which proceed radiating pencils or tufts of long, straight, stiff, elastic filaments, like threads of spun glass, varying greatly in length, and each terminated by a little knob of the same material. The *tout ensemble* of this object is very attractive and beautiful, and its history is a tale of marvels.

No wonder that Ehrenberg, supposing this form to

be an independent animal, gave it a generic and specific name. He called it *Acineta mystacina.* For

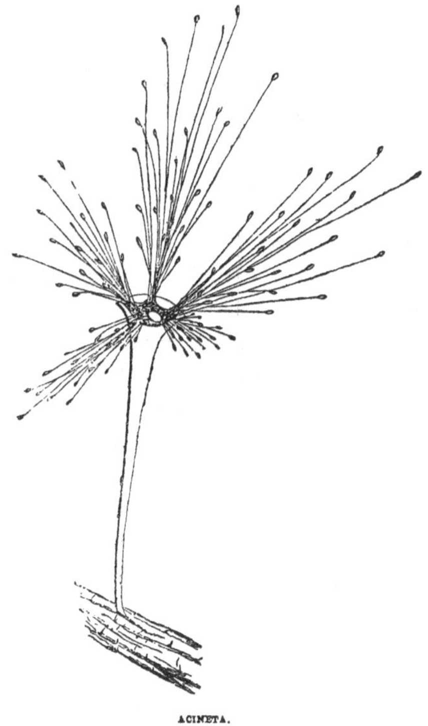

ACINETA.

who would have suspected that this stiff and motionless object, with its tufts of flexible but inanimate threads, had any connexion with the sprightly vases which we have been examining? Yet it is the same animalcule, in what we may, with a certain liberty of phrase, call its chrysalis condition!

The history of the *Vorticella*, as it has been elaborately worked out by Dr. Stein, exhibits phenomena analogous to those marvellous changes which we lately considered under the appellation of the Alternation of Generations. Large individuals withdraw their circle of cilia, close up the mouth, and become globular, and then secrete from their whole surface a gummy substance, which hardens into a spherical transparent shell, inclosing the *Vorticella* in its cavity, in the form of a simple vesicle. Within this vesicle is seen the band-shaped nucleus, unchanged, and what was the contractile bladder, which, however, no longer contracts.

By and by this torpid *Vorticella* enlarges itself irregularly, pushing out its substance in tufts of threads, and frequently protruding from one side a larger mass, which becomes an adhering stalk. Thus it has become an *Acineta*, such as we now behold.

From this condition two widely different results may proceed. In the one case, the encysted *Vorticella* separates itself from the walls of the *Acineta*, contracts into an oval body, furnished at one end with a circle of vibratory cilia, by whose movements it rotates vigorously in its prison, while the more obtuse end is perforated by a mouth leading into an internal cavity. In the interior of this active oval body there are seen the band-like nucleus, and a cavity which has again begun to contract and to expand at regular intervals. It is, in fact, in every respect like a *Vorticella* vase, which has just freed itself from its stalk. Presently, the perpetual ciliary action so far thins away the walls of the *Acineta* that they burst at some point or other, and the little *Vorticella* breaks out of prison, and commences life afresh. The *Acineta*, meanwhile, soon heals its wound, and after a while develops a new nucleus, which passes through the same stages as I have described, and bursts out a second *Vorticella*.

But the cycle of changes may be quite different from this. For sometimes the nucleus within the *Acineta*, instead of forming a *Vorticella*, breaks itself up into a great number of tiny clear bodies, resembling Monads, which soon acquire independent motion, and glide rapidly about the cell formed by the inclosed *Vorticella*-body as in a little sea. But by and by, this body, together with the *Acineta* wall, suddenly bursts, and the whole group of Monad-like embryos are shot out, to the number of thirty or upwards. The *Acineta* now collapses and disappears, having done its office, while the embryos shoot hither and thither in newly acquired freedom. It is assumed, on pretty good grounds, that these embryos soon become fixed, develop stalks, which are at first not contractile, and gradually grow into perfect *Vorticellæ*, small at the beginning, but capable of self-division, and of passing into the *Acineta* stage, and gradually attaining the full size of the race.

Some forms of the same family, *Vorticelladæ*, are interesting as dwelling in beautiful crystalline houses, of various shapes, always elegant. All these have been ascertained to pass through the same or similar *Acineta* stages. *Cothurnia imberbis* is one of the prettiest of these. The cell is of an elegant ampulla-like form, perfectly transparent and colourless, set on a stiff foot, or short pedicle, which shows many transverse folds, like those of leather. From the mouth of the vase projects the animal, whose form may be distinctly traced through the clear walls of the cell, attached to its bottom, whence it stretches upward when seeking prey, or to which it shrinks when alarmed.

In the former condition the body resembles a much elongated *Vorticella*, with a similar circular orifice, set round with cilia. Often the animal performs its ciliary vibrations within the shelter of its house, not venturing to protrude beyond its rim. If carmine be mixed with the water, the atoms are seen in the customary vortex, and some are occasionally drawn into the cell nearly half-way down its cavity, and then swiftly driven out again. On a slight tap upon the table the animal withdraws, and in the same moment the urn bends down upon its leathery pedicle, at a point where there is always an angle, until the rim of the cell is in contact with the plant to which it is attached. This action is instantaneous. Presently, however, it rises, and resumes its former position, and then the mouth of the cell slowly opens, and the animal again protrudes, the cilia appearing first, and finally the head or front part of the animal, which is then opened and begins to rotate.

Very similar to this are the *Vaginicolæ*, but the cells which they inhabit are not stalked, but are immoveably affixed to plants. In *V. crystallina*, the cell is a tall goblet, standing erect, perfectly colourless; while in *V. decumbens*, it is slipper-shaped, attached along its side, and of a golden-brown hue, but still quite transparent. Here is, fortunately, a group of the latter species, scattered about the leaves of the *Nitella.*

Though, in general, both in form and habits, closely like the *Cothurnia,* yet the *Vaginicola* has

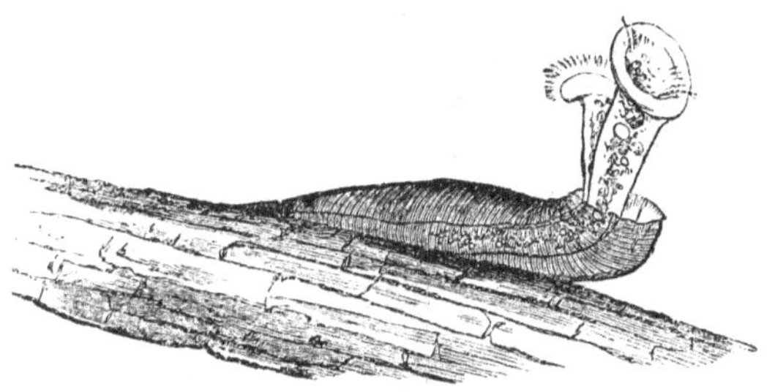

VAGINICOLA.

some peculiarities of interest. The cilia are more developed, and can be more distinctly seen than in either *Cothurnia* or *Vorticella,* forming, when in swift action, a filmy ring above the margin, along which, as if upon a wheel, one or more dark points are frequently seen to run swiftly round; the optical expression, as I presume, of a momentary slackening in the speed of the wave. The act of self-division takes place in this animal, as in the *Vorticella;* and it is curious to see two *Vaginicolæ,* exactly alike, lovingly inhabiting the same cell. One of the cells which we are now examining is in this doubly tenanted condition.

I will now exhibit to you some examples of the most highly organized forms of this class of animals, in which we discern a marked superiority over any that we have yet looked at, and a distinct approach to those animals whose more precise movements are performed by means of special limbs. These creatures are excessively common, both in fresh and sea water, wherever vegetable matter is in process of decomposition; and hence their presence can at all times be commanded by keeping infusions. In this old infusion of sage leaves, for instance, they occur in vast multitudes, past all imagination; as you may see with a lens in this drop.

This group belongs to the genus *Stylonychia*, and I believe to the species *S. pustulata*. It presents the form of an oval disk, which, when seen sideways, is found to be flat beneath and convex above. It commonly swims with the belly upwards, and when exhibited on the stage of the microscope, in almost every case, this surface is presented to the eye. It darts about very irregularly, with a bobbing motion, rarely going far in one direction, but shooting a little distance, and then instantly receding, turning short round, and starting hither and thither, so fitfully that it is very difficult to obtain a fair sight of its structure. Its margin, however, is surrounded by short cilia; the mouth, which is a long opening on the front part, and at the left side (as to the animal) of the ventral surface, is fringed with long cilia, which are continually vibrating. These are the organs of the darting motion; but the creature crawls like a mouse, along the stems of *conferva*, &c., which it performs by means of curved spines, called *uncini*, near the front part, the points of which are applied to the stem, and also by long stiff styles, or bristles, which project backward and downward from the hinder part. Sometimes the animalcules crawl for a moment back-downward, on the inner surface of the glass cover, when the bases of the anterior curved spines appear dilated like large spots. The spines are not capable of much action, but they are rapidly used. The general appearance of the creature reminds us of the little Wood-louse or Armadillo of our gardens. The interior of the body is occupied with a granular substance, in which are scattered many globular vesicles of different sizes. The animal is very transparent, and almost colourless. They increase very fast by transverse division, which is performed under the microscope, so as greatly to increase the number under examination, even in an hour or two. A constriction forms in the middle of one, which quickly deepens, dividing the oblong creature into two of circular figure. The mouth of the new one, with its vibratile cilia, is formed long before separation is complete, and at the same end and side as in the parent. The styles and bristles then form, and the creatures are held together for a few seconds by these organs, even when the bodies are distinctly severed. When separated, they retain the round form for some time.

When a drop of water is examined between two plates of glass, it is amusing to observe the numbers that congregate in the sinuosities left by the gradual drying of the fluid. This probably becomes unfit for respiration, for the motion of the cilia becomes more and more languid, and the creatures die before the water is dry. They not only die but *vanish*, so that where there were scores, so close that in moving they indented each other's sides and crawled over one another,—if we look away for a few minutes, and again look, we see nothing but a few loose granules. This puzzled me, till I watched some dying, and I found that each one burst and as it were dissolved. The cilia moved up to the very last moment, especially the strong ones in front, until, from some point in the outline, the edge became invisible, and immediately the animal became shapeless, and from the part which had dissolved the interior parts seemed to escape, or rather the skin, so to speak, seemed to dissolve, leaving only the loose viscera. From the midst of these then pressed, as if by the force of an elastic fluid within, several vesicles of a pearly appearance, varying in number and size, and then the whole became evanescent.

You will have observed that the admixture of carmine to the water, while the animalcules were active, shows the direction of the ciliary motion with great distinctness. The particles form two vortices, one on each side of the front, which meet in the centre in a strong current, and pass off behind the mouth on each side. We do not perceive that any of them swallow the particles of carmine, for the internal vessels remain colourless.

I have found that if a drop of water containing these animals be placed on a slip of glass exposed to the open air, they do not burst as the water dries away; but dry flat on the glass, their bodies broader, but shorter than when alive, and quite entire. Their cilia are then very manifest. On being again wetted, though after only a few minutes' desiccation, I have never been able to revive them, nor any other *Infusoria* in like circumstances, notwithstanding what is stated in books.

Here is another species in equally amazing profusion, *S. mytilus.* Its form is oblong, with rounded extremities, the anterior obliquely dilated. This species affords a good example of the various organs of locomotion. A transparent oblong shield, which is quite soft and flexible, is spread over the back, which does not prevent the eye discerning all the organs through it, though much more commonly the animal, when under the microscope, crawls belly-upward, beneath the glass cover of the live-box. Around the anterior part, which is broadened, are placed cilia, which are vibratile; these are continued round the mouth, a sort of fold on the side. Towards the posterior extremity on each side are other rows of cilia, which being large are well displayed. On the ventral surface, chiefly towards the front part, are seen several thick pointed processes, shaped like the prickles of a rose, but flexible, and capable of being turned every way. These are the *uncini*, and are evidently used as feet, the tips being applied to the glass. The optical effect of the throwing about of these *uncini*, when the place which they touch is in focus, is very curious. They are rapidly moved, but without regularity; the tips bend as they touch the surface of the glass; some of them seem to have accessary hairs, equally long, but slender, proceeding from the same base. On the hinder quarter of the ventral surface are several thick pointed spines; these are inflexible nearly straight, placed side by side, but not in regular order, some reaching beyond others. I have not seen these used, but they commonly remain sticking out in a horizontal direction. These organs are termed *styles*. Besides these, there are three slender bristles, called *setæ*, placed at the hinder extremity, the central one in the line of the body, the others radiating at an angle. These are distinguished from the cilia, not only by their length, but by not being vibratile. The motions of these animals are powerful, but irregular, and fitful, very much like those of the former species. They dart hither and thither, backward as well as forward, occasionally shooting round and round in a circle, with many gyrations, much like the pretty little polished beetles (*Gyrinus*) that play in mazy dances on the surface of a pool. The two extremities seem covered with minute pits or stipplings, but colourless; the central part is occupied with yellowish granules of different sizes.

I once witnessed the dissolution of one of these animals under peculiar circumstances. Two or three stems of an aquatic plant had become crossed in the live-box so as to form an area, into which the *Stylonychia* had somehow introduced himself. There was just room for him to move backward and forward without turning, and the space was about three times his own length. Within this narrow limit he impatiently continued crawling to and fro, moving his uncini with great rapidity and showing their extreme flexibility, for as he applied them now to the stem, now to the surface of the glass, these whip-like uncini were sometimes bent double. The so-called styles at the posterior extremity, though less frequently used so, were yet occasionally bent and applied to the surface as feet, so that they are certainly not inflexible as supposed, nor do I see any essential difference between them and the uncini. The whole body was flexible, taking the form of any passage or nook into which it was thrust, yet recovering its elasticity immediately the pressure was removed. Its proper form appeared to be convex above and concave beneath, rather than flat. After having been thus employed about half an hour under my observation, it became still, moving only its cilia, when I left it a little while, and on my return found that it was dissolved; the outline having entirely disappeared, and nothing being left but the granules, and globular vesicles, that had constituted its viscera, some of which still contained the carmine which had been very perceptible in the living animal. This was the more remarkable as there was plenty of water. It looked like suicide, a spontaneous choosing of death rather than hopeless captivity.

Common as these *Stylonychiæ* are, and abundant beyond all calculation, where they do occur, from their tendency to self-division, they are not so universally met with as their cousins, of the genus *Euplotes*. These are still more highly organized, and will please you by their activity and sprightly intelligence, I am sure. Here are several individuals in the live-box at this moment.

They differ from the *Stylonychiæ*, in having the soft body covered with a plate of crystal mail, hard and inflexible, much like the shield of a Tortoise.

Several species have this glassy shield marked with delicate lines running lengthwise; sometimes in the form of parallel ridges, as in a little species found in infusions (perhaps *E. charon*); at others forming rows of minute round knobs, as *E. truncatus*, the species now before us. The shield is ample, considerably overlapping the soft body; it rises into an arched form in the centre; and is more or less round or oval. The mouth is oblique, and extends a long way down the under surface; it is set with strong and fine cilia, which also spread over the front. The organs of motion are, as before, long styles, pointed and rather stiff processes, which project from beneath the shell backwards and downwards, and soft hook-like uncini which are set chiefly near the forepart of the inferior surface. In the species before us these are about six or seven in number, but in *E. charon* they are more numerous. The twinkling rapidity with which these little feet are applied to the surface in crawling affords a pleasing sight; particularly when the animal is running back-downwards on the upper glass plate of the live-box. Some species have bristles (or *setæ*), affixed to the hinder part of the shell, from which they diverge. In *E. truncatus* these are four, but they are wanting in *E. charon*. The body displays a mass of granules, vacuoles, and vesicles of different sizes.

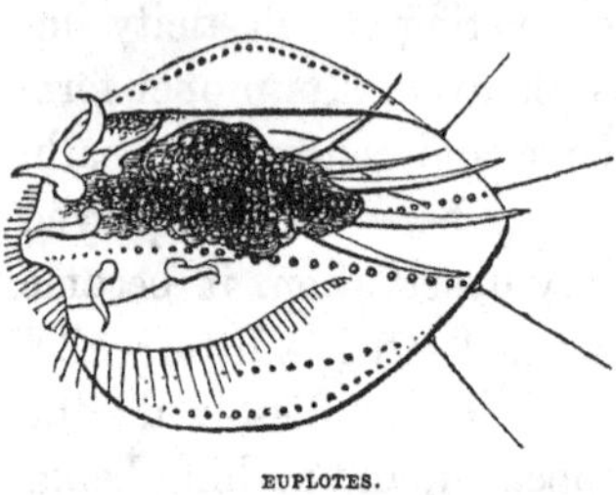
EUPLOTES.

These are very beautiful objects; and their sprightly motions, and apparent intelligence, give them an additional interest. They crawl more than they swim, running with great swiftness hither and thither, frequently taking short starts, and suddenly stopping. The specimens which we are examining are taken from water which had been kept in a jar for several weeks. The vegetable matters are decaying, and among the stems and filaments this pretty species crawls and dodges about. It seems reluctant to leave the shelter of the decaying solution; sometimes one will creep out a little way into the open water; but in an instant it darts back, and settles in among the stems and flocculent matter. Any attempt by turning the glass cover to bring it out into view only makes it dive deeper into the mass, as if seeking concealment. This is about $\frac{1}{250}$th of an inch in length of lorica; and the *E. charon* is not more than one-fourth of this size. These creatures remind one of an *Oniscus*, especially when in profile.

There is an animal very closely allied to these, but much more beautiful, being of a clear greenish translucency, with several vesicles filled with a rose-coloured or purple fluid of much brilliancy. This creature, which bears the name of *Chlamidodon*, has the peculiarity of a set of wand-like teeth arranged in a hollow cylinder.

And with these we dismiss the *Infusoria*, a class of animals, which, from their minuteness, the number and variety of their species, their exceeding abundance, the readiness with which they may be procured, and, as it were, made to our hand (by simply steeping vegetable matter in water), and the uncertainty which still prevails as to many parts of their structure and economy; and therefore, as to their true affinities in the great Plan of creation,—offer one of the most promising fields of research which a young microscopist could cultivate.

These are thy glorious works, Parent of good,
Almighty; thine this universal frame;
Thus wondrous fair; Thyself how wondrous then!
Unspeakable, who sitt'st above these heav'ns,
To us invisible, or dimly seen
In these thy lowest works; yet *these* declare,
Thy goodness beyond thought, and power divine.

INDEX.

THE END.

Printed in the United States
By Bookmasters